MICROBIOLOGY RESEARCH ADVANCES

MYCOTOXINS

OCCURRENCE, TOXICOLOGY AND MANAGEMENT STRATEGIES

MICROBIOLOGY RESEARCH ADVANCES

Additional books in this series can be found on Nova's website
under the Series tab.

Additional e-books in this series can be found on Nova's website
under the e-book tab.

MICROBIOLOGY RESEARCH ADVANCES

MYCOTOXINS

OCCURRENCE, TOXICOLOGY AND MANAGEMENT STRATEGIES

CESAR RIOS
EDITOR

nova publishers
New York

Library of Congress Cataloging-in-Publication Data

ISBN: 978-1-63483-544-2

Published by Nova Science Publishers, Inc. † New York

CONTENTS

PREFACE

Exposure to certain types of mycotoxins, the bioactive secondary metabolites of filamentous fungi, significantly impact the animal industry as well as human health. The contamination of food with mycotoxins is a worldwide problem in animal production and direct consequences are the reducing of food intake and production. In this book, the authors increase the public awareness of the implications of certain types of mycotoxins exposure to promote the health of livestock as well as the general public; an analysis of trichothecens (TCEs), a large group of mycotoxins, and their impact on lifestock health and production are also examined, as well as the occurrence of mycotoxins in animal products such as goat milk. Different strategies implemented for the analysis of a wide spectrum of mycotoxins, as well as their advantages and disadvantages are provided by the authors. In the next few chapters, the occurrence of mycotoxins in cereals and cereal products and in particular, their toxic properties are looked at; a discussion on deoxynivalenol (DON) and related 8-ketotrichothecene mycotoxins, which are extensively distributed in cereal-based foods and feed stuffs worldwide are reviewed; potential toxigenic fungi from diverse habitats are identified with special emphasis on the methods as well as on the genetic markers employed; the importance of performing exposure assessments to Aflatoxin B and some of the most important aspects to consider in the risk assessment process, including the simultaneous presence of other mycotoxins and the challenge of choosing the most suitable method to perform exposure assessment are analyzed; and finally, a review of the occurrence of dietary mycotoxins in Africa as well as the advances in analytical methods of mycotoxin extraction and detection over the last decade is provided.

Chapter 1 – Exposure to certain types of mycotoxins, the bioactive secondary metabolites of filamentous fungi, significantly impacts the animal industry as well as human health. Mycotoxins from *Fusarium* or *Aspergillus* molds have been studied extensively because of the frequent contamination by these mold species and their various toxic effects. On the contrary, *Penicillium* mycotoxins such as ochratoxin A (OTA), citrinin (CIT), patulin (PAT), penicillic acid (PA), mycophenolic acid (MPA) and roquefortine C (RQC) are less well studied. Though they are present endemically in wide varieties of crops, fruits, and food stuffs from many different climate regions. A common characteristic of many *Penicillium* mycotoxins is their immunomodulatory and immunotoxic properties that could affect livestock productivity and human health by pre-disposing the host to secondary diseases. *Penicillium* mycotoxins can directly interfere with the activities of both innate and adaptive immune cells, and some appear to affect immune-associated gene expression through epigenetic mechanisms. The intent of this review paper is to increase the public awareness of the implications of *Penicillium* mycotoxins exposure and to promote health of livestock as well as the general public at large.

Chapter 2 – Trichothecene (TCE) contamination of common agricultural grains such as wheat, barley and maize is an increasingly common problem for the livestock industry because it can cause mycotoxicosis in animals affecting their production and survival. Many questions remain as to how TCE are metabolized by different livestock species, what their effects on health and production are, and what levels of intake are tolerable to ruminants without affecting health and productivity. The following TCEs: T-2, deoxynivalenol (DON), nivalenol (NIV), fusarenon-X (FUS-X), diacetoxyscirpenol (DAS) and neosolaniol (NEO) are reviewed herein to assess their effects on livestock health and production. These TCEs commonly affect the immune system, reproduction, and feed intake and weight gain across species. However, further research investigating the effects of these TCEs needs to be conducted in ruminants. Research on the combined effects of mixed-TCE exposure is especially necessary since this type of exposure typically occurs in nature.

Chapter 3 – Mycotoxins are toxic secondary metabolites produced by certain fungi in agricultural commodities. Among mycotoxigenic fungi, *Aspergillus, Penicillium* and *Fusarium* are the most important genera. The contamination of feed with mycotoxins is a worldwide problem in animal production and the direct consequences are the reducing of feed intake, production, reproduction and weight gain. The complex diet of ruminants consisting of forages and concentrates can be a source of diverse mixtures of mycotoxins. The mycotoxins can be transferred to animal products mainly in milk. Once milk and milk products are a good source of many nutrients such as proteins and calcium, mycotoxin residues in milk can pose a hazard to human health. The cow milk is the most consumed; however, goat milk could substitute it because there is a percentage of the population that has allergies and gastro-intestinal disorders with cow milk.

Researchers demonstrate the beneficial effects of goat milk, but there is few studies about the incidence of mycotoxins in goat milk. Several countries have established regulations for mycotoxins in feeds, milk and its products due the risk for consumers and the adverse effects on animal health. This chapter describes the occurrence of mycotoxins in goat milk and its dairy products.

Chapter 4 – This chapter presents an overview of current chemometric tools utilized for the evaluation of several mycotoxins often found in foodstuff, such as cereals, peanuts, pistachio nuts and condiments. Mycotoxins are compounds produced as a result of the secondary metabolism of fungus, such as *Aspergillus*, *Fusarium* and *Penicillium*. Because of their harmful effect in animals and humans, their levels in food are strictly controlled.

Discussions are focused on the analysis of first and second-order data generated through several techniques such as near infrared spectroscopy, fluorescence spectroscopy and high performance liquid chromatography with diode array detector. This review manuscript describes the algorithms frequently used to model this type of data, including Partial Least Squares (PLS) regression, Parallel Factor Analysis (PARAFAC) and Multivariate Curve Resolution Alternating Least Squares (MCR-ALS). These multivariate calibration methods allow the quantification of multiple analytes in complex food samples with the application of minimum pre-treatment techniques. Another important chemometric tool commented in this review is the Response Surface Method (RSM), commonly applied in the optimization of experimental procedures and chemical measurements.

This chapter discusses the different strategies implemented for the analysis of a wide spectrum of mycotoxins, as well as their advantages and disadvantages.

Chapter 5 – Mycotoxins as a metabolic product of fungi (moulds) may be formed in many crops at various stages, starting from the plant growth phase through harvest time and up to crop storage. The most common mycotoxins are related to three species: *Aspergillus*, *Penicillium* and *Fusarium*. The first two are generally recognised as contaminants of food during drying and later storage, while *Fusarium* species are pathogens damaging plant and producing mycotoxins during vegetation or just after harvest. Mycotoxin content is an important indicator of cereal grain hygiene status and their quality. Due to the broad spectrum of their toxic properties and mode of action, followed – in general – by high resistance to temperature, the presence of mycotoxins in food and feeds generates a serious health hazard both for humans and animals. The consequence is of prime social concern. Since mycotoxicoses – poisonings and diseases caused by fungal toxins – are worldwide and agency problems are observed, international organisations emphasise the pressing need of continuous quality assessment for plant and animal food in terms of toxic fungal metabolites. Symptoms of mycotoxicoses depend on the type and concentration of consumed mycotoxins, time of exposure as well as the age, health conditions and sex of the organism with significant influence of diet.

Chapter6 – Deoxynivalenol (DON) and related 8-ketotrichothecene mycotoxins are extensively distributed in the cereal-based food and feed stuffs worldwide. Monitoring the oral exposures is crucial in assessing acute and chronic toxicities of the mixed type of 8-ketotrichothecenes mainly important in grain-based staple. For the efficient mucosal biomonitoring, development of effective gut-derived biomarkers is a very important procedure of the toxicological risk assessment of 8-ketotrichothecenes. In the luminal parts of mucosa, the 8-ketotrichothecenes are mixed with dietary components and gut microbiota, all of which interact together, leading to complex events in the body. For instance, de-epoxidation of DON by gut microbes can lead to more fecal secretion of less toxic DOM-1 (metabolic inactivation). By contrast, most of plant-derived DON-glucosides are also easily catabolized to free DON by gut microbes, which produces more burden to the body (metabolic activation). We can use these 8-ketotrichothecene metabolites as efficient mucosal exposure biomarkers from luminal parts of the exposed host. Moreover, the diversity and presence of the 8-ketotrichothecene-metabolizing microbes in the gut are another important determinant of the susceptibility to the toxicity, provided as useful susceptibility biomarkers. In terms of the mucosal biological responses to 8-ketotrichothecenes in food, mucosal effector biomarkers are important in estimating the gut-related diseases including gastroenteritis and igA nephritis in human and animal population. Enterocyte chemokines as the early effector mucosal biomarker and mucosal igA as the chronic biomarker have been investigated. In particular, mucosal igA is very sensitive effector biomarker which can be prominently correlated with exposure at dose as low as the regulatory limit in most countries. Moreover, anorexia monitoring using murine models are recently developed to monitor 8-ketotrichothcene-induced feed refusal at the gut-brain axis. These behavioral responses are mechanistically linked to alteration of eating disorder-linked hormones from the intestinal neuroendocrine cells, which can be useful biomarkers in some species. Using these gut-derived biomarkers, reporter readout systems and more realistic gut mimetic models have been developed and would be useful in monitoring the potent risk of 8-ketotrichothecene mycotoxins with minimal use of experimental animals.

Chapter 7 – Mycotoxins are harmful secondary metabolites mainly produced by members of the genera *Fusarium*, *Aspergillus* and *Penicillium*. The identification of these fungi has been traditionally carried out through procedures that require their growth under standard laboratory conditions. Nevertheless, in the recent years a significant progress has been made in the development of metagenomics, a culture independent approach that aims at studying the overall microbial genomes present in a particular habitat. Metagenomic analysis of uncultured fungal communities can be used for the identification of potential toxigenic fungi and also to assess their distribution, which may have important implications for their management. Thus, molecular methods based on the polymerase chain reaction (PCR) have been used to access the uncultured fraction of fungal communities in different environments as diverse as soils and air masses. Here, different studies using a sequence-based metagenomics approach and identifying potential toxigenic fungi from diverse habitats are reviewed with special emphasis on the methods as well as on the genetic markers employed.

Chapter 8 – Mycotoxins are harmful secondary metabolites mainly produced by members of the genera *Fusarium*, *Aspergillus* and *Penicillium*. The identification of these fungi has been traditionally carried out through procedures that require their growth under standard laboratory conditions. Nevertheless, in the recent years a significant progress has been made in the development of metagenomics, a culture independent approach that aims at studying the overall microbial genomes present in a particular habitat. Metagenomic analysis of uncultured fungal communities can be used for the identification of potential toxigenic fungi and also to assess their distribution, which may have important implications for their management. Thus, molecular methods based on the polymerase chain reaction (PCR) have been used to access the uncultured fraction of fungal communities in different environments as diverse as soils and air masses. Here, different studies using a sequence-based metagenomics approach and identifying potential toxigenic fungi from diverse habitats are reviewed with special emphasis on the methods as well as on the genetic markers employed.

Chapter 9 – Mycotoxins are prevalently toxic compounds produced as secondary metabolites by various fungi and excreted on agricultural crops at pre and post-harvest stages of food and feed production under a wide range of climatic conditions. Majority of mycotoxins are produced by fungi that belong to the genera *Aspergillus*, *Fusarium* and *Penicillium*. Most of the known hosts include maize, peanut, wheat, rice, sorghum, barley and various tree nuts which are staple foods in most parts of Africa. Exposure to mycotoxins is mostly by direct ingestion of contaminated agricultural crops or by a carry-over of mycotoxins into animal tissues, milk and eggs and can also be inhaled or absorbed through the skin. Mycotoxins exert various toxic effects from relatively mild G.I disturbances to severe organ damage on exposed humans and animals. Risks of mycotoxin contamination of food and feed in Africa is increasing due to climatic conditions, agronomic practices and the socio-economic and food security status of the public (inhabitants).

Climatic conditions such as hot and humid environment favour the proliferation of fungal and mycotoxin contamination of food and feeds. Farming practices in Africa sustain fungal and mycotoxin contamination of food as well as feeds. The socio-economic status of greater parts of the public predisposes them to the consumption of contaminated food either directly or at various points in the food chain. The resulting health risk of mycotoxin contamination depends on the toxicological property of the particular compound (carcinogenicity, mutagenicity, teratogenicity, genotoxicity, immunosuppression, nephrotoxicity and hepatotoxicity) as well as the extent of exposure. Mycotoxins have assumed great economic importance worldwide because of their influence on human health and livestock and many countries have set regulatory limits in order to reduce their intake. In Africa, fifteen countries which accounts for about 59% of the population of the continent, were considered to have mycotoxin regulations in 2003.

Mycotoxin control strategies should also include the adoption of appropriate agronomic practices at the pre and post-harvest prevention stages and also the physical, biological and chemical methods of decontamination at the post-harvest stage. Strategies for the diversion of contaminated products to lower risk uses and laboratory safety measures to protect researchers from the toxic effects of mycotoxin should also be included. There should also be producer /consumer education and accurate and affordable detection methods of mycotoxin in developing African countries. The aim of this review is to highlight the occurrence of dietary mycotoxins in Africa as well as the advances in analytical methods of mycotoxin extraction and detection over the last decade (2003-2013).

In: Mycotoxins
Editor: Cesar Rios
ISBN: 978-1-63483-544-2

Chapter 1

Silage *Penicillium* Mycotoxins: Hidden Modulators of the Immune System

Se-Young Oh[1,*], Rebecca E. Fisher[2], H. V. L. N. Swamy[3], Herman J. Boermans[2], Alexandros Yiannikouris[4] and Niel A. Karrow[1,*]

[1]Department of Animal and Poultry Science (APS), Ontario Agriculture College (OAC), University of Guelph, Guelph, ON, Canada

[2]Department of Biomedical Science, Ontario Veterinary College (OVC), University of Guelph, Guelph, ON, Canada

[3]Devenish Nutrition Ltd, Bengaluru, Karnataka State, India

[4]Alltech Inc., Nicholasville, KY, US

Abstract

Exposure to certain types of mycotoxins, the bioactive secondary metabolites of filamentous fungi, significantly impacts the animal industry as well as human health. Mycotoxins from *Fusarium* or *Aspergillus* molds have been studied extensively because of the frequent contamination by these mold species and their various toxic effects. On the contrary, *Penicillium* mycotoxins such as ochratoxin A (OTA), citrinin (CIT), patulin (PAT), penicillic acid (PA), mycophenolic acid (MPA) and roquefortine C (RQC) are less well studied. Though they are present endemically in wide varieties of crops, fruits, and food stuffs from many different climate regions. A common characteristic of many *Penicillium* mycotoxins is their immunomodulatory and immunotoxic properties that could affect livestock productivity and human health by pre-disposing the host to secondary diseases. *Penicillium* mycotoxins can directly interfere with the activities of both innate and adaptive immune cells, and some appear to affect immune-associated gene expression through epigenetic mechanisms. The intent of this review paper is to

* Corresponding Authors: Se-Young Oh, Department of Animal& Poultry Science, University of Guelph, Guelph, Ont. N1G2W1. Email: ugohs530@gmail.com. Tel: (519) 824-4120 ext. 58365.

Niel A. Karrow, Department of Animal& Poultry Science, University of Guelph, Guelph, Ont. N1G2W1. Email: nkarrow@uoguelph.ca. Tel: (519) 824-4120 ext. 53646.

increase the public awareness of the implications of *Penicillium* mycotoxins exposure and to promote health of livestock as well as the general public at large.

Keywords: *Penicillium* mycotoxins, immunomodulation, proliferation, cell death, inflammation, epigenetic

LIST OF ABBREVIATION

AbMIR	Antibody-mediated immune response
BALF	Bronchioaleolar lavage fluid
B-cell	B-lymphocytes
Bmi-1	Polycomb ring finger oncogene
BoMacs	Bovine macrophage cell line
Caco-2	Human epithelial colorectal adenocarcinoma cells
CCR	Chemokine receptors
CD	Cluster of Differentiation
CIT	Citrinin
CMIR	Cell-mediated immune response
CO_2	Carbon Dioxide
Con A	Concanavalin A
COX-2	Cyclooxygenase-2
CTLs	Cytotoxic T-cells
DC	Dendritic cells
DNA	Deoxyribonucleic acid
DNMT	DNA methyltransferases
DON	Deoxynivalenol
E. coli	*Escherichia coli*
ELISA	Enzyme-linked immunosorbent assay
ERK	Extracellular signal-regulated kinases
FAO	Food and Agriculture Organization
Foxp3	Forkhead Box P3
GALT	Gut-associated lymphoid tissues
GDP	Guanosine Diphosphate
GSH	Glutathione
GSSG	GSH reductase
GTP	Guanosine triphosphate
H_2O_2	Hydrogen Peroxide
H3K27	Histone H3 lysine 27
H3K4	Histone H3 lysine 4
HDAC	Histone deacetylases
HPLC	High performance Lipid Chromatography
hsp72	Heat shock protein 72
Iκβ-α	Inhibitor of NF-κβ
IFN	Interferon
IgA	Immunoglobulin A

IgG	Immunogloublin G
IgM	Immunoglobulin M
IL	Interleukin
IMPDH	Inosine monophosphate dehydrogenase
iNOS	Inducible nitric oxide synthase
iv	Intravenous injection
ip	Intraperitoneal injection
JAK	Janus kinase
JMJD3	Jumonji domain-containing protein 3
JNK	Jun N-terminal kinase
LC	Liquid Chromatography
LD50	Lethal dose that kills 50% of a group of test animals
LPS	Lipopolysaccharide
MAPK	Mitogen-activated-protein kinase
MIP	Macrophage inflammatory protein
MMF	Mycophenolate mofetil
Mono-Mac-6	Human monocytic cell line
MPA	Mycophenolic acid
MS	Mass spectrometry
NF-κB	Nuclear factor-κB
NK cell	Natural Killer cell
NO	Nitric oxide
Po	Oral administration
OTA	Ochratoxin A
PA	Penicillic acid
PAT	Patulin
PBL	Peripheral blood lymphocytes
PBMC	Peripheral blood mononuclear cell
PHA	Phytohaemagglutinin
PRRs	Pattern recognition receptors
PWM	Pokeweed mitogen
RBCs	Red blood cells
RNA	Ribonucleic acid
ROS	Reactive oxygen species
RQC	Roquefortine C
Spp	Species
STAT	Sgnal transducers and activators of transcription
T-2 toxin	Trichothecene mycotoxin
T-cell	T-lymphocytes
TER	Transepithelial resistance
TGF-β	Transforming growth factor
Th	T helper cell type
TLC	Thin Layer Chromatography
TNFR-1	TNF-α receptors-1
TNF-α	Tumor necrosis factor-α
TRADD	TNF receptor-1-associated death domain

T-reg cell	Regulatory T-cells
UTX	Tetratricopeptide-repeat X
VCAM-1	Vascular cell adhesion molecule 1

1. INTRODUCTION

1.1. General Overview

Mycotoxins are biologically active secondary metabolites of filamentous fungi that commonly contaminate a wide variety of cereal and livestock feed crops worldwide. The Food and Agriculture Organization (FAO) estimates that approximately 25% of food and food stuffs around the world are contaminated by fungal mycotoxins [1]. Increasing humidity and temperature accompanying global climate change will likely favour more fungal contamination of feedstuffs, affecting the safety of both human and animal feeds [2]. Economic losses from mycotoxin contamination occur due to reduced crop yield and value, as well as decreased livestock productivity and health from the consumption of contaminated feed [3].

Penicillium molds are one of the common fungal species found in crops, fruits and animal feeds of Northern Europe, North and South America as well as East Asia [4–7]. Poor storage conditions for most grains and food products often provide favourable environmental conditions, such as moderate temperature with high humidity, for mycotoxin production from *Penicillium* mold species [8, 9]. Due to their dramatic increase during the ensilaging of animal feed, the *Penicillium* mycotoxins are also refereed as 'silage mycotoxins.'

A number of *Penicillium* mycotoxins can be detected in stored feed or food such as ochratoxin (OTA), citrinin (CIT), patulin (PAT), penicillic acid (PA), mycophilic acid (MPA), and roquefortine C (RQC) (Table 1) [10, 11]. These *Penicillium* mycotoxins are known to cause a variety of toxic effects in animals (Table 2). For example, OTA, CIT and PA are nephrotoxic [12, 13] and hepatotoxic [14] in monogastric species, while PAT is a potent hepatotoxicant in both monogastric and ruminant species [15–17]. These *Penicillium* mycotoxins also affect the host immune response, which could predispose animals to secondary diseases [18–20].

Regardless of their abundance in animal feed and their potential to affect the immune system, the agricultural and economic importance of *Penicillium* mycotoxins are often under-estimated because it is generally assumed that *Penicillium* mycotoxins are readily metabolized by gut microbiota and hepatic enzymes [21, 22]. However, considering that most of these *Penicillium* mycotoxins have anti-microbial properties [23] and are found to be hepatotoxic to animals [13–15], it is possible that detoxification of the *Penicillium* mycotoxins is attenuated by these properties. RQC for example, was found to inhibit cytochrome-p450 activity by reducing nicotinamide adenine dinucleotide phosphate (NADPH) consumption [24], and p450 is an important enzyme for detoxifying mycotoxins [25]. Furthermore, mycotoxins can also enter the circulation and target tissues without passing though the digestive tract. For example, it has been reported that some *Fusarium* and/or *Penicillium* mycotoxins, such as OTA, can enter the circulation during inhalation [26,

27] and dermal exposure [28]. Since entry via these routes bypasses initial detoxification provided by gut microbiota and the liver, exposure via inhalation and dermal contact could be a potential health risk for farmers and livestock [29–31].

1.2. The General Occurrence and Toxicity of *Penicillium* Mycotoxins

1.2.1. Ochratoxin A (OTA)

OTA is a common *Penicillium* mycotoxin found in stored cereal grains, nuts as well as some brewed liquors, because temperature and humidity during storage and brewing favour OTA production [11, 32]. OTA was initially isolated from *Aspergillius ochraeus* in 1965, but was later also isolated from *P. citrinin* and *P. verrucosum* [33]. The lethal doses that kill 50% (LD50) of mice via intraperitoneal and oral exposure were estimated to be 29.4 mg/kg and 46.0 mg/kg, respectively [34], which are much higher than typical OTA concentrations found in silage.

Nephrotoxicity is the most common pathology associated with OTA exposure at a concentration greater than 0.05 mg/kg in monogastric species, such as rats and pigs [35, 36]. However, a high dose of 13 mg/kg, or cumulative exposure to a concentration as low as 3 mg/kg was also reported to cause teratogenicity to embryos of rats and chicks [37–39] as well as goats [40]. A few studies have also shown that OTA significantly reduced the overall performance of both laying hens and broiler chickens [41, 42]. OTA was also found to have synergistic toxicity in poultry species when mixed with either CIT or PA [43, 44].

1.2.2. Citrinin (CIT)

CIT is a common mycotoxin isolated along with the natural yellow biopigment produced by *Monascus* spp. [45]. CIT was first isolated from *P. citrinin* [46], and was subsequently detected from *P. verrucosum* and *P. expansum* that frequently contaminate cereal grains, rotten fruits, and juice [47, 48].

Several animal studies have reported significant toxicity mediated by CIT, and the oral LD50 to the rat was estimated to be 50 mg/kg [49]. The LD50s in response to intraperitoneal exposure have been estimated to be 35-58 mg/kg for the mouse and 19 mg/kg for the rabbit [49]. CIT appears to be a nephrotoxin based on its toxicity to porcine kidney cells [12], while teratogenicity was also observed in rats that were exposed to 35 mg/kg during gestation [50] and in chicken embryos exposed to CIT concentrations between 1.0 to 10.0 μg [44]. The toxicity of CIT was shown to be synergistically enhanced in poultry species when co-exposure occurred with OTA [44].

1.2.3. Patulin (PAT)

PAT is one of the most potent mycotoxins found in food [51]. *P. expansum* is the most common mold species that produces PAT [16]; however, other *Penicillium* and *Aspergillius* mold species including *P. urticae, P. melinii, P. claviforme, A. clavatus* and, *A. terreus* can also produce PAT [52]. PAT can be found in fruits, flour, cereal grains, and meat [47, 52]; however, it is most often detected in the fruit-based products, such as apple and grape juices [53, 54].

The LD50s of acute PAT exposure for mice, neonatal rats, and weaning rats were determined to be 17, 6.7 and 108-118 mg/kg BW, respectively, with the following pathological signs: focal hepatic necrosis, atelectasis, alveolar septal congestion and intra-alveolar haemorrhaging [15]. In 1954, it was suggested that PAT caused haemorrhagic syndrome and the death of over 100 Japanese dairy cattle that had been fed malt contaminated with PAT produced from *P. urticae* [17]. A similar incident occurred in Germany and France as a result of PAT produced from *A. clavatus* contamination of wheat [52]. Chronic exposure to 0.1 mg/kg of PAT in rat feed resulted in various pathological signs, including haemorrhage, plasma cell hyperplasia as well as cell destruction and abnormalities of the nucleus and organelles of rat liver and kidneys [55]. No evidence of either teratogenicity or mutagenicity has been observed in mice [56].

1.2.4. Penicillic Acid (PA)

PA is often detected in various grains, tobacco, and has even been found in moldy sausages [52, 57–59]. Its potential toxicity has been known since 1896; however, it was first isolated from *P. puberulum* in 1913 [60]. Several different *Penicillium* spp. including *P. puberulum, P.martensii, P. palitans*, and *P. cyclopium* are the major producers of PA [57, 61], but it is also produced by a few *Aspergillius* spp. such as *A. ochraceus* [62].

The LD50s for PA for rats via subcutaneous, intravenous and oral exposure were determined to be 110, 250 and 600 mg/kg, respectively [63]. PA has been shown to be hepatotoxic and hemorrhagic in dogs [64], and potential carcinogenicity was reported in early rat and mice PA studies [14]. Similar to CIT, PA has been reported to be nephrotoxic and hepatotoxic to mice, and its toxic effects can be synergistically enhanced when mixed with OTA [13].

1.2.5. Mycophenolic Acid (MPA)

MPA is a common contaminant found in ensilaged and grain feeds [65], as well as some fermented dairy products [66]. MPA is produced by *P. stoloniferum, P. brevicompactum, P.roqueforti* and *P. carneurn* [67–69].

LD50s for MPA by oral exposure were determined to be 352 mg/kg for the rat, 1000 mg/kg for the mouse, and over 6000 mg/kg for the rabbit [70], which indicates that MPA has relatively low toxicity in animals compared to other *Penicillium* mycotoxins. Although low in toxicity, one of the most distinctive characteristics of MPA is its anti-proliferative effect on specific cell types. MPA has been shown to inhibit the proliferation of smooth muscle cells [71], endothelial cells [72], as well as some immune cells, such as lymphocytes [73, 74]. Ironically, MPA's low toxicity and immunomodulatory properties have resulted in many human pharmaceutical studies, but few animal studies have been carried out despite the fact that it is frequently found at high concentrations in ensiled animal feed [9]. MPA has been used therapeutically for the treatment of arteriosclerosis and rheumatoid arthritis [72, 75], and to prevent organ rejection following kidney and heart transplantation [76].

1.2.6. Roquefortine C (RQC)

RQC is a major mycotoxin produced from *P.Roqueforti*, which is used to manufacture 'Roquefort' or 'blue' cheese. Other *Penicillium* spp. found in blue cheese or milk, including *P commune, P crustosum, P. chrysogenum, P. cyclopium* and *P. glabrum* are also capable of producing RQC [77–79]. The presence of RQC is very common in blue cheese products, yet it is considered to be safe for human consumption due to the low exposure range (0.05 - 10 mg/kg food) and its low toxicity [80]. However, RQC is found at relatively higher concentrations (up to 17 mg/kg) in moldy animal feed grains and silage, especially in the maize-based silage [77, 81, 82].

Despite its relatively low toxicity, it has been reported that natural RQC exposure concentrations may be high enough to cause potential toxicity to both monogastrics [83] and ruminants [84]. The LD50 for RQC in the mouse by intraperitoneal exposure was previously estimated to be 20 mg/kg [85], but Tiwary et al. (2009) later re-determined the LD50 via intraperitoneal exposure to be between 169-184 mg/kg for mice [86]. The pathology of RQC toxicity includes paralysis, trembling and lethargy in canine and murine species [78, 79, 87]. There are no reports of lethality in ruminant species, and a single study has reported various clinical signs, including paresis, indigestion, diarrhea as well as increased rate of mastitis and abortion for dairy cows fed 25 mg/kg of RQC [84].

1.3. Effect of *Penicillium* Mycotoxins on Physical Barriers

Epithelial cells and commensal microbes create an important physical barrier at mucosal surfaces that protect the host against pathogenic microbes and toxins. Epithelial cells form a mechanical and bioactive barrier providing attachment surfaces for commensal microbes, while the commensal microbes maintain a symbiotic relationship with the host to construct a protective barrier between the outside gut environment and host immune system [88, 89]. *Penicillium* mycotoxins could potentially disrupt these physical barriers by altering commensal microbial populations and destabilizing the epithelial cell barrier.

1.3.1. Disruption of Commensal Microbial Populations

Hundreds of different microbial species reside within the GIT of monogastrics and ruminants where they maintain a symbiotic relationship with the host. These commensal microbes form a barrier on epithelial surfaces preventing attachment and colonization of pathogenic microbes [90]. Likewise, the biodiversity of these commensal populations is crucial for maintaining gut homeostasis, integrity and normal function of immune system [91, 92] (Figure 1A). For example, colonization of the gut lumen with *Escherichia coli*, *Klebsiella pnemoniae*, and *Streptococcus viridans* increases paracellular permeability, whereas, *Lactobacillus brevis* has the opposite effect [93]. It is hypothesized that when commensal microbial populations are altered, protective and homeostatic interaction between host and commensals is impaired, and this can lead to disease, such as inflammatory bowel disease [94, 95].

The biodiversity of commensal microbiota is especially important for the development and maturation of the neonatal immune system. Rapid colonization of the neonatal gut by commensal bacteria for example, leads to hypertrophy of Peyer's patches, and attenuates the responsiveness of B cells to produce secretory IgA against commensal microbes [96]. Furthermore, specific commensal bacterial species promote the development of mucous-associated lymphoid tissues [97], as well as immune system maturation to facilitate immunological tolerance to mucosal commensals and effector function against pathogenic microbes [98]. Interestingly, both delayed and inappropriate colonization of mucosal surfaces appears to be a risk factor for neonatal disease [98].

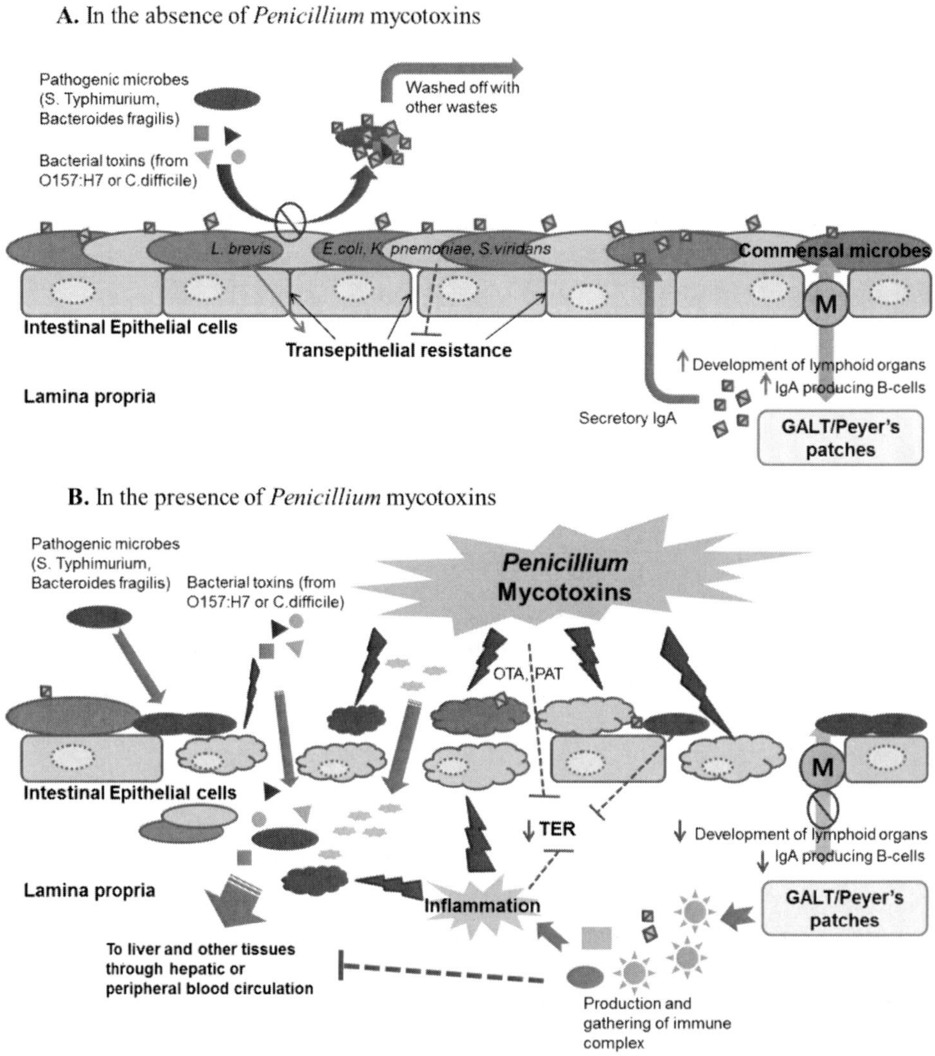

Figure 1. Simple schematic diagram of how the antimicrobial properties of *Penicillium* mycotoxins could disrupt the homeostasis of intestinal immune system. **(A)** In the absence of *Penicillium* mycotoxins, commensal microbes and secretory IgA act together to create physical barrier to protect the epithelial cells against the colonization of pathogenic microbes, such as E.coli O157:H7, *C. difficile* and *S. typhimurium*, as well as endotoxins [90, 101, 102]. Furthermore, rapid colonization of commensal bacteria leads to hypertrophy of Peyer's patches, but attenuates the responsiveness of B cells to produce secretory IgA [96]. Specific commensal bacteria species also promote the development of gut-associated lymphoid tissues (GALT) [97]. **(B)** In the presence of *Penicillium* mycotoxins, some *Penicillium* mycotoxins, such as CIT, PAT, MPA, PA and RQC, possess antimicrobial properties [23, 24], which could disrupt commensal biodiversity and population numbers. Disruption of the commensal microbes could compromise the protection against the pathogenic microbes [90, 101] and endotoxins [102, 103]. Development of GALT and immune tolerance could also be compromised [96, 97].

Several *Penicillium* mycotoxins, such as CIT, PAT, MPA, PA and RQC, possess antimicrobial properties [23, 24] that could disrupt commensal biodiversity and population numbers (Figure 1B). For example, CIT has an antimicrobial effect against several pathogenic strains of bacteria and fungi [99], while PA has antimicrobial activities against Gram-positive bacterial species [23]. The antimicrobial properties of *Penicillium* mycotoxins are probably an evolutionary strategy by which *Penicillium* molds protect themselves against other microbes and reduce competition with other molds within their surrounding environment. Although it is generally assumed that many *Penicillium* mycotoxins are rapidly metabolized by commensal microbes [21, 22, 100], their antimicrobial properties could possibly attenuate microbial detoxifying mechanisms. Furthermore, *Penicillium* mycotoxins may also disrupt the normal colonization or maintenance of commensal microbes, which could compromise host protection against pathogenic microbes [90, 101] and toxins [102, 103]. In support of this hypothesis, enteropathogenic *Salmonella Typhimurium* was shown to colonize the murine intestine more efficiently when it was not competing with commensal microbiota [101]. Additionally, when commensal microbes are disrupted, bacterial toxins from *E. coli* O157:H7 [103] and endotoxins from *Clostridium difficile* [102] can cause systemic pathology.

1.3.2. Physical Disruption of Epithelial Surfaces

Penicillium mycotoxins can also directly disrupt the integrity of epithelial surfaces increasing risk of secondary disease. For example, Maresca et al. (2001) reported a significant reduction in the transepithelial resistance (TER) of HT-29-D4 and Caco-2 cells, which is an indicator of epithelial barrier damage, at OTA concentrations greater than 1 μM [104]. Another study also observed that PAT destabilized epithelial cell membrane cysteine thiol and sulfhydryl groups that compromised HT-29-D4 and Caco-2 cell integrity and reduced TER [105]. This epithelial barrier damage increases potential uptake of foreign substances such as *Penicillium* mycotoxins and pathogenic microbes into the host systemic circulation, which could predispose animals to disease [106].

2. *PENICILLIUM* MYOTOXINS: IMMUNOMODULATION AND IMMUNOTOXICITY

In general, the paradigm in immunology categorises the immune system into two branches: innate and adaptive immunity for ease of understanding [107], even though some common cells and molecules are involved in both types of immunity. Innate immunity is characterized as the first-line of defence during microbial infection after physical barriers have been penetrated.

It generally involves immune surveillance, inflammation and immune regulation, all of which are considered non-specific immune responses. Innate immunity is triggered by the recognition of pathogens by humoral and cellular pattern recognition receptors (PRRs) that recognize pathogen-associated molecular patterns (PAMP) with limited specificity and affinity.

Table 1. The occurrence of *Penicillium* mycotoxins in animal feed and food products

Mycotoxins	Location or source: natural condition (nc), artificial condition (ac)	Sample material	Method of analysis	Mean concentration	Converted concentration (mg/kg)	References
OTA	Mexico, conditions created (nc)	corn silage	ELISA	5.1 ppb	0.0051	[226]
	All regions from Portugal (nc)	swine feed	HPLC	11.0 μg/kg	0.011	[227]
		laying hen feed		4.0 μg/kg	0.0040	
CIT	Calvados, France (nc)	corn silage	HPLC-MS	15.0 ppb	0.015	[228]
	Normandy, France (nc)			32.0 μg/kg	0.032	[229]
	Local store, Portugal (nc)	rotten apples	TLC	0.55 mg/kg	0.55	[47]
PAT	Pennsylvania, USA (nc)	maize silage	HPLC-MS	0.08 μg/g	0.080	[9]
	Germany (ac)		HPLC; 2D-TLC	15.10 mg/kg	15.10	[58]
	Local store, Protugal (nc)	rotten apples	TLC	20.45 mg/kg	20.45	[47]
	Local store, Japan (nc)	apple juice	LC-MS	4.74 ng/g	0.0050	[53]
PA	Germany (ac)	maize silage	HPLC; 2D-TLC	3.06 mg/kg	3.06	[58]
MPA	Pennsylvania, USA (nc)	maize silage	HPLC-MS	0.16 μg/g	0.16	[9]
	Bavaria, Germany (nc)		LC-MS	690 μg/kg	0.69	[230]
		grass silage		2200 μg/kg	2.20	
	Germany (ac)	maize silage	HPLC; 2D-TLC	3.56 mg/kg	3.56	[58]
RQC	Pennsylvania, USA (nc)	maize silage	HPLC-MS	0.38 μg/g	0.38	[9]
	Northern Germany (nc)		HPLC	17.0 mg/kg	17.00	[81]
	Tochigi prefecture, Japan (ac)		TLC	1.66 mg/kg	1.66	[82]
	Local store, Germany (nc)	Blue cheese	HPLC	1.47 mg/kg	1.47	[80]

Table 2. Brief toxicological information of *Penicillium* mycotoxins other than immunotoxicity (all references provided in the text)

Mycotoxins	*Penicillium* spp. Producing mycotoxins	Potential antimicrobial property	Previous reported toxicological signs other than immunotoxicity	LD50s intraperitoneal (ip), intravenous (iv), subcutaeous (sc), oral (po)
OTA	*P. citrinin* *P. verrucosum*	No	nephrotoxic (rats, pigs, chicken) hepatotoxic (goat, pigs, chicken) teratogenic (rats)	ip 29.4 mg/kg (mice) po 46.0 mg/kg (mice)
CIT	*P. citrinin* *P. verrucosum* *P. expansum*	Yes	nephrotoxic (pigs) teratogenic (rats) genotoxic (rats)	po 50 mg/kg (mice) ip 35-58 mg/kg (mice) ip 19 mg/kg (rabbits)
PAT	*P. expansum* *P. urticae* *P. melinii* *P. claviforme*	Yes	haemorrhagic syndrome (rats, cattles) teratogenicity/mutagenicity (mice) hepatotoxic (rats) destruction of various cell types (rats)	po 17 mg/kg (mice) po 108-118 mg/kg (rats)
PA	*P. puberulum* *P. martensii* *P. palitans* *P.cyclopium*	Yes	hepatotoxic (dogs, rats, mice) nephorotoxic (rats, mice) haemorrhagic syndrome (dogs) carcinogenic (rats, mice)	sc 110 mg/kg (rats) iv 250 mg/kg (rats) po 600 mg/kg (rats)
MPA	*P. stoloniferm* *P. brevicompactum* *P. roqueforti* *P. carneurn*	Yes	No report of other significant toxic effects than immunotoxicity	po 352 mg/kg (rats) po 1000 mg/kg (mice) po 6000 mg/kg (rabbits)
RQC	*P. roqueforti* *P commune* *P crustosum* *P chrysogenum* *P cyclopium*	Yes	mastitis, abortion, diarrhea (cattles) paralysis, trembling, fatigue (dogs, mice)	ip 20 mg/kg (mice) ip re-determined 169-184 mg/kg (mice)

Table 3. Summary of immunomodulatory effect of *Penicillium* mycotoxins (all references provided in the text; NA-not available)

	OTA	CIT	PAT	PA	MPA	RQC
Target leukocytes	Neutrophils Basophils Monocytes Macrophages NK cells Lymphocytes	Macrophages Lymphocytes	Monocytes Macrophages DCs CTLs	Macrophages Lymphocytes	Macrophages DCs NK cells activated T-cells naïve B-cells	Macrophages Neutrophils
Proliferation	↓ (*in vitro* and *in vivo*)	↓ (*in vitro*) ↑ (*in vivo*; mitogenic activity)	↓ (*in vitro* and *in vivo*)	↓ (*in vitro* and *in vivo*)	↓ (*in vitro* and *in vivo*)	↓ (*in vitro*)
Cell death	↑ caspase 3/9 apoptotic pathways: activated through TRADD ↑ necrosis due to oxidative stress	↑ caspase 3/9 apoptotic pathways ↑ necrosis mediated by oxidative stress due to ↓GSH-based antioxidants		↓ caspase 3/8/9 mediated apoptotic pathways ↑ necrosis due to oxidative stress at concentration higher than 100 μM (*in vitro*)	Severely affect the cells (activated T-cells & naïve B-cells) utilizing type II IMPDH for GTP synthesis ↑ caspase 3/9 apoptotic pathways ↑ necrosis due to ↓ GTP/GDP synthesis	NA
Inflammation	↑ NF-κβ and MAPK Pathways ↑ COX-2, iNOS, hsp72, TNF-α	↓ NF-κβ and MAPK Pathways ↓ IL-1β, IL-6, TNF-α, iNOS, NO	↓ IL-1β, IL-6, TNF-α, –β, IL-10, ROS production	NA	↓ MAPK pathways ↓ ROS production ↑ trafficking of DCs to inflammatory sites by ↑ CCR-1 ↑ IL-10, TGF-β	May act as an irritant ↑ MIP-2, TNF- α in BALF
Cell-mediated immune response	↓ IFN-α, -β, -γ ↓ phagocytosis, anti-tumor activity of NK cells, basophil hypersensitivity, delayed type hypersensivitiy	↓ IFN-γ ↑ infectivity of intracellular pathogens (*T. gondii*)	↓ IFN-γ, IL-12, IL-2 ↓ phagocytosis and microbicidal activity to *S. cerevisiae*	NA	↓ activation of NK cells, DCs, T-cells ↑ T-reg cell number ↑ intracellular infection	NA
Antibody-mediated immune response	↓ IgM and IgG (due to ↓ cell number)	↓ IL-4 (less sensitive compared to IFN-γ) ↓ antibody response to antigens (sheep RBC, ovalbumin)	↓ IL-4 (less sensitive compared to IFN-γ) ↓ IgM, IgA, IgG ↑ IgG1/IgG2 ratio	NA	↓ activation B-cells ↓ IgM, IgA, IgG ↑ IL-5 No effect on plasma cells	NA

Adaptive immunity, on the other hand, recognises and targets pathogens with higher specificity and affinity. In this case, pathogen antigen recognition is mediated by either T-cell receptors on T-lymphocytes, or membrane-bound immunoglobulins on B-lymphocytes and antibodies secreted by plasma cells. Adaptive immunity is most often distinguishable from innate immunity by its ability to induce a memory response that is characterized as having a more rapid response with higher specificity and affinity during subsequent pathogen exposure.

Penicillium mycotoxins can potentially modulate molecules and cells involved in both innate and adaptive immune responses in many different ways (Table 3). For example, most of the *Penicillium* mycotoxins that will be discussed in this review affect leukocyte cell numbers that could be due to cell death and changes in cell proliferation, and cell differentiation and trafficking throughout the body. Furthermore, some of these *Penicillium* mycotoxins may also alter the function and expression of cytokines, inflammatory molecules, as well as the expression of immunoglobulins, which will be described in more detail in the following sections. *Penicillium* mycotoxins have also been reported to alter the morphological structure and function of lymphoid organs such as thymus, spleen and lymph nodes, which are important for the development, and maintenance, and activation of adaptive immune system.

2.1. Ochratoxin A (OTA)

2.1.1. Leukocyte Number

OTA is a potent immunomodulatory mycotoxin that can reduce circulating numbers of leukocytes leading to leukopenia [39, 108]. For example, OTA decreased the total number of blood monocytes and lymphocytes in a dose-dependent manner in broiler chickens [108], and a significant decrease in the number of blood neutrophils, basophils, and monocytes was also observed in mice fed 5 mg/kg of OTA in feed [109]. This decrease in the blood leukocyte number was partially due to the reduced production of leukocytes in lymphoid organs, as OTA targets the primary lymphoid organ, such as bone marrow, thymus and bursa of Fabricius, as well as secondary lymphoid organs, including spleen and lymph node of various species [39, 43, 110, 111]. For instance, OTA decreased the number of bone marrow pluripotent stem cells that are required for generating leukocytes such as granulocyte macrophage progenitors, as well as both $CD4^+$ and $CD8^+$ T-cell subpopulations in spleen of mice [110, 112]. OTA reduces proliferation and viability of these immune cells, leading to the overall decrease in the number of blood and tissue leukocytes [113]. In support of this, OTA prevented the lymphoblastogenesis of porcine and human lymphocytes [114, 115]. Charoenpornsook et al. (1998) also reported that 1.5 µl/ml of OTA inhibited concanavalin A (Con A)-, phytohaemagglutinin (PHA)- and pokeweed mitogen (PWM)-induced bovine peripheral blood mononuclear cell (PBMC) proliferation by 50% (IC50) [113], and similar IC50s have been reported for bovine lymphocytes [73, 116]. Lea et al. (1989) looked more specifically at the effect of OTA on B cell proliferation, which was induced with polyclonal rabbit anti-human IgM antibodies in the presence of bovine B-cell growth factor (BCGF-1) [114]; they reported a significant inhibitory effect of OTA (50 µM) on B-cell proliferation. We have also observed an inhibitory effect of OTA on the proliferation of a bovine macrophage cell line (BoMacs) at concentration of 5.2 µM and higher [74]. In contrast, OTA concentrations between 0.5 and 20 μM failed to inhibit the proliferation of murine lymphocytes induced by either Con A or lipopolysaccharide (LPS) [112]. These contradictory results may be due to species variation.

2.1.2. Inflammatory Response and Cell Death

Several studies have demonstrated the OTA as a potential inducer of inflammatory molecule production by macrophages. Ferrante et al. (2006, 2008) for example, observed a significant increase of pro-inflammatory molecules such as cyclooxygenase-2 (COX-2), inducible nitric oxide synthase (iNOS) and heat shock protein 72 (hsp72) from murine macrophages and the macrophage cell line J774A [19, 117]. These inflammatory molecules, especially COX-2, stimulate macrophage tumor necrosis factor (TNF)-α production, which is a potent pro-inflammatory cytokine [118]. These studies have also found a strong positive association between the OTA-induced pro-inflammatory mediators and the activity of nuclear factor (NF)-κB, a transcription factor that stimulates a cellular pro-inflammatory response [19, 118, 119]. For example, OTA induced p65-NF-κβ protein expression in the macrophage cell line [118], and these authors also showed that inhibition of NF-κβ blocked TNF-α release that was induced by OTA from perfused rat liver. Ferrante et al. (2008) later demonstrated that OTA induced the degradation of cytosolic inhibitor of NF-κβ (Iκβ-α) [19], which allows NF-κβ for translocation into nucleus and increased expression of pro-inflammatory molecules (Figure 2). In addition to this, TNF-α also activates the Jun N-terminal kinase (JNK)-p38 mitogen-activated-protein kinase (MAPK) signaling pathway [120], which further enhances the pro-inflammatory response [121], and this was also previously observed with both canine renal epithelial cells and rat proximal tubular cells [122, 123].

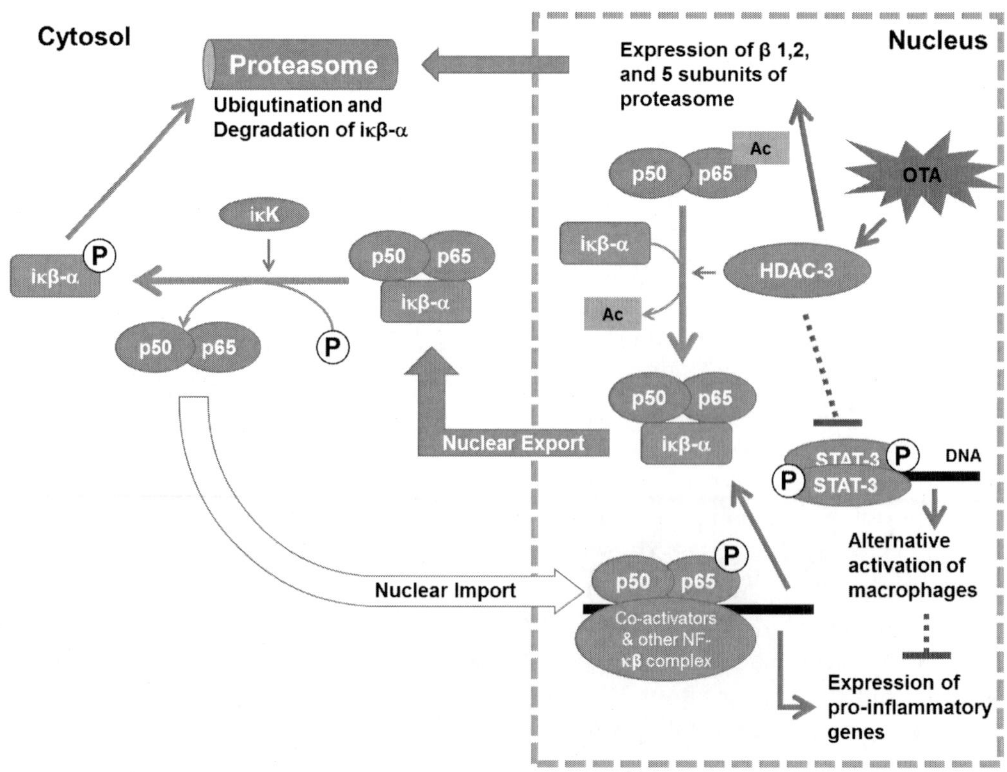

Figure 2. The Schematic diagram of how *Penicillium* mycotoxins alter inflammatory response of macrophages by altering HDAC-3 [213]. Solid arrows represent 'stimulation,' and dotted lines indicate 'inhibition.' OTA induces the expression by favoring HDAC-3 activity [118, 209]. HDAC-3 induces the export of NF-κβ molecules out of nucleus, which then reduces the inflammatory response [212]. However, it also induces the synthesis of catalytic part of proteasome. The increased proteasome catalytic activity eventually increases the degradation of Iκβ-α, leading to nuclear translocation of NF-κβ into nucleus [19, 214] and induction of the pro-inflammatory response. The deacetylation of lysine residues in STAT-3 by Type-1 HDAC, such as HDAC-3, destabilizes the dimerization of STAT-3 [218]. The inhibition of HDACs on the other hand, stabilizes STAT-3 dimers, leading to alternative activation of macrophages that have anti-inflammatory phenotypes [217].

Increased TNF-α production by OTA not only stimulates pro-inflammatory response, but also affects cell viability. Interaction between TNF-α and membrane-bound TNF-α receptors-1 (TNFR-1) for example, activates TNF receptor-1-associated death domain (TRADD), which then triggers caspase-dependent apoptosis as indicated by a decrease in Bcl:Bax family ratio, the activation of caspases, and the release of mitochondrial protein cytochrome C [124, 125]. It has been proposed that OTA reduces the expression of anti-apoptotic protein Bcl-x_L, which results in the release of cytochrome C as well as the activation of caspase-3 and -9. These activated caspases trigger a loss in mitochondrial trans-membrane potential leading to apoptotic cell death [126–128]. Activation of the JNK-p38-MAPK pathway by TNF-α further enhances the cell death process [120, 122]. Lastly, OTA was also shown to induce apoptosis, independent of interaction between TNF-α and TNFR-1 in hepatocytes, which appears to be due to intracellular oxidative stress as indicated by increased production of H_2O_2, which then triggers apoptosis [129, 130].

2.1.3. Effect on Adaptive Immunity

In addition to stimulating the pro-inflammatory response and death of leukocytes, OTA also modulates other immune cell functions, specifically down-regulating the cell-mediated immune response (CMIR). Harvey et al. (1992) reported that feeding 2.5 mg of OTA to growing gilts during a one-month period decreased the CMIR, including macrophage phagocytic activity, cutaneous basophil hypersensitivity, as well as delay-type hypersensitivity response [115].

The decreased CMIR may be partially attributed to the decreased total number of T-cells, since OTA inhibits T-cell proliferation and viability [73, 108, 112]. However, several other studies also suggested that the selective suppression of the CMIR by OTA exposure might have also occurred via the inhibition of both type-I and type-II interferon (IFN) production [111, 131]. The type-I IFNs IFN-α and IFN-β are potent activators of CMIR, and their suppression decreases NK cell anti-viral and anti-tumor activity [132]. Type-2 IFN-γ is a crucial cytokine for promoting cytotoxic T-cells (CTLs) and macrophage activation [133, 134]. Luster et al. (1987) reported that oral exposure of mice to OTA at doses between 3.4 to 13.4 mg/kg decreased the serum concentration of total IFNs in a dose-dependent manner [131]. They also observed selective inhibition of natural killer (NK) cell anti-tumor activity against murine mastocytoma P815 tumor cells, but not anti-tumor activity of either blood CTLs or macrophages. This finding indicates that OTA may specifically inhibit the production of type-I IFNs in mice. However, a much lower concentration of 0.25 mg/kg OTA, when co-exposed with T-2 toxin, led to a significant decrease in serum IFN-γ concentration as well as a slight reduction in the splenic gene expression of IFN-γ in broiler chickens [111]. Furthermore, *in vitro* exposure of OTA significantly inhibited not only the cytotoxic activity of NK cells but also the cytotoxic activity of CTLs against mouse lymphoma Yac-1 target cells [112]. Oral administration of 0.05-0.45 mg/kg of OTA to male rats also decreased the cytotoxic activity of CTL and splenocytes against sheep red blood cells (RBC) [135]. Based on these studies, it can be suggested that OTA suppresses not only type-I IFN, but may also suppress IFN-γ that is important for stimulating the cytotoxic activity of CTL [134].

Some studies have reported that OTA also modulates the antibody-mediated immune response (AbMIR). For example, intraperitoneal injection of 1 μg/kg of OTA significantly decreased sheep RBC-specific IgG and IgM responses by 90% in mice [136]. This decrease may be attributed to a decrease in the total number of B-cells, since OTA was shown to decrease lymphocyte numbers by inhibiting cell proliferation as discussed earlier [114]. In contrast to the previous study, the oral administration of 2.5 mg/kg OTA to pigs did not affect total immunoglobulin concentrations, or the specific AbMIR to chicken RBCs [115]. It is possible that intraperitoneal injection of OTA, used in the Creppy et al. (1983) murine study, might have bypassed microbial and hepatic metabolism [136], which were previously shown to detoxify mycotoxins [50, 137]

2.2. Citrinin (CIT)

2.2.1. Leukocyte Number

Similar to OTA, CIT has significant effects on leukocyte numbers. For example, CIT exposure reduced the proliferation of porcine and bovine T-lymphocytes, *in vitro* [73, 116]. CIT doses of 5 or 10 mg/kg BW via oral gavage also decreased the number of murine splenic macrophages, $CD8^+$ CTLs and $CD4^+$ T-helper cells, while the total number of B cells was reduced in intestinal intra-epithelium, lamina propria and Peyer's patches of BALB/c mice [138]. In contrast, these authors also observed that the CIT exposure increased the number of CTLs, T-helper cells, and $CD4^+CD25^+Foxp3^+$ regulatory T-cells in mesenteric lymph nodes, as well as intestinal CTLs in the intra-epithelium, lamina propria and Peyer's patches of mice. Another *in vivo* study reported similar results, where they observed T-cell blastogenesis was stimulated in the spleens of the mice that received CIT at the doses greater than 0.12 mg/kg by ip injection [139]. Islam et al. (2012) suggested that the increase in T-cell population may be due to CIT acting as a T-cell-specific mitogen, activating NF-κβ or MAPK. However, they also stated that CIT could also stimulate the migration of T-cells from spleen to other immune tissues, which would explain the increased T-cell numbers that they observed in secondary lymphoid organs, and reduced numbers of T-cells in the spleen. This seems to be a suitable explanation for the *in vivo* studies; however, it does not explain the suppressed T-cell proliferation observed *in vitro*, since CIT should have induced the T-cell proliferation if CIT was indeed acting as a T-cell mitogen. Therefore, there seems to be more complex interaction among T-cells and other cell types when animals are exposed to CIT as compared to *in vitro* exposure. In support of this, Dziarski (1985) previously reported a higher murine lymphocyte mitogenic response *in vivo* than *in vitro* following exposure to various types of mitogens [140]. Furthermore, the effect observed in the *in vivo* studies may not be due to CIT in its original form, but rather the effect of its metabolites. CIT is metabolized in the liver and possibly other organs, and these metabolites may have a different potency and toxic effects in animals [50, 137]. It is also likely that the mitogenic or inhibitory response is dose-dependent, where a low concentration of CIT could have resulted in hormetic effect *in vivo*. Hormesis is a biphasic concentration-response relationship manifesting as a stimulating response at low concentrations, followed by an inhibitory effect at high concentrations [141].

2.2.2. Oxidative Stress and Cell Death

Regardless of potential mitogenic activity of CIT on lymphocyte proliferation, CIT can also decrease total leukocyte numbers by inducing cell death. Similar to OTA, CIT triggers the apoptosis of leukocytes by inducing caspase pathways [142], and this is probably initiated by increased intracellular oxidative stress.

CIT induces oxidative stress by decreasing the level of intracellular anti-oxidant, glutathione (GSH) [142], and this in turn reduces other GSH-dependent anti-oxidant enzymes such as GSH reductase (GSSG) and trans-hydrogenases [143]. Intracellular glutathione peroxidase (GSH-Px) functions as an anti-oxidant enzyme that inhibits the production of reactive oxygen species (ROS) molecules, such as superoxide anion and hydrogen peroxide [144]. Oxidative stress subsequently stimulates the production of superoxide anion and ROS via the mitochondrial respiratory chain pathway [145], while at the same time arrests the G0 to G1 phase of cell cycle, reduces Bcl-2/Bax activity ratio, and induces cytochrome C release from mitochondria; all of which eventually stimulate and sensitize cells to the p53-mediated caspase-3 and -9 apoptotic death signal [138, 142, 146]. CIT has also been shown to reduce the gene and protein expression of inflammatory interleukin (IL)-1β, IL-6 and TNF-α in human Mono-Mac-6 cells [147] and murine macrophages [138]. In contrast, Liu et al. (2010) observed CIT inhibited the Janus kinase (JAK)/signal transducers and activators of transcription (STAT)-1α and NF-κB signaling pathways, leading to the inhibition of iNOS and nitric oxide (NO) production in a glomerular mesangial cell line in response to LPS and IFN-γ exposure [148]. Therefore, CIT appears to induce leukocyte oxidative stress [142, 143] independent of NO production, which in turn activates the intrinsic apoptosis pathway that involves stimulation of p53-mediated caspase-3 and caspase-9 [138, 142, 149].

2.2.3. Effect on Adaptive Immunity

CIT appears to affect the balance of T helper cell type (Th)1 and Th2 cytokines that support CMIR and AbMIR, respectively. An *in vitro* exposure study for example, reported that CIT suppressed both IFN-γ and IL-4 production from CD3- and CD28-stimulated human blood mononuclear cells in a dose-dependent manner [150]; the IC50 for IL-4 cytokine production was 21.6 μg/ml CIT, however, it was only 8.3 μg/ml CIT for IFN-γ production. This suggests CIT affects IFN-γ production more severely than IL-4 production thereby favouring AbMIR over CMIR at low exposure concentrations. Taken together, the suppression of iNOS and IFN-γ production by CIT has the potential to compromise host defence against intracellular pathogens [148]. In support of this, Herzog-Soares and Freire (2004) demonstrated that when murine macrophages were pre-exposed to 10 μg/ml of CIT, the infectivity of the cells by *Toxoplasma gondii* tachyzoites (*T. gondii*) increased by 18%, and also enhanced the replication of the tachyzoites within the cells [20].

At higher concentrations, CIT also appears to suppress AbMIR, since CIT was shown to decrease total B-cell number as well as IL-4 production at higher concentrations [138, 150]. Furthermore, 2.5 mg/kg of CIT by ip injection decreased both sheep RBC-specific antibody and complement activity by 75-87.5% and 87.5-93.8%, respectively, in albino mice [45]. Similar results were also observed in Wistar rats that were fed 10 mg/kg of CIT when sheep RBC- and ovalbumin-specific antibody responses were measured [151].

2.3. Patulin (PAT)

2.3.1. Leukocyte Number

PAT is a potent immunotoxin that can severely alter the number of leukocytes in blood and tissues, and modulate host immune function. *In vitro* studies have demonstrated that PAT reduces the proliferation and viability of leukocytes, specifically T-cells and macrophages [73, 74, 152]. For example, Keblys et al. (2004) observed dose-dependent decrease in the proliferation of blood lymphocytes when porcine T-cells were exposed to PAT *in vitro* [73].

In vivo studies have also demonstrated similar effects on the number of blood leukocytes from mice exposed orally to PAT, by both acute and chronic exposure. Escoula et al. (1988) for example, reported a significant decrease in total blood lymphocyte numbers in Swiss mice fed 10 mg/kg of PAT for 5 to 10 days [153]. A similar observation was also reported with B6C3F mice that consumed PAT at concentrations greater than 1.28 mg/kg for 28 days, where PAT decreased the number of total blood leukocytes by 30% [154]. These authors also reported that PAT increased the number of certain subsets of immune cells at a predicted human exposure level of 0.08 – 2.56 mg/kg. For example, PAT after a 28-day exposure period, increased the number of splenic CTLs, NK cells and monocytes by 50%, 30% and 24%, respectively, while the number of $CD3^{+}$T-cells, $CD4^{+}$T-cells, and MAC-3^{+} monocytes remained unchanged [154, 153]. Similar to the 28-day chronic exposure study, acute 4-day exposure of mice to 10 mg/kg PAT via stomach tubing, and ip injection of 2.5 mg/kg PAT in rabbits did not affect the number of either splenic neutrophils or CTLs, but increased the number of splenic T-cells even though the mitogenic response of T-cells and the number of splenic B cells were suppressed [153]. Furthermore, another *in vivo* study reported that 0.1 mg/kg of PAT for 90 days significantly increased apoptosis of rat thymic dendritic cells (DCs) that are required for normal T-cell development [155].

2.3.2. Oxidative Stress and Cell Death

In addition to the above reported cytotoxic effects of PAT on rat thymic DCs [155], *in vitro* studies have also shown PAT-induced cytotoxicity to rat alveolar macrophages [152], and a bovine macrophage cell line (BoMacs) [74]. Similar to OTA and CIT, PAT appears to induce p53-mediated apoptotic cell death triggered by caspase-3 and -9 pathways, which was observed in both murine skin cells [156], and human leukemia HL60 cells [157]. PAT acted as a strong inhibitor of inflammatory cytokines, suppressing the production of IL-1β, IL-6 as well as TNF-α and TNF-β from human mononuclear cells and T-cells [158]. However, PAT also induced ROS production leading to intracellular oxidative stress, which then stimulated the caspase pathways to induce cell apoptosis [144, 159].

The mechanism by which PAT induces oxidative stress appears to resemble that of CIT, where both mycotoxins have been shown to inhibit anti-oxidant enzymes, specifically targeting the intracellular GSH-dependent enzymes, such as GSSG [143, 144, 158]. Although it is still unclear for CIT, it appears that PAT destabilizes the disulfide bonding of cysteine within GSH [94, 144]. PAT-induced ROS production damages DNA which arrests the cell cycle and eventually initiates caspase apoptotic death pathway by inducing Bax pro-apoptotic protein expression as well as p21 and p53 [156, 157].

2.3.3. Effect on Adaptive Immunity

Intracellular GSH levels have also been explored as a marker to determine the polarization of DCs, with reduced GSH levels being a marker of DCs of the Th2 phenotype that support AbMIR [160]. Studies have reported that PAT decreased intracellular GSH levels in human PBMCs and T-cells as well as murine DCs [94, 158]. These cells, at the same time, expressed a Th2 dominant phenotype in terms of their cytokine production. These studies have also shown that PAT decreased the production of both IFN-γ and IL-12. These cytokines are crucial for promoting a Th1 response, and stimulate the CMIR. Impaired production of these cytokines could increase host susceptibility to intracellular pathogens, such as *T. gondii* [161] or *Mycobacterium* [95]. In support of this, PAT at a concentration greater than 0.5 μg/ml severely suppressed the phagocytic and microbicidal activity of murine peritoneal macrophages challenged *in vitro* with yeast *Saccharomyces cerevisiae* [162].

In contrast to its effect on the CMIR, PAT had less effect on the expression of Th2 cytokines, such as IL-5 and IL-13, by murine splenocytes and mediastinal lymph node cells [94]. The production of IL-4 by human PBMCs on the other hand, was inhibited at 0.050 μg/ml of PAT [158], but this study also observed greater suppression of IFN-γ production as compared to that of IL-4. In fact, a low concentration of 0.003 μg/ml PAT, rather increased human PBMC IL-4 production, but suppressed IFN-γ production suggesting a possible quadratic effect of PAT on IL-4 production [150]. PAT also suppressed the production of other cytokines, including IL-2 and IL-10, all of which are important for the survival and regulation of leukocyte function [163, 164].

Similar to CIT, PAT at high doses may also suppress the AbMIR by down-regulating plasma cell (PC) antibody production. For example, intake of 10 mg/kg of PAT by oral gavage and ip administration of 2.5 mg/kg dramatically decreased total serum IgG, IgA and IgM levels without altering levels of serum albumin and complement component C3 [153]. The reduction of these immunoglobulins may be attributed to B-cell apoptosis. However, other studies using lower concentrations of PAT reported that PAT did not necessarily decrease immunoglobulin production. In fact, 28 days of oral exposure to 0.08 – 2.56 mg/kg of PAT, unlike CIT, did not alter the serum IgM response to sheep RBC in B6C3F1 mice [45, 154]. Furthermore, oral and nasal exposure to 5 – 20 μg of PAT increased ovalbumin-specific IgE and IgG1 in BALF/C mice without altering IgG2a levels [94]. The increased ratio of IgG1/IgG2a is also an indication of a Th2 response, suggesting that PAT shifts the immune response toward AbMIR [94].

2.4. Penicillic Acid (PA)

2.4.1. Leukocyte Number

There are only few studies available on the subject of PA immunotoxicity. PA may not necessarily alter immune function, yet it was shown to reduce the proliferation and viability of immune cells, decreasing the overall number of immune cells *in vivo*. For example, feeding a diet containing 2 mg/kg of PA to broiler chicks led to a severe decrease in the weight of both primary (thymus and bursa of Fabricus) and secondary lymphoid organs (spleen), and this effect was amplified by co-exposure with OTA [43]. A reduction in lymphoid organ size may be due to suppressed proliferation of immune cells such as macrophages and lymphocytes rather than the reduced viability, because a much higher PA concentration is required to induce cell death [43, 73, 74]. Stoev et al. (2004) also observed a concentration-dependent decrease in the proliferation of poultry PBMCs, but they were unable to reproduce these results using human PBMCs in a later study [165]. The IC50s for PA based on cell proliferation have been determined to be 18.0 and 30.0 μM for porcine T-cells and bovine macrophages, respectively [73, 74].

2.4.2. Cell Death via Necrosis

PA can also decrease the viability of immune cells, yet this occurs at much higher concentrations. Sorenson and Simpson (1986) for example, reported that concentrations of PA greater than 1.0 mM caused significant cytotoxicity to rat alveolar macrophages after 2 hours of exposure [166]. PA concentrations greater than 160 μM also significantly decreased the viability of bovine macrophages when exposed for 24 hours, and the LC50 was determined to be 176.0 μM [74].

In contrast to the *Penicillium* mycotoxins discussed earlier, the cytotoxicity induced by PA does not occur via apoptosis but rather necrosis, whereby PA reduces cell membrane integrity, leading to cytolysis of immune cells [167]. Actually, concentrations of PA between 100 and 200 μM effectively blocked the activity and self-activation of caspase-3, -8, and -9 in a lymphoma cell line, and inhibition of these caspase pathways resulted in decreased DNA fragmentation and apoptotic cell bodies. However, these concentrations at the same time caused unfavourable cell necrosis that reduced the overall cell viability [167].

PA could also affect other parameters of immune system, even though this has not widely been studied. Sorenson and Simpson (1986) reported reduced phagocytic capacity along with suppressed RNA and protein synthesis in rat alveolar macrophages that were treated with 0.3 mM PA for 2 hours [166]. The metabolic activity of the human PBMC, on the other hand, was induced by PA [165]. Changes in RNA and protein synthesis as well as metabolic activity could potentially affect the production of anti-microbial molecules produced by immune cells, however, we are unaware of any studies that have been carried out to assess the potential effect of PA related to these functional aspects.

2.5. Mycophenolic Acid (MPA)

MPA is a well-known potent immunosuppressive *Penicillium* mycotoxin that has been effectively used to prevent organ rejection following transplantation [168], as well as treat inflammatory disorders such as arteriosclerosis and rheumatoid arthritis [72, 75]. For these reasons, MPA and its derivative mycophenolate mofetil (MMF) are by far the most widely studied *Penicillium* mycotoxins in terms of their immunomodulatory properties. Both functional and transcriptomic studies have revealed that MPA most frequently mediates changes in cell proliferation, viability, as well as the expression of co-receptors required for migration and activation of immune cells [169–172].

2.5.1. Innate Immune Cell Proliferation

The effect of MPA on cell proliferation and viability has been thoroughly studied in human medicine [75, 168]. A study performed by Borrow and colleagues (2006) reported that when a high oral dose of MPA, up to 1.95 g/day, was given to patients following renal transplantation, there was a significant decrease in the number of blood leukocytes in comparison to patients who received a low dose of MPA [168]. MPA and MMF likely cause a decrease in the proliferation of leukocytes resulting in reduced leukocyte numbers [169, 172, 173]. For, example, MPA and MMF at clinically effective concentrations between 0.6 and 10 μM, have been shown to inhibit the proliferation of rat and human mesangial cells *in vitro* without inducing cellular necrosis or apoptosis [169, 171]. Other studies have demonstrated that MPA also inhibited the proliferation of various immune cell types including bovine macrophages [74], human monocyte-derived dendritic cells [172], polymorphonuclear cells [172] and NK cells [173] without causing apoptosis or necrosis. The decreased proliferation of these innate immune cells may partially explain why there was reduced rejection in transplant recipients following treatment with MPA.

2.5.2. Inhibition of NK Cell Activation

NK cells are important mediators of the immune response that limit the growth of tumors and spread of microbial infection [173], and also contribute to allograft rejection during the first month of post-transplantation [174, 175]. For these reasons, NK cells are one of the innate immune cell groups that have most widely been studied in the context of MPA immunosuppression. Many studies have shown that MPA decreases the activity of NK cells by inhibiting their activation [173, 176]. For example, when human NK cells were incubated with leukemic cells and exposed to 10 μg/ml of MPA *in vitro* for 7 days, there was a significant decrease in their activation markers, NKG2D, NKp30, NKp44 and NKp46 [173]. These NK cells also had reduced cytotoxic activity to the leukemic cells K562 and Daudi [173]. Eissen et al. (2010) also reported that MPA decreased the shift in NK cell receptor repertoire, such as NKG2A, which might have partially contributed in the decreased cytotoxic activity of NK cells against K562 [174]. A decrease in NK cell activation markers reported from the studies indicates that MPA has inhibitory effects on both the maturation and activation of NK cells, thereby reducing their cytotoxic activity. This is another reason why MPA has been used as an effective post-transplantation drug.

2.5.3. Changes in Dendritic Cell (DC) Surface Proteins

DCs are important regulators of the immune system as they are considered to be key mediators of naïve lymphocyte activation [177]. These cells are highly susceptible to immunomodulation by MPA. It is thought that MPA affects the expression of membrane proteins on DCs that are required for the activation and function of the host immune response [171, 178]. For example, one study demonstrated that *in vivo* exposure of MPA reduced the trafficking of human blood myeloid DCs to lymph nodes by down-regulating chemokine receptor (CCR)-7 expression, but increased the migration of these cells to sites of inflammation by up-regulating CCR-1 expression [178]. MPA also induced the expression of endothelial VCAM-1 and E-selectin, both of which are required for the trans-migration of DCs and other leukocytes across endothelial cells during inflammation [171]. Therefore, these studies suggest that MPA stimulates the recruitment of myeloid DCs towards sites of inflammation.

MPA also inhibits the maturation of DCs thereby limiting their ability to stimulate other immune mediators, such as T and B cells [171, 178]. Surface proteins on DCs play an essential role in lymphocyte activation by acting as co-receptors during antigen presentation to T and B cells [177]. Therefore, down-regulation of these co-receptors would lead to incomplete activation of lymphocytes and reduced function [179, 180]. For example, a study performed by Cicinnati and colleagues (2009) demonstrated that MPA inhibited the maturation of human myeloid DCs by impairing TLR-3 ligation to polyinosinic-polycytidylic acid, a synthetic double stranded RNA used as a viral antigen to activate innate cells [178]. Other *in vitro* studies also reported down-regulation of human DC maturation markers, such as CD80 and CD86, as well as other co-stimulatory and adhesion molecules including CD1a, CD40, CD54, CD83 and CD 205 [163, 172, 181]. The *in vitro* exposure of MMF also down-regulated the expression of co-receptors, CD40, CD80/86, and ICAM-1, on murine DCs [182]. DCs lacking the above co-receptors would not be able to efficiently activate lymphocytes, and therefore would affect the activation, maturation and function of both B- and T-cells [179, 180, 183]. The absence of CD40 signal from DCs for example, impairs the proliferation as well as maturation of B-cells [181, 183]. Additionally, incomplete activation of T-cells can either stimulate apoptosis of the T-cells, or favor their differentiation into T-reg cells [179, 180]. In support of this, when human DCs were pre-treated with MPA, they displayed increased production of IL-5 and the anti-inflammatory cytokines IL-10 and TGF-β, and induced the differentiation of naïve T-cells to T-reg cells [163, 184]; IL-10 and TGF-β are strong immunosuppressive cytokines that are required for the differentiation of T-reg cells [185]. Interestingly, direct exposure of naive and memory human T-cells with 1 μM MPA also induced the differentiation to T-reg cells, as indicated by the increased expression of the T-reg cell markers, Foxp3, CTLA4 and PD-1 [186, 187]. Based on these studies, it is speculated that MPA increases the T-reg cell population, via DC-dependent and independent signals. Since T-reg cells have strong immunosuppressive phenotypes [188], an increase in their number could lead to immunosuppression, favoring allograft acceptance [72, 168].

2.5.4. Proliferation and Viability of T-Cells

The immunosuppressive effect of MPA may also be due in part to the reduced number of T-cells [73, 189, 190]. For example, a patient treated with 2 g/day of MPA for a year following organ transplantation had reduced numbers of total T-cells in their peripheral blood [176]. This decrease in lymphocyte number is partially due to MPA's ability to inhibit cell proliferation as well as its cytotoxicity to T-cells [189].

It has been proposed that the selective effect of MPA on proliferation and viability of activated T-cells is closely linked to down-regulation of type II inosine monophosphate dehydrogenase (IMPDH-II) activity, which inhibits *de-novo* purine biosynthesis [170, 171, 191]. There are two isotypes of IMPDH, type I and II. Both isotypes have similar enzymatic function for purine biosynthesis, yet they are differentially expressed in different cell types. For example, Nagai and colleagues (1992) demonstrated that activated T-cells and HL60 leukemic cells predominantly express type II IMPDH, while type I IMPDH was more constitutively conserved across most other cell types [191]. Regardless of their similar functionality, the activity of type I IMPDH cannot be substituted for type II IMPDH [192, 193]. Activated T-cells and leukemic cells rely heavily on type II IMPDH for GTP synthesis, while naïve T-cells and other cell types, such as macrophages or epithelial cells, actively utilize type I IMPDH [191, 194]. Interestingly, MPA has been shown to specifically down-regulate type II IMPDH, but had no effect on Type I IMPDH [171, 191]. This would allow MPA to selectively target the proliferation and viability of activated lymphocytes as well as leukemic cells. Inhibited purine biosynthesis compromises the cell's ability to synthesize GTP/GDP, which is essential for normal immune cell functions such as differentiation, proliferation and apoptosis [191, 195]. The decreased expression of GTP following MPA treatment partially led to the reduction in ERK 1/2 and MAPK signalling, which could reduce the cell viability by activating capase-dependent apoptotic death signals [196, 197]. For example, MPA induced the cell death of the MOLT-4 T-lymphocytic cell line as well as PHA-activated $CD4^+$ T-cells, but not resting naïve T-cells [75, 189]. Other studies have reported that MPA induced apoptosis of the Jurkat human T-cell line and gastric cancer cell line, as indicated by increased caspase-3 and -9 activity [170, 198].

The depletion of cellular GTP/GDP not only trigger apoptosis, but also causes necrotic cell death in some immune cell types [199, 200]. Chaigne-Delalande and colleagues (2008) demonstrated that MPA caused cytotoxicity to B- and T-lymphoma cell lines as well as mitogen-activated human PBLs by inducing necrosis and caspase-dependent apoptotic pathways [199]. MPA did not necessarily induce either apoptosis or necrosis to naïve PBLs, but rather caused them to become sensitized to MPA-mediated necrosis upon activation [199]. Necrotic cell death was also demonstrated using the $Jurkat^{DoxR}$ and CEM^{DoxR} lymphocytic leukemia cell lines, which are resistant to doxorubicin-induced apoptosis [200]. Therefore, MPA specifically targets activated lymphocytes that heavily rely on type II IMPDH activity for GTP/GDP synthesis, causing either apoptosis or necrosis.

2.5.5. Effect on B-Cell Activation

Some studies have also investigated the potential effect of MPA and MMF on the activities of B-cells [194, 201]. Weigel and colleagues (2002) reported that MMF treatment resulted in almost a 50% reduction in B-cell numbers in patients who received MMF for a year [176]. The study also reported that these B-cells from MMF-treated patients were lacking the B-cell activation marker CD69. Karnell et al. (2011) also noted that *in vitro* MPA exposure to concentrations between 1.0 and 3.0 μM selectively impaired the early activation of human B-cells [194]. Similarly, another study reported significant decreases in the activation as well as differentiation of human B-cells when these cells were exposed to 5 μM MPA *in vitro* [201]. This decreased activation and differentiation of B-cells could decrease the number of plasma cells (PCs). In support of this, Eickenberg et al. (2012) reported that the patients who received an oral dose of MMF between 0.3 and 0.5 μM had significantly lower serum IgG concentrations compared to those from other treatment groups [201]. In addition to this, Karnell et al. also observed that *in vitro* exposure to MMF significantly decreased the production of IgG, IgM and IgA from plasma cells.

Regardless of the significant effect of MPA on B-cell activation and differentiation [176, 194, 201], MPA does not seem to affect cell viability and/or the responsiveness of terminally differentiated human plasma cells. Karnell et al. (2011) discussed that the selective effect of MPA on B-cells but not plasma cells is probably due to the differential utilization of IMPDH enzymes [194]. B cells, like activated T-cells, are predominantly dependent on type II IMPDH activity for GTP/GDP, while terminally differentiated plasma cells use type I IMPDH [194].

2.6. Roquefortine C (RQC)

2.6.1. Immunomodulatory Effect of RQC

Similar to the other *Penicillium* mycotoxins, RQC can also inhibit the proliferation of lymphocytes. For example, *in vitro* RQC exposure suppressed the proliferation of porcine lymphocytes, but to a much lesser extent when compared to OTA, CIT or PAT [73]. In addition to this, Rand et al. (2005) observed that intra-tracheal instillation of RQC dissolved in a saline solution resulted in a significant increase in the numbers of macrophages and neutrophils in the bronchoalveolar lavage fluid (BALF) of mice after 6 hours of exposure [78]. There was also a significant increase in BALF inflammatory molecules including MIP-2 and TNF-α. Rand and colleagues in their later study suggested that fungal compounds, such as RQC, with low molecular weight may elevate inflammatory gene transcription in alveolar macrophages [202], inducing pro-inflammatory molecules, such as MIP-2 and TNF-α [78]. Based on these studies, it can be speculated that RQC irritated the lung tissues to induce the infiltration of immune complexes, such as neutrophils and macrophages, and the elevation of other inflammatory mediators, such as TNF-α, in the mice BALF [78, 203]. This however, would require more evidence to support the speculation at this point.

Considering that high concentrations of RQC (25 mg/kg) was shown to cause various toxicological signs in ruminant species [84] and the concentration of RQC in animal silage reported ranged from 0.4 to 17 mg/kg [9, 81], it would be important to further investigate the potential for RQC immunotoxicity.

3. Epigenetic Changes and Its Relation to the Inflammatory Response

Epigenetics are heritable changes in gene expression caused by mechanisms other than changes in the underlying DNA sequence, and these tend to follow environmental cues [204]. These changes are often regulated through DNA and/or histone covalent modifications, such as methylation or acetylation [205, 206], and are crucial for optimal cellular activity. The regulation of epigenetic modifications is important for the proper functioning of immune cells [207]. For example, macrophage-specific genes that characterize the M2 macrophage subset are regulated by reciprocal changes in the methylation status of lysine 4 and 27 on histone H3 (H3K4 and H3K27), both of which are regulated by epigenetic enzymes [208].

As previously discussed, the exposure of *Penicillium* mycotoxins could lead to cellular and functional changes of the immune response [74, 152, 166]. It has been proposed that some of these functional changes may be influenced by alterations in epigenetic patterns and the activities of the enzymes regulating them [209, 210]. Given this, it seems reasonable that exposure to mycotoxins may also alter the functioning of immune cells through epigenetic modifications [207, 210]. In support of this, Marin-Kuan et al. (2007) reported that chronic OTA exposure resulted in an increased concentration of histone deacetylase (HDAC)-3 protein in mice kidneys [209]. HDAC-3 is a class-1 (HDAC) that regulates cell functions by deacetylating lysine residues in histone 3 and/or 4. The class-1 HDACs stimulate inflammatory activities of cells by interacting with signalling molecules involved in NF-κβ transcriptional pathway (Figure 2) [211–213]. The inhibition of HDACs suppresses proteasome activity, and reduces the degradation iκβ-α [118, 214], nuclear translocation and overall activity of NF-κβ [19]. OTA was shown to reduce the level of cytosolic iκβ-α expression and this in turn elevated of nuclear p65- NF-κβ expression in rat liver [118]. Additionally, decreases in the activity of HDACs demonstrated impairment of NF-κβ signalling pathways and decreased inflammatory response in human myeloid dendritic cells [215, 216]. All these studies indicate that OTA can induce the pro-inflammatory response by up-regulating HDAC-3 activity, which activates NF-κβ inflammatory pathways [19, 118, 119, 209].

Recently, many studies have shown that HDAC-3 can alter the stability of the STAT-3 dimer [217, 218]. The STAT-3 dimer functions as a transcription molecule for differentiating M2 macrophages that exhibit anti-inflammatory phenotypes [219, 220]. Upon the stimulation of macrophages with anti-inflammatory cytokines such as IL-4, the cytoplasmic portion of the IL-4 receptor acts as a kinase to phosphorylate cytoplasmic STAT-3 [198]. Phosphorylated STAT-3 subsequently translocates into the nucleus with the help of importin-3 [221].

These phosphorylated STATs then spontaneously bind to each other to form STAT-3 dimers [221]. However, Type-1 HDACs such as HDAC-3 deacetylate lysine^{685} of STAT-3 to destabilize its dimerization, which prevents M2 macrophage differentiation [218]. In support of this, Mullican et al. (2011) reported that the macrophages lacking HDAC-3 are hyper-responsive to IL-4 stimulation, resulting in IL-4-induced alternative activation of macrophages both *in vivo* and *in vitro* [219]. From these studies, it is speculated that the increased HDAC-3 activity by OTA could prevent differentiation of macrophages into M2 macrophages [220].

Other epigenetic enzymes, such as DNA methyltransferases, histone demethylase and histone acetylase (Bmi-1), also play an important role in regulating cellular activities [222, 223]. However, there is not enough information available on these enzymes in relation to *Penicillium* mycotoxicoses. Therefore, it would be interesting to look at the effect of *Penicillium* mycotoxins on these enzymes, because changes in cell viability and proliferation are one of the most common mechanisms of *Penicillium* mycotoxicoses as was previously discussed in this review [73, 74, 224].

SUMMARY AND CONCLUSION

Overall, this review provides insight into the potential immunomodulatory effects of *Penicillium* mycotoxins through alterations in immune cell viability, proliferation, inflammatory response, activation as well as both cytokine and antibody production [73, 74, 111]. *Penicillium* mycotoxins differentially affect these parameters by directly altering cellular activities, such as caspase apoptotic pathways and/or NF-κβ transcriptional pathways [117, 130, 142]. Some researchers have also proposed that *Penicillium* mycotoxins affect the cellular activity of immune cells indirectly as a result of alteration of the enzymes regulating epigenetic modifications [209, 210]. Many *in vitro* studies and also some *in vivo* studies have provided the evidence of their immunomodulatory effects in poultry, rodents and humans [43, 73, 140, 155, 168]. Regardless of the fact that ruminants are constantly exposed to *Penicillium* mycotoxins via consumption of ensiled feed [9], the potential toxic effect of

Penicillium mycotoxins, including immunomodulation, have been disregarded under a general assumption that *Penicillium* mycotoxins are rapidly degraded by commensal microbes in the gut [22, 100]. However, there are previously reported cases of various toxicological signs associated with the consumption of *Penicillium* mycotoxin-contaminated silage in ruminant species [17, 52, 77]. This suggests that *Penicillium* mycotoxins could potentially break through the protection from commensal microbes, potentially leading to immunomodulation, and therefore contribute to secondary diseases, in ruminant species. In addition to this, most studies discussed herein have focussed on the risk observed from individual *Penicillium* mycotoxin exposure, which may not necessarily reflect the true risk due to their potential interactions [44, 74, 165, 225], because most *Penicillium* mycotoxins are often found in combinations of two or more in animal feed [9]. Therefore, in order to gain a greater perspective of how to prevent mycotoxins in the agricultural industry, combinational mycotoxicoses should be considered in future ruminant studies.

Conflict of Interest

The authors declare that there are no conflicts of interest.

Acknowledgment

This review was funded by Alltech Inc., Nicholasville, Kentucky.

References

[1] Boutrif E, Canet C: Mycotoxin prevention and control: FAO programmes. *Rev Med Vet* 1998, 149:681–694.

[2] Miraglia M, Marvin HJP, Kleter GA, Battilani P, Brera C, Coni E, Cubadda F, Croci L, De Santis B, Dekkers S, Filippi L, Hutjes RWA, Noordam MY, Pisante M, Piva G, Prandini A, Toti L, van den Born GJ, Vespermann A: Climate change and food safety: an emerging issue with special focus on Europe. *Food Chem Toxicol Int J Publ Br Ind Biol Res Assoc* 2009, 47:1009–1021.

[3] Mycotoxins in Crops: A Threat to Human and Domestic Animal Health[http://www.apsnet.org/education/IntroPlantPath/Topics/mycotoxins/Pages/impact.html]

[4] Almeida AP, Corrêa B, Mallozzi MAB, Sawazaki E, Soares LMV: Mycoflora and aflatoxin/fumonisin production by fungal isolates from freshly harvested corn hybrids. *Braz J Microbiol* 2000, 31.

[5] Creppy EE: Update of survey, regulation and toxic effects of mycotoxins in Europe. *Toxicol Lett* 2002, 127:19–28.

[6] Duong TA: Infection due to Penicillium marneffei, an emerging pathogen: review of 155 reported cases. *Clin Infect Dis Off Publ Infect Dis Soc Am* 1996, 23:125–130.

[7] Xiao CL, Kim YK, Boal RJ: First Report of Occurrence of Pyrimethanil Resistance in *Penicillium expansum* from Stored Apples in Washington State. *Plant Dis* 2011, 95:72–72.

[8] Dalcero A, Magnoli C, Luna M, Ancasi G, Reynoso MM, Chiacchiera S, Miazzo R, Palacio G: Mycoflora and naturally occurring mycotoxins in poultry feeds in Argentina. *Mycopathologia* 1998, 141:37–43.

[9] Mansfield MA, Jones AD, Kuldau GA: Contamination of fresh and ensiled maize by multiple penicillium mycotoxins. *Phytopathology* 2008, 98:330–336.

[10] Lennox CL, Spotts RA, Cervantes LA: Populations of *Botrytis cinerea* and *Penicillium* spp. on Pear Fruit, and in Orchards and Packinghouses, and Their Relationship to Postharvest Decay. *Plant Dis* 2003, 87:639–644.

[11] Mateo R, Medina A, Mateo EM, Mateo F, Jiménez M: An overview of ochratoxin A in beer and wine. *Int J Food Microbiol* 2007, 119:79–83.

[12] Braunberg RC, Gantt O, Barton C, Friedman L: In vitro effects of the nephrotoxins ochratoxin A and citrinin upon biochemical function of porcine kidney. *Arch Environ Contam Toxicol* 1992, 22:464–470.

[13] Sansing GA, Lillehoj EB, Detroy RW, Miller MA: Synergistic toxic effects of citrinin, ochratoxin A and penicillic acid in mice. *Toxicon Off J Int Soc Toxinology* 1976, 14:213–220.

[14] Dickens F, Jones HEH: Further Studies on the Carcinogenic Action of Certain Lactones and Related Substances in the Rat and Mouse. *Br J Cancer* 1965, 19:392–403.

[15] Hayes AW, Phillips TD, Williams WL, Ciegler A: Acute toxicity of patulin in mice and rats. *Toxicology* 1979, 13:91–100.

[16] Sommer NF, Buchanan JR, Fortlage RJ: Production of patulin by Penicillium expansum. *Appl Microbiol* 1974, 28:589–593.

[17] Ukai T, Yamamoto Y, Yamamoto T: Substance from a strain of Penicillium. II. Culture method of Hori-Yamamoto strain and chemical structure of its poisonous substance. *J Pharm Soc Jpn* 1954, 74:450.

[18] Al-Anati L, Petzinger E: Immunotoxic activity of ochratoxin A. *J Vet Pharmacol Ther* 2006, 29:79–90.

[19] Ferrante MC, Raso GM, Bilancione M, Esposito E, Iacono A, Meli R: Differential modification of inflammatory enzymes in J774A.1 macrophages by ochratoxin A alone or in combination with lipopolysaccharide. *Toxicol Lett* 2008, 181:40–46.

[20] Herzog-Soares JDA, Freire RB: Effect of citrinin and in association with aflatoxin B(1) on the infectivity and proliferation of Toxoplasma gondii in vitro. *Braz J Infect Dis Off Publ Braz Soc Infect Dis* 2004, 8:101–108.

[21] Abrunhosa L, Paterson RRM, Venâncio A: Biodegradation of Ochratoxin A for Food and Feed Decontamination. *Toxins* 2010, 2:1078–1099.

[22] Fuchs S, Sontag G, Stidl R, Ehrlich V, Kundi M, Knasmüller S: Detoxification of patulin and ochratoxin A, two abundant mycotoxins, by lactic acid bacteria. *Food Chem Toxicol Int J Publ Br Ind Biol Res Assoc* 2008, 46:1398–1407.

[23] Kavanagh F: Activities of Twenty-two Antibacterial Substances against Nine Species of Bacteria. *J Bacteriol* 1947, 54:761–766.

[24] Aninat C, Hayashi Y, André F, Delaforge M: Molecular requirements for inhibition of cytochrome p450 activities by roquefortine. *Chem Res Toxicol* 2001, 14:1259–1265.

[25] Smith JE, Harran G: Microbial degradation of mycotoxins. *Int Biodeterior Biodegrad* 1993, 32:205–211.

[26] Halstensen AS, Nordby K-C, Elen O, Eduard W: Ochratoxin A in grain dust--estimated exposure and relations to agricultural practices in grain production. *Ann Agric Environ Med AAEM* 2004, 11:245–254.

[27] Skaug MA, Eduard W, Størmer FC: Ochratoxin A in airborne dust and fungal conidia. *Mycopathologia* 2001, 151:93–98.

[28] Panda A, Sharma N, Das G, Kumar N, Satpathy G: Mycotic keratitis in children: epidemiologic and microbiologic evaluation. *Cornea* 1997, 16:295–299.

[29] EFSA: Scientific Opinion on the risks for public and animal health related to the presence of citrinin in food and feed. *Volume 10*. EFSA; 2012:2605.

[30] Reijula K, Tuomi T: Mycotoxins of aspergilli: exposure and health effects. *Front Biosci J Virtual Libr* 2003, 8:s232–235.

[31] Smith JE, Solomons G, Lewis C, Anderson JG: Role of mycotoxins in human and animal nutrition and health. *Nat Toxins* 1995, 3:187–192; discussion 221.

[32] Visconti A, Pascale M, Centonze G: Determination of ochratoxin A in wine by means of immunoaffinity column clean-up and high-performance liquid chromatography. *J Chromatogr A* 1999, 864:89–101.

[33] Pitt JI: Penicillium viridicatum, Penicillium verrucosum, and production of ochratoxin A. *Appl Environ Microbiol* 1987, 53:266–269.

[34] Moroi K, Suzuki S, Kuga T, Yamazaki M, Kanisawa M: Reduction of ochratoxin A toxicity in mice treated with phenylalanine and phenobarbital. *Toxicol Lett* 1985, 25:1–5.

[35] Baudrimont I, Betbeder AM, Gharbi A, Pfohl-Leszkowicz A, Dirheimer G, Creppy EE: Effect of superoxide dismutase and catalase on the nephrotoxicity induced by subchronical administration of ochratoxin A in rats. *Toxicology* 1994, 89:101–111.

[36] Milićević D, Jurić V, Stefanović S, Jovanović M, Janković S: Survey of slaughtered pigs for occurrence of ochratoxin A and porcine nephropathy in Serbia. *Int J Mol Sci* 2008, 9:2169–2183.

[37] Brown MH, Szczech GM, Purmalis BP: Teratogenic and toxic effects of ochratoxin A in rats. *Toxicol Appl Pharmacol* 1976, 37:331–338.

[38] O'Brien E, Prietz A, Dietrich DR: Investigation of the teratogenic potential of ochratoxin A and B using the FETAX system. *Birth Defects Res B Dev Reprod Toxicol* 2005, 74:417–423.

[39] Stoev SD: Studies on carcinogenic and toxic effects of ochratoxin A in chicks. *Toxins* 2010, 2:649–664.

[40] Ribelin WE, Fukushima K, Still PE: The toxicity of ochratoxin to ruminants. *Can J Comp Med Rev Can Médecine Comparée* 1978, 42:172–176.

[41] Niemiec J, Borzemska W: The effect of Ochratoxin A on egg quality development of embryos and the level of toxin in egg and tissue of hens and chicks. *J Anim Feed Sci* 1994, 3:309–316.

[42] Prior MG, O'Neil JB, Sisodia CS: Effects of ochratoxin A on growth response and residues in broilers. *Poult Sci* 1980, 59:1254–1257.

[43] Stoev SD, Stefanov M, Denev S, Radic B, Domijan AM, Peraica M: Experimental mycotoxicosis in chickens induced by ochratoxin A and penicillic acid and intervention with natural plant extracts. *Vet Res Commun* 2004, 28:727–746.

[44] Veselá D, Veselý D, Jelínek R: Toxic effects of ochratoxin A and citrinin, alone and in combination, on chicken embryos. *Appl Environ Microbiol* 1983, 45:91–93.

[45] Carvalho J, Oishi B, Pandey A, Soccol C: Biopigments from Monascus: strains selection, citrinin production and color stability. *Braz Arch Biol Technol* 2005, 48:885–894.

[46] Hetherington A, Raistrick H: Studies in the biochemistry of micro-organisms. XI. On the production and chemical constitution of a new yellow colouring matter, citrinin, produced from glucose by Penicillium citrinum Thom. *Philos Trans R Soc Lond Ser* 1931, B220:269–297.

[47] Martins ML, Gimeno A, Martins HM, Bernardo F: Co-occurrence of patulin and citrinin in Portuguese apples with rotten spots. *Food Addit Contam* 2002, 19:568–574.

[48] Scott PM, Van Walbeek W, Kennedy B, Anyeti D: Mycotoxins (ochratoxin A, citrinin, and sterigmatocystin) and toxigenic fungi in grains and other agricultural products. *J Agric Food Chem* 1972, 20:1103–1109.

[49] Citrinin fact sheet [http://services.leatherheadfood.com/eman/FactSheet.aspx?ID=14]

[50] Reddy RV, Mayura K, Hayes AW, Berndt WO: Embryocidal, teratogenic and fetotoxic effects of citrinin in rats. *Toxicology* 1982, 25:151–160.

[51] Puel O, Galtier P, Oswald IP: Biosynthesis and toxicological effects of patulin. *Toxins* 2010, 2:613–631.

[52] Ciegler A: Patulin. In *Mycotoxins in human and animal health.* Park Forest South, Ill.: Pathotox Publishers Inc; 1977:609–24.

[53] Kataoka H, Itano M, Ishizaki A, Saito K: Determination of patulin in fruit juice and dried fruit samples by in-tube solid-phase microextraction coupled with liquid chromatography-mass spectrometry. *J Chromatogr A* 2009, 1216:3746–3750.

[54] Scott PM, Somers E: Stability of patulin and penicillic acid in fruit juices and flour. *J Agric Food Chem* 1968, 16:483–485.

[55] Arzu Koçkaya E, Selmanoğlu G, Ozsoy N, Gül N: Evaluation of patulin toxicity in the thymus of growing male rats. *Arh Hig Rada Toksikol* 2009, 60:411–418.

[56] Reddy CS, Chan PK, Hayes AW: Teratogenic and dominant lethal studies of patulin in mice. *Toxicology* 1978, 11:219–223.

[57] Ciegler A, Kurtzman CP: Penicillic acid production by blue-eye fungi on various agricultural commodities. *Appl Microbiol* 1970, 20:761–764.

[58] Müller HM, Amend R: Formation and disappearance of mycophenolic acid, patulin, penicillic acid and PR toxin in maize silage inoculated with Penicillium roqueforti. *Arch Für Tierernähr* 1997, 50:213–225.

[59] Snow JP, Lucas GB, Harvan D, Pero RW, Owens RG: Analysis of tobacco and smoke condensate for penicillic acid. *Appl Microbiol* 1972, 24:34–36.

[60] Ciegler A, Detroy RW, Lillehoj EB: Patulin, penicillic acid and other carcinogenic lactones. *Microb Toxin* 1971, Ch6:409–434.

[61] Alsberg C, Black O: Contributions to the study of maize deterioration; biochemical and toxicological investigations of Penicillium puberulum and Penicillium stoloniferum (1913) US Dept Agr Bur Plant Ind Bull. *Dept Agr Bur Plant Ind Bull* 1913, 270:1–47.

[62] Ciegler A, Mintzlaff HJ, Weisleder D, Leistner L: Potential production and detoxification of penicillic acid in mold-fermented sausage (salami). *Appl Microbiol* 1972, 24:114–119.

[63] Murnaghan MF: The pharmacology of penicillic acid. *J Pharmacol Exp Ther* 1946, 88:119–132.

[64] Hayes AW, Under PD, Williams WL: Acute toxicity of penicillic acid and rubratoxin B in dogs. *Ann Nutr Aliment* 1977, 31:711–721.

[65] Burkin AA, Kononenko GP: Producers of mycophenolic acid in ensiled and grain feeds. *Appl Biochem Microbiol* 2010, 46:545–550.

[66] Usleber E, Dade M, Schneider E, Dietrich R, Bauer J, Märtlbauer E: Enzyme immunoassay for mycophenolic acid in milk and cheese. *J Agric Food Chem* 2008, 56:6857–6862.

[67] Bartman CD, Doerfler DL, Bird BA, Remaley AT, Peace JN, Campbell IM: Mycophenolic Acid Production by Penicillium brevicompactum on Solid Media. *Appl Environ Microbiol* 1981, 41:729–736.

[68] Boysen M, Skouboe P, Frisvad J, Rossen L: Reclassification of the Penicillium roqueforti group into three species on the basis of molecular genetic and biochemical profiles. *Microbiol Read Engl* 1996, 142 (Pt 3):541–549.

[69] Lafont P, Debeaupuis J, Gaillardin M, Payen J: Production of mycophenolic acid by Penicillium roqueforti strains. *Appl Envrion Microbiol* 1979, 37:365–368.

[70] Menon SK, Modi NR, Mistry B, Joshi K: Improvement of some pharmaceutical properties of mycophenolate mofetil (MMF) by para sulphonatocalix[4]resorcinarene inclusion complex. *J Incl Phenom Macrocycl Chem* 2011, 70:121–128.

[71] Ahn HJ, Park J, Song JS, Ju MK, Kim MS, Ha H, Song KH, Kim YS: Mycophenolic acid inhibits oleic acid-induced vascular smooth muscle cell activation by inhibiting cellular reactive oxygen species. *Transplantation* 2007, 84:634–638.

[72] Mohacsi PJ, Tüller D, Hulliger B, Wijngaard PL: Different inhibitory effects of immunosuppressive drugs on human and rat aortic smooth muscle and endothelial cell proliferation stimulated by platelet-derived growth factor or endothelial cell growth factor. *J Heart Lung Transplant Off Publ Int Soc Heart Transplant* 1997, 16:484–492.

[73] Keblys M, Bernhoft A, Höfer CC, Morrison E, Larsen HJS, Flåøyen A: The effects of the Penicillium mycotoxins citrinin, cyclopiazonic acid, ochratoxin A, patulin, penicillic acid, and roquefortine C on in vitro proliferation of porcine lymphocytes. *Mycopathologia* 2004, 158:317–324.

[74] Oh S-Y, Boermans HJ, Swamy HVLN, Sharma BS, Karrow NA: Immunotoxicity of Penicillium Mycotoxins on Viability and Proliferation of Bovine Macrophage Cell Line (BOMACs). *Open Mycol J* 2012, 6:11–16.

[75] Cohn RG, Mirkovich A, Dunlap B, Burton P, Chiu SH, Eugui E, Caulfield JP: Mycophenolic acid increases apoptosis, lysosomes and lipid droplets in human lymphoid and monocytic cell lines. *Transplantation* 1999, 68:411–418.

[76] Weber LT, Shipkova M, Lamersdorf T, Niedmann PD, Wiesel M, Mandelbaum A, Zimmerhackl LB, Schütz E, Mehls O, Oellerich M, Armstrong VW, Tönshoff B: Pharmacokinetics of mycophenolic acid (MPA) and determinants of MPA free fraction in pediatric and adult renal transplant recipients. German Study group on Mycophenolate Mofetil Therapy in Pediatric Renal Transplant Recipients. *J Am Soc Nephrol JASN* 1998, 9:1511–1520.

[77] Häggblom P: Isolation of roquefortine C from feed grain. *Appl Environ Microbiol* 1990, 56:2924–2926.

[78] Rand TG, Giles S, Flemming J, Miller JD, Puniani E: Inflammatory and cytotoxic responses in mouse lungs exposed to purified toxins from building isolated Penicillium brevicompactum Dierckx and P. chrysogenum Thom. *Toxicol Sci Off J Soc Toxicol* 2005, 87:213–222.

[79] Wagener RE, Davis ND, Diener UL: Penitrem A and Roquefortine Production by Penicillium commune. *Appl Environ Microbiol* 1980, 39:882–887.

[80] Finoli C, Vecchio A, Galli A, Dragoni I: Roquefortine C occurrence in blue cheese. *J Food Prot* 2001, 64:246–251.

[81] Auerbach H, Oldenburg E, Weissbach F: Incidence ofPenicillium roqueforti and roquefortine C in silages. *J Sci Food Agric* 1998, 76:565–572.

[82] Ohmomo S, Kitamoto HK, Nakajima T: Detection of roquefortines in penicillium roqueforti isolated from moulded maize silage. *J Sci Food Agric* 1994, 64:211–215.

[83] Puls R, Ladyman E: Roquefortine Toxicity in a Dog. *Can Vet J* 1988, 29:569.

[84] Häggblom P, Persson I, Holmberg T: A mycotoxin with neurological effects isolated from grain used for feed. *Sven Veterinärtidning* 1990, 42:619–620.

[85] Ohmomo S, Sato T, Utagawa T, Abe M: Isolation of Festuclavine and Three New Indole Alkaloids, Roquefortine A, B and C from the Cultures of Penicillium roqueforti (Production of Alkaloids and Related Substances by Fungi Part XII). *J Antibact Antifung Agents* 1975, 49:615–623.

[86] Tiwary AK, Puschner B, Poppenga RH: Using roquefortine C as a biomarker for penitrem A intoxication. *J Vet Diagn Investig Off Publ Am Assoc Vet Lab Diagn Inc* 2009, 21:237–239.

[87] Scott PM, Kennedy PC: Analysis of blue cheese for roquefortine and other alkaloids from Penicillium roqueforti. *J Agric Food Chem* 1976, 24:865–868.

[88] Yan F, Polk DB: Commensal bacteria in the gut: learning who our friends are. *Curr Opin Gastroenterol* 2004, 20:565–571.

[89] Yu LC-H, Wang J-T, Wei S-C, Ni Y-H: Host-microbial interactions and regulation of intestinal epithelial barrier function: From physiology to pathology. *World J Gastrointest Pathophysiol* 2012, 3:27–43.

[90] Croswell A, Amir E, Teggatz P, Barman M, Salzman NH: Prolonged impact of antibiotics on intestinal microbial ecology and susceptibility to enteric Salmonella infection. *Infect Immun* 2009, 77:2741–2753.

[91] Magalhaes JG, Tattoli I, Girardin SE: The intestinal epithelial barrier: how to distinguish between the microbial flora and pathogens. *Semin Immunol* 2007, 19:106–115.

[92] Tappenden KA, Deutsch AS: The physiological relevance of the intestinal microbiota--contributions to human health. *J Am Coll Nutr* 2007, 26:679S–83S.

[93] García-Lafuente A, Antolín M, Guarner F, Crespo E, Malagelada JR: Modulation of colonic barrier function by the composition of the commensal flora in the rat. *Gut* 2001, 48:503–507.

[94] Schütze N, Lehmann I, Bönisch U, Simon JC, Polte T: Exposure to mycotoxins increases the allergic immune response in a murine asthma model. *Am J Respir Crit Care Med* 2010, 181:1188–1199.

[95] Wagner D, Sangari FJ, Kim S, Petrofsky M, Bermudez LE: Mycobacterium avium infection of macrophages results in progressive suppression of interleukin-12 production in vitro and in vivo. *J Leukoc Biol* 2002, 71:80–88.

[96] Shroff KE, Meslin K, Cebra JJ: Commensal enteric bacteria engender a self-limiting humoral mucosal immune response while permanently colonizing the gut. *Infect Immun* 1995, 63:3904–3913.

[97] Rhee K-J, Sethupathi P, Driks A, Lanning DK, Knight KL: Role of commensal bacteria in development of gut-associated lymphoid tissues and preimmune antibody repertoire. *J Immunol Baltim Md 1950* 2004, 172:1118–1124.

[98] Tourneur E, Chassin C: Neonatal immune adaptation of the gut and its role during infections. *Clin Dev Immunol* 2013, 2013:270301.

[99] Devi P, D'Souza L, Kamat T, Rodrigues C, Naik C: Batch culture fermentation of Penicillium chrysogenum and a report on the isolation, purification, identification and antibiotic activity of citrinin. *IJMS* 2009, 38:38–44.

[100] Kiessling KH, Pettersson H, Sandholm K, Olsen M: Metabolism of aflatoxin, ochratoxin, zearalenone, and three trichothecenes by intact rumen fluid, rumen protozoa, and rumen bacteria. *Appl Environ Microbiol* 1984, 47:1070–1073.

[101] Stecher B, Robbiani R, Walker AW, Westendorf AM, Barthel M, Kremer M, Chaffron S, Macpherson AJ, Buer J, Parkhill J, Dougan G, von Mering C, Hardt W-D: Salmonella enterica serovar typhimurium exploits inflammation to compete with the intestinal microbiota. *PLoS Biol* 2007, 5:2177–2189.

[102] Pamer EG: Immune responses to commensal and environmental microbes. *Nat Immunol* 2007, 8:1173–1178.

[103] Tesh VL, Ramegowda B, Samuel JE: Purified Shiga-like toxins induce expression of proinflammatory cytokines from murine peritoneal macrophages. *Infect Immun* 1994, 62:5085–5094.

[104] Maresca M, Mahfoud R, Pfohl-Leszkowicz A, Fantini J: The mycotoxin ochratoxin A alters intestinal barrier and absorption functions but has no effect on chloride secretion. *Toxicol Appl Pharmacol* 2001, 176:54–63.

[105] Mahfoud R, Maresca M, Garmy N, Fantini J: The mycotoxin patulin alters the barrier function of the intestinal epithelium: mechanism of action of the toxin and protective effects of glutathione. *Toxicol Appl Pharmacol* 2002, 181:209–218.

[106] Maresca M, Yahi N, Younès-Sakr L, Boyron M, Caporiccio B, Fantini J: Both direct and indirect effects account for the pro-inflammatory activity of enteropathogenic mycotoxins on the human intestinal epithelium: stimulation of interleukin-8 secretion, potentiation of interleukin-1beta effect and increase in the transepithelial passage of commensal bacteria. *Toxicol Appl Pharmacol* 2008, 228:84–92.

[107] Murphy K, Paul T, Walport M: *Janeway's Immunobiology*. LLC, UK: Garland Science, Taylor & Francis Group; 2008 [7th Ed].

[108] Chang CF, Huff WE, Hamilton PB: A leucocytopenia induced in chickens by dietary ochratoxin A. *Poult Sci* 1979, 58:555–558.

[109] Gupta M, Bandopadhyay S, Paul B, Majumder SK: Hematological changes produced in mice by ochratoxin A. *Toxicology* 1979, 14:95–98.

[110] Boorman GA, Hong HL, Dieter MP, Hayes HT, Pohland AE, Stack M, Luster MI: Myelotoxicity and macrophage alteration in mice exposed to ochratoxin A. *Toxicol Appl Pharmacol* 1984, 72:304–312.

[111] Xue CY, Wang GH, Chen F, Zhang XB, Bi YZ, Cao YC: Immunopathological effects of ochratoxin A and T-2 toxin combination on broilers. *Poult Sci* 2010, 89:1162–1166.

[112] Alvarez-Erviti L, Leache C, González-Peñas E, de Cerain AL: Alterations induced in vitro by ochratoxin A in rat lymphoid cells. *Hum Exp Toxicol* 2005, 24:459–466.
[113] Charoenpornsook K, Fitzpatrick JL, Smith JE: The effects of four mycotoxins on the mitogen stimulated proliferation of bovine peripheral blood mononuclear cells in vitro. *Mycopathologia* 1998, 143:105–111.
[114] Lea T, Steien K, Størmer FC: Mechanism of ochratoxin A-induced immunosuppression. *Mycopathologia* 1989, 107:153–159.
[115] Harvey RB, Elissalde MH, Kubena LF, Weaver EA, Corrier DE, Clement BA: Immunotoxicity of ochratoxin A to growing gilts. *Am J Vet Res* 1992, 53:1966–1970.
[116] Stec J, Rachubik J, Szczotka M, Kuzmak J: Effects of Penicillium Mycotoxins: Ctrinin, Ochratoxin A, and Patulin on In Vitro Proliferation of Bovine Lymphocytes. *Bulletin of the Vet Insti in Pulaway* 2008, 52:163–167.
[117] Ferrante MC, Bilancione M, Raso GM, Esposito E, Iacono A, Zaccaroni A, Meli R: Expression of COX-2 and hsp72 in peritoneal macrophages after an acute ochratoxin A treatment in mice. *Life Sci* 2006, 79:1242–1247.
[118] Al-Anati L, Katz N, Petzinger E: Interference of arachidonic acid and its metabolites with TNF-alpha release by ochratoxin A from rat liver. *Toxicology* 2005, 208:335–346.
[119] Lawrence T: The Nuclear Factor NF- B Pathway in Inflammation. *Cold Spring Harb Perspect Biol* 2009, 1:a001651–a001651.
[120] Deng Y, Ren X, Yang L, Lin Y, Wu X: A JNK-dependent pathway is required for TNFalpha-induced apoptosis. *Cell* 2003, 115:61–70.
[121] Elsea CR, Roberts DA, Druker BJ, Wood LJ: Inhibition of p38 MAPK suppresses inflammatory cytokine induction by etoposide, 5-fluorouracil, and doxorubicin without affecting tumoricidal activity. *PloS One* 2008, 3:e2355.
[122] Gekle M, Schwerdt G, Freudinger R, Mildenberger S, Wilflingseder D, Pollack V, Dander M, Schramek H: Ochratoxin A induces JNK activation and apoptosis in MDCK-C7 cells at nanomolar concentrations. *J Pharmacol Exp Ther* 2000, 293:837–844.
[123] Sauvant C, Holzinger H, Gekle M: The nephrotoxin ochratoxin A induces key parameters of chronic interstitial nephropathy in renal proximal tubular cells. *Cell Physiol Biochem Int J Exp Cell Physiol Biochem Pharmacol* 2005, 15:125–134.
[124] Aggarwal S, Gollapudi S, Yel L, Gupta AS, Gupta S: TNF-alpha-induced apoptosis in neonatal lymphocytes: TNFRp55 expression and downstream pathways of apoptosis. *Genes Immun* 2000, 1:271–279.
[125] George J, Banik NL, Ray SK: Genistein induces receptor and mitochondrial pathways and increases apoptosis during BCL-2 knockdown in human malignant neuroblastoma SK-N-DZ cells. *J Neurosci Res* 2010, 88:877–886.
[126] Assaf H, Azouri H, Pallardy M: Ochratoxin A induces apoptosis in human lymphocytes through down regulation of Bcl-xL. *Toxicol Sci Off J Soc Toxicol* 2004, 79:335–344.
[127] Chao DT, Linette GP, Boise LH, White LS, Thompson CB, Korsmeyer SJ: Bcl-XL and Bcl-2 repress a common pathway of cell death. *J Exp Med* 1995, 182:821–828.
[128] Chen M, Guerrero AD, Huang L, Shabier Z, Pan M, Tan T-H, Wang J: Caspase-9-induced mitochondrial disruption through cleavage of anti-apoptotic BCL-2 family members. *J Biol Chem* 2007, 282:33888–33895.
[129] Buttke TM, Sandstrom PA: Oxidative stress as a mediator of apoptosis. *Immunol Today* 1994, 15:7–10.

[130] Essid E, Dernawi Y, Petzinger E: Apoptosis induction by OTA and TNF-α in cultured primary rat hepatocytes and prevention by silibinin. *Toxins* 2012, 4:1139–1156.
[131] Luster MI, Germolec DR, Burleson GR, Jameson CW, Ackermann MF, Lamm KR, Hayes HT: Selective immunosuppression in mice of natural killer cell activity by ochratoxin A. *Cancer Res* 1987, 47:2259–2263.
[132] Swann JB, Hayakawa Y, Zerafa N, Sheehan KCF, Scott B, Schreiber RD, Hertzog P, Smyth MJ: Type I IFN contributes to NK cell homeostasis, activation, and antitumor function. *J Immunol Baltim Md 1950* 2007, 178:7540–7549.
[133] Schroder K, Hertzog PJ, Ravasi T, Hume DA: Interferon-gamma: an overview of signals, mechanisms and functions. *J Leukoc Biol* 2004, 75:163–189.
[134] Wigginton JM, Gruys E, Geiselhart L, Subleski J, Komschlies KL, Park JW, Wiltrout TA, Nagashima K, Back TC, Wiltrout RH: IFN-gamma and Fas/FasL are required for the antitumor and antiangiogenic effects of IL-12/pulse IL-2 therapy. *J Clin Invest* 2001, 108:51–62.
[135] Alvarez L, Gil AG, Ezpeleta O, García-Jalón JA, López de Cerain A: Immunotoxic effects of Ochratoxin A in Wistar rats after oral administration. *Food Chem Toxicol Int J Publ Br Ind Biol Res Assoc* 2004, 42:825–834.
[136] Creppy EE, Størmer FC, Röschenthaler R, Dirheimer G: Effects of two metabolites of ochratoxin A, (4R)-4-hydroxyochratoxin A and ochratoxin alpha, on immune response in mice. *Infect Immun* 1983, 39:1015–1018.
[137] Thust R, Kneist S: Activity of citrinin metabolized by rat and human microsome fractions in clastogenicity and SCE assays on Chinese hamster V79-E cells. *Mutat Res* 1979, 67:321–330.
[138] Islam MR, Roh Y-S, Cho A, Kim J, Kim J-H, Eo S, Lim C-W, Kim B: Immune modulatory effects of the foodborne contaminant citrinin in mice. *Food Chem Toxicol Int J Publ Br Ind Biol Res Assoc* 2012, 50:3537–3547.
[139] Reddy RV, Taylor MJ, Sharma RP: Evaluation of citrinin toxicity on the immune function of mice. *J Food Prot* 1988, 51:32–36.
[140] Dziarski R: Comparison of in vitro and in vivo mitogenic and polyclonal antibody and autoantibody responses to peptidoglycan, LPS, protein A, PWM, PHA and Con A in normal and autoimmune mice. *J Clin Lab Immunol* 1985, 16:93–109.
[141] Calabrese EJ, Baldwin LA: H ORMESIS: The Dose-Response Revolution. *Annu Rev Pharmacol Toxicol* 2003, 43:175–197.
[142] Kumar R, Dwivedi PD, Dhawan A, Das M, Ansari KM: Citrinin-generated reactive oxygen species cause cell cycle arrest leading to apoptosis via the intrinsic mitochondrial pathway in mouse skin. *Toxicol Sci Off J Soc Toxicol* 2011, 122:557–566.
[143] Ribeiro SM, Chagas GM, Campello AP, Klüppel ML: Mechanism of citrinin-induced dysfunction of mitochondria. V. Effect on the homeostasis of the reactive oxygen species. *Cell Biochem Funct* 1997, 15:203–209.
[144] Papp G, Horváth E, Mike N, Gazdag Z, Belágyi J, Gyöngyi Z, Bánfalvi G, Hornok L, Pesti M: Regulation of patulin-induced oxidative stress processes in the fission yeast Schizosaccharomyces pombe. *Food Chem Toxicol Int J Publ Br Ind Biol Res Assoc* 2012, 50:3792–3798.

[145] Chagas GM, Campello AP, Klüppel MLW: Mechanism of citrinin-induced dysfunction of mitochondria. I. Effects on respiration, enzyme activities and membrane potential of renal cortical mitochondria. *J Appl Toxicol* 1992, 12:123–129.

[146] Basu A, Haldar S: The relationship between BcI2, Bax and p53: consequences for cell cycle progression and cell death. *Mol Hum Reprod* 1998, 4:1099–1109.

[147] Johannessen LN, Nilsen AM, Løvik M: The mycotoxins citrinin and gliotoxin differentially affect production of the pro-inflammatory cytokines tumour necrosis factor-alpha and interleukin-6, and the anti-inflammatory cytokine interleukin-10. *Clin Exp Allergy J Br Soc Allergy Clin Immunol* 2005, 35:782–789.

[148] Liu B-H, Chi J-Y, Hsiao Y-W, Tsai K-D, Lee Y-J, Lin C-C, Hsu S-C, Yang S-M, Lin T-H: The fungal metabolite, citrinin, inhibits lipopolysaccharide/interferon-γ-induced nitric oxide production in glomerular mesangial cells. *Int Immunopharmacol* 2010, 10:1608–1615.

[149] Chan W-H: Citrinin induces apoptosis via a mitochondria-dependent pathway and inhibition of survival signals in embryonic stem cells, and causes developmental injury in blastocysts. *Biochem J* 2007, 404:317–326.

[150] Wichmann G, Herbarth O, Lehmann I: The mycotoxins citrinin, gliotoxin, and patulin affect interferon-gamma rather than interleukin-4 production in human blood cells. *Environ Toxicol* 2002, 17:211–218.

[151] Singh ND, Sharma AK, Dwivedi PD, Kumar M, Patil R: Immunosuppressive effect of combined citrinin and endosulfan toxicity in pregnant Wistar rats. *Vet Arh* 2011, 81:751–763.

[152] Sorenson WG, Simpson J, Castranova V: Toxicity of the mycotoxin patulin for rat alveolar macrophages. *Environ Res* 1985, 38:407–416.

[153] Escoula L, Bourdiol D, Linas MD, Recco P, Seguela JP: Enhancing resistance and modulation of humoral immune response to experimental Candida albicans infection by patulin. *Mycopathologia* 1988, 103:153–156.

[154] Llewellyn GC, McCay JA, Brown RD, Musgrove DL, Butterworth LF, Munson AE, White KL: Immunological evaluation of the mycotoxin patulin in female B6C3F1 mice. *Food Chem Toxicol Int J Publ Br Ind Biol Res Assoc* 1998, 36:1107–1115.

[155] Ozsoy N, Selmanoğlu G, Koçkaya EA, Gül N, Cebesoy S: Effect of patulin on the interdigitating dendritic cells (IDCs) of rat thymus. *Cell Biochem Funct* 2008, 26:192–196.

[156] Saxena N, Ansari KM, Kumar R, Dhawan A, Dwivedi PD, Das M: Patulin causes DNA damage leading to cell cycle arrest and apoptosis through modulation of Bax, p(53) and p(21/WAF1) proteins in skin of mice. *Toxicol Appl Pharmacol* 2009, 234:192–201.

[157] Wu T-S, Liao Y-C, Yu F-Y, Chang C-H, Liu B-H: Mechanism of patulin-induced apoptosis in human leukemia cells (HL-60). *Toxicol Lett* 2008, 183:105–111.

[158] Luft P, Oostingh GJ, Gruijthuijsen Y, Horejs-Hoeck J, Lehmann I, Duschl A: Patulin influences the expression of Th1/Th2 cytokines by activated peripheral blood mononuclear cells and T cells through depletion of intracellular glutathione. *Environ Toxicol* 2008, 23:84–95.

[159] Liu B-H, Wu T-S, Yu F-Y, Su C-C: Induction of oxidative stress response by the mycotoxin patulin in mammalian cells. *Toxicol Sci Off J Soc Toxicol* 2007, 95:340–347.

[160] Peterson JD, Herzenberg LA, Vasquez K, Waltenbaugh C: Glutathione levels in antigen-presenting cells modulate Th1 versus Th2 response patterns. *Proc Natl Acad Sci U S A* 1998, 95:3071–3076.

[161] Hunter CA, Subauste CS, Van Cleave VH, Remington JS: Production of gamma interferon by natural killer cells from Toxoplasma gondii-infected SCID mice: regulation by interleukin-10, interleukin-12, and tumor necrosis factor alpha. *Infect Immun* 1994, 62:2818–2824.

[162] Bourdiol D, Escoula L, Salvayre R: Effect of patulin on microbicidal activity of mouse peritoneal macrophages. *Food Chem Toxicol Int J Publ Br Ind Biol Res Assoc* 1990, 28:29–33.

[163] Colic M, Stojic-Vukanic Z, Pavlovic B, Jandric D, Stefanoska I: Mycophenolate mofetil inhibits differentiation, maturation and allostimulatory function of human monocyte-derived dendritic cells. *Clin Exp Immunol* 2003, 134:63–69.

[164] Dooms H, Kahn E, Knoechel B, Abbas AK: IL-2 induces a competitive survival advantage in T lymphocytes. *J Immunol Baltim Md 1950* 2004, 172:5973–5979.

[165] Stoev S, Denev S, Dutton M, Nkosi B: Cytotoxic Effect of Some Mycotoxins and their Combinations on Human Peripheral Blood Mononuclear Cells as Measured by the MTT Assay. *Open Toxinology J* 2009, 2:1–8.

[166] Sorenson WG, Simpson J: Toxicity of penicillic acid for rat alveolar macrophages in vitro. *Environ Res* 1986, 41:505–513.

[167] Bando M, Hasegawa M, Tsuboi Y, Miyake Y, Shiina M, Ito M, Handa H, Nagai K, Kataoka T: The mycotoxin penicillic acid inhibits Fas ligand-induced apoptosis by blocking self-processing of caspase-8 in death-inducing signaling complex. *J Biol Chem* 2003, 278:5786–5793.

[168] Borrows R, Chusney G, Loucaidou M, James A, Lee J, Tromp JV, Owen J, Cairns T, Griffith M, Hakim N, McLean A, Palmer A, Papalois V, Taube D: Mycophenolic acid 12-h trough level monitoring in renal transplantation: association with acute rejection and toxicity. *Am J Transplant Off J Am Soc Transplant Am Soc Transpl Surg* 2006, 6:121–128.

[169] Dubus I, Vendrely B, Christophe I, Labouyrie J-P, Delmas Y, Bonnet J, Combe C: Mycophenolic acid antagonizes the activation of cultured human mesangial cells. *Kidney Int* 2002, 62:857–867.

[170] Dun B, Sharma A, Xu H, Liu H, Bai S, Zeng L, She J-X: Transcriptomic changes induced by mycophenolic acid in gastric cancer cells. *Am J Transl Res* 2013, 6:28–42.

[171] Hauser IA, Johnson DR, Thévenod F, Goppelt-Strübe M: Effect of mycophenolic acid on TNF alpha-induced expression of cell adhesion molecules in human venous endothelial cells in vitro. *Br J Pharmacol* 1997, 122:1315–1322.

[172] Mezger M, Wozniok I, Blockhaus C, Kurzai O, Hebart H, Einsele H, Loeffler J: Impact of mycophenolic acid on the functionality of human polymorphonuclear neutrophils and dendritic cells during interaction with Aspergillus fumigatus. *Antimicrob Agents Chemother* 2008, 52:2644–2646.

[173] Ohata K, Espinoza JL, Lu X, Kondo Y, Nakao S: Mycophenolic acid inhibits natural killer cell proliferation and cytotoxic function: a possible disadvantage of including mycophenolate mofetil in the graft-versus-host disease prophylaxis regimen. *Biol Blood Marrow Transplant J Am Soc Blood Marrow Transplant* 2011, 17:205–213.

[174] Eissens DN, Van Der Meer A, Van Cranenbroek B, Preijers FWMB, Joosten I: Rapamycin and MPA, but not CsA, impair human NK cell cytotoxicity due to differential effects on NK cell phenotype. *Am J Transplant Off J Am Soc Transplant Am Soc Transpl Surg* 2010, 10:1981–1990.

[175] Kroemer A, Xiao X, Degauque N, Edtinger K, Wei H, Demirci G, Li XC: The Innate NK Cells, Allograft Rejection, and a Key Role for IL-15. *J Immunol* 2008, 180:7818–7826.

[176] Weigel G, Griesmacher A, Karimi A, Zuckermann AO, Grimm M, Mueller MM: Effect of mycophenolate mofetil therapy on lymphocyte activation in heart transplant recipients. *J Heart Lung Transplant Off Publ Int Soc Heart Transplant* 2002, 21:1074–1079.

[177] Novak N, Koch S, Allam J-P, Bieber T: Dendritic cells: bridging innate and adaptive immunity in atopic dermatitis. *J Allergy Clin Immunol* 2010, 125:50–59.

[178] Cicinnati VR, Hou J, Lindemann M, Horn PA, Sotiropoulos GC, Paul A, Gerken G, Beckebaum S: Mycophenolic acid impedes the antigen presenting and lymph node homing capacities of human blood myeloid dendritic cells. *Transplantation* 2009, 88:504–513.

[179] Daniel PT, Kroidl A, Cayeux S, Bargou R, Blankenstein T, Dörken B: Costimulatory signals through B7.1/CD28 prevent T cell apoptosis during target cell lysis. *J Immunol Baltim Md 1950* 1997, 159:3808–3815.

[180] Vogel I, Verbinnen B, Maes W, Boon L, Van Gool SW, Ceuppens JL: Foxp3+ regulatory T cells are activated in spite of B7-CD28 and CD40-CD40L blockade. *Eur J Immunol* 2013, 43:1013–1023.

[181] Wadia PP, Herrera ND, Abecassis MM, Tambur AR: Mycophenolic acid inhibits maturation and function of human dendritic cells and B cells. *Hum Immunol* 2009, 70:692–700.

[182] Mehling A, Grabbe S, Voskort M, Schwarz T, Luger TA, Beissert S: Mycophenolate mofetil impairs the maturation and function of murine dendritic cells. *J Immunol Baltim Md 1950* 2000, 165:2374–2381.

[183] Wykes M, Macpherson G: Dendritic cell-B-cell interaction: dendritic cells provide B cells with CD40-independent proliferation signals and CD40-dependent survival signals. *Immunology* 2000, 100:1–3.

[184] Lagaraine C, Lemoine R, Baron C, Nivet H, Velge-Roussel F, Lebranchu Y: Induction of human CD4+ regulatory T cells by mycophenolic acid-treated dendritic cells. *J Leukoc Biol* 2008, 84:1057–1064.

[185] Taylor A, Verhagen J, Blaser K, Akdis M, Akdis CA: Mechanisms of immune suppression by interleukin-10 and transforming growth factor-beta: the role of T regulatory cells. *Immunology* 2006, 117:433–442.

[186] Deng G, Xiao Y, Zhou Z, Nagai Y, Zhang H, Li B, Greene MI: Molecular and biological role of the FOXP3 N-terminal domain in immune regulation by T regulatory/suppressor cells. *Exp Mol Pathol* 2012, 93:334–338.

[187] He X, Smeets RL, Koenen HJPM, Vink PM, Wagenaars J, Boots AMH, Joosten I: Mycophenolic acid-mediated suppression of human CD4+ T cells: more than mere guanine nucleotide deprivation. *Am J Transplant Off J Am Soc Transplant Am Soc Transpl Surg* 2011, 11:439–449.

[188] Sakaguchi S: Naturally arising CD4+ regulatory t cells for immunologic self-tolerance and negative control of immune responses. *Annu Rev Immunol* 2004, 22:531–562.

[189] Chapuis AG, Paolo Rizzardi G, D'Agostino C, Attinger A, Knabenhans C, Fleury S, Acha-Orbea H, Pantaleo G: Effects of mycophenolic acid on human immunodeficiency virus infection in vitro and in vivo. *Nat Med* 2000, 6:762–768.

[190] Quéméneur L, Flacher M, Gerland L-M, Ffrench M, Revillard J-P, Bonnefoy-Berard N: Mycophenolic acid inhibits IL-2-dependent T cell proliferation, but not IL-2-dependent survival and sensitization to apoptosis. *J Immunol Baltim Md 1950* 2002, 169:2747–2755.

[191] Nagai M, Natsumeda Y, Weber G: Proliferation-linked regulation of type II IMP dehydrogenase gene in human normal lymphocytes and HL-60 leukemic cells. *Cancer Res* 1992, 52:258–261.

[192] Gu JJ, Stegmann S, Gathy K, Murray R, Laliberte J, Ayscue L, Mitchell BS: Inhibition of T lymphocyte activation in mice heterozygous for loss of the IMPDH II gene. *J Clin Invest* 2000, 106:599–606.

[193] Jain J, Almquist SJ, Ford PJ, Shlyakhter D, Wang Y, Nimmesgern E, Germann UA: Regulation of inosine monophosphate dehydrogenase type I and type II isoforms in human lymphocytes. *Biochem Pharmacol* 2004, 67:767–776.

[194] Karnell JL, Karnell FG, Stephens GL, Rajan B, Morehouse C, Li Y, Swerdlow B, Wilson M, Goldbach-Mansky R, Groves C, Coyle AJ, Herbst R, Ettinger R: Mycophenolic acid differentially impacts B cell function depending on the stage of differentiation. *J Immunol Baltim Md 1950* 2011, 187:3603–3612.

[195] Yalowitz JA, Jayaram HN: Molecular targets of guanine nucleotides in differentiation, proliferation and apoptosis. *Anticancer Res* 2000, 20:2329–2338.

[196] Allan LA, Morrice N, Brady S, Magee G, Pathak S, Clarke PR: Inhibition of caspase-9 through phosphorylation at Thr 125 by ERK MAPK. *Nat Cell Biol* 2003, 5:647–654.

[197] Ha H, Kim MS, Park J, Huh JY, Huh KH, Ahn HJ, Kim YS: Mycophenolic acid inhibits mesangial cell activation through p38 MAPK inhibition. *Life Sci* 2006, 79:1561–1567.

[198] Choi H, Choi H, Han J, Jin SH, Park J-Y, Shin DW, Lee TR, Kim K, Lee A-Y, Noh M: IL-4 Inhibits the Melanogenesis of Normal Human Melanocytes through the JAK2–STAT6 Signaling Pathway. *J Invest Dermatol* 2013, 133:528–536.

[199] Chaigne-Delalande B, Guidicelli G, Couzi L, Merville P, Mahfouf W, Bouchet S, Molimard M, Pinson B, Moreau J-F, Legembre P: The immunosuppressor mycophenolic acid kills activated lymphocytes by inducing a nonclassical actin-dependent necrotic signal. *J Immunol Baltim Md 1950* 2008, 181:7630–7638.

[200] Guidicelli G, Chaigne-Delalande B, Dilhuydy M-S, Pinson B, Mahfouf W, Pasquet J-M, Mahon F-X, Pourquier P, Moreau J-F, Legembre P: The necrotic signal induced by mycophenolic acid overcomes apoptosis-resistance in tumor cells. *PloS One* 2009, 4:e5493.

[201] Eickenberg S, Mickholz E, Jung E, Nofer J-R, Pavenstadt HJ, Jacobi AM: Mycophenolic acid counteracts B cell proliferation and plasmablast formation in patients with systemic lupus erythematosus. *Arthritis Res Ther* 2012, 14:R110.

[202] Rand TG, Dipenta J, Robbins C, Miller JD: Effects of low molecular weight fungal compounds on inflammatory gene transcription and expression in mouse alveolar macrophages. *Chem Biol Interact* 2011, 190:139–147.

[203] Anderson SE, Franko J, Jackson LG, Wells JR, Ham JE, Meade BJ: Irritancy and Allergic Responses Induced by Exposure to the Indoor Air Chemical 4-Oxopentanal. *Toxicol Sci* 2012, 127:371–381.

[204] Delcuve GP, Rastegar M, Davie JR: Epigenetic control. *J Cell Physiol* 2009, 219:243–250.

[205] Fuks F, Hurd PJ, Wolf D, Nan X, Bird AP, Kouzarides T: The methyl-CpG-binding protein MeCP2 links DNA methylation to histone methylation. *J Biol Chem* 2003, 278:4035–4040.

[206] Rice JC, Allis CD: Histone methylation versus histone acetylation: new insights into epigenetic regulation. *Curr Opin Cell Biol* 2001, 13:263–273.

[207] Bowdridge S, Gause WC: Regulation of alternative macrophage activation by chromatin remodeling. *Nat Immunol* 2010, 11:879–881.

[208] Biel M, Wascholowski V, Giannis A: Epigenetics--an epicenter of gene regulation: histones and histone-modifying enzymes. *Angew Chem Int Ed Engl* 2005, 44:3186–3216.

[209] Marin-Kuan M, Nestler S, Verguet C, Bezençon C, Piguet D, Delatour T, Mantle P, Cavin C, Schilter B: MAPK-ERK activation in kidney of male rats chronically fed ochratoxin A at a dose causing a significant incidence of renal carcinoma. *Toxicol Appl Pharmacol* 2007, 224:174–181.

[210] Marin-Kuan M, Cavin C, Delatour T, Schilter B: Ochratoxin A carcinogenicity involves a complex network of epigenetic mechanisms. *Toxicon Off J Int Soc Toxinology* 2008, 52:195–202.

[211] Ashburner BP, Westerheide SD, Baldwin AS: The p65 (RelA) subunit of NF-kappaB interacts with the histone deacetylase (HDAC) corepressors HDAC1 and HDAC2 to negatively regulate gene expression. *Mol Cell Biol* 2001, 21:7065–7077.

[212] Chen L, Fischle W, Verdin E, Greene WC: Duration of nuclear NF-kappaB action regulated by reversible acetylation. *Science* 2001, 293:1653–1657.

[213] Rahman I, Marwick J, Kirkham P: Redox modulation of chromatin remodeling: impact on histone acetylation and deacetylation, NF-kappaB and pro-inflammatory gene expression. *Biochem Pharmacol* 2004, 68:1255–1267.

[214] Place RF, Noonan EJ, Giardina C: HDAC inhibition prevents NF-kappa B activation by suppressing proteasome activity: down-regulation of proteasome subunit expression stabilizes I kappa B alpha. *Biochem Pharmacol* 2005, 70:394–406.

[215] Song W, Tai Y-T, Tian Z, Hideshima T, Chauhan D, Nanjappa P, Exley MA, Anderson KC, Munshi NC: HDAC inhibition by LBH589 affects the phenotype and function of human myeloid dendritic cells. *Leukemia* 2011, 25:161–168.

[216] Zhu H, Shan L, Schiller PW, Mai A, Peng T: Histone deacetylase-3 activation promotes tumor necrosis factor-alpha (TNF-alpha) expression in cardiomyocytes during lipopolysaccharide stimulation. *J Biol Chem* 2010, 285:9429–9436.

[217] Sica A, Bronte V: Altered macrophage differentiation and immune dysfunction in tumor development. *J Clin Invest* 2007, 117:1155–1166.

[218] Yuan Z-L, Guan Y-J, Chatterjee D, Chin YE: Stat3 dimerization regulated by reversible acetylation of a single lysine residue. *Science* 2005, 307:269–273.

[219] Mullican SE, Gaddis CA, Alenghat T, Nair MG, Giacomin PR, Everett LJ, Feng D, Steger DJ, Schug J, Artis D, Lazar MA: Histone deacetylase 3 is an epigenomic brake in macrophage alternative activation. *Genes Dev* 2011, 25:2480–2488.

[220] Stout RD, Jiang C, Matta B, Tietzel I, Watkins SK, Suttles J: Macrophages sequentially change their functional phenotype in response to changes in microenvironmental influences. *J Immunol Baltim Md 1950* 2005, 175:342–349.

[221] Liu L, McBride KM, Reich NC: STAT3 nuclear import is independent of tyrosine phosphorylation and mediated by importin- 3. *Proc Natl Acad Sci* 2005, 102:8150–8155.

[222] Fatemi M, Hermann A, Gowher H, Jeltsch A: Dnmt3a and Dnmt1 functionally cooperate during de novo methylation of DNA. *Eur J Biochem FEBS* 2002, 269:4981–4984.

[223] Subkhankulova T, Zhang X, Leung C, Marino S: Bmi1 directly represses p21Waf1/Cip1 in Shh-induced proliferation of cerebellar granule cell progenitors. *Mol Cell Neurosci* 2010, 45:151–162.

[224] Stec J, Zmudzki J, Szczotka M, Rachubik J: Effects of aflatoxin B1, ochratoxin A, patulin, citrinin, and zearalenone on the in vitro proliferation of pig blood lymphocytes. *Bulletin of the Vet Insti in Pulaway* 2009, 53:129–34.

[225] Stork LG, Gennings C, Carter WH, Johnson RE, Mays DP, Simmons JE, Wagner ED, Plewa MJ: Testing for additivity in chemical mixtures using a fixed-ratio ray design and statistical equivalence testing methods. *J Agric Biol Environ Stat* 2007, 12:514–533.

[226] Reyes-Velázquez WP, Isaías Espinoza VH, Rojo F, Jiménez-Plasencia C, de Lucas Palacios E, Hernández-Góbora J, Ramírez-Alvarez A: Occurrence of fungi and mycotoxins in corn silage, Jalisco State, Mexico. *Rev Iberoam Micol* 2008, 25:182–185.

[227] Martins HM, Almeida I, Camacho C, Costa JM, Bernardo F: A survey on the occurrence of ochratoxin A in feeds for swine and laying hens. *Mycotoxin Res* 2012, 28:107–110.

[228] Garon D, Richard E, Sage L, Bouchart V, Pottier D, Lebailly P: Mycoflora and multimycotoxin detection in corn silage: experimental study. *J Agric Food Chem* 2006, 54:3479–3484.

[229] Richard E, Heutte N, Sage L, Pottier D, Bouchart V, Lebailly P, Garon D: Toxigenic fungi and mycotoxins in mature corn silage. *Food Chem Toxicol Int J Publ Br Ind Biol Res Assoc* 2007, 45:2420–2425.

[230] Schneweis I, Meyer K, Hörmansdorfer S, Bauer J: Mycophenolic acid in silage. *Appl Environ Microbiol* 2000, 66:3639–3641.

In: Mycotoxins
Editor: Cesar Rios

ISBN: 978-1-63483-544-2

Chapter 2

THE EFFECTS OF THE TRICHOTHECENES: T-2, DEOXYNIVALENOL, DIACETOXYSCIRPENOL, FUSARENON-X, NIVALENOL AND NEOSOLANIOL, ON LIVESTOCK HEALTH AND PRODUCTION WITH EMPHASIS ON RUMINANTS

Mikayla Ross[1], Se-Young Oh[1,*],
Alexandros Yiannikouris[2] and Niel A. Karrow[1,*]
[1]Department of Animal and Poultry Science (APS),
Ontario Agriculture College (OAC), University of Guelph,
Guelph, ON, Canada
[2]Alltech Inc., Nicholasville, KY, US

ABSTRACT

Trichothecene (TCE) contamination of common agricultural grains such as wheat, barley and maize is an increasingly common problem for the livestock industry because it can cause mycotoxicosis in animals affecting their production and survival. Many questions remain as to how TCE are metabolized by different livestock species, what their effects on health and production are, and what levels of intake are tolerable to ruminants without affecting health and productivity. The following TCEs: T-2, deoxynivalenol (DON), nivalenol (NIV), fusarenon-X (FUS-X), diacetoxyscirpenol (DAS) and neosolaniol (NEO) are reviewed herein to assess their effects on livestock health and production. These TCEs commonly affect the immune system, reproduction, and feed intake and weight gain across species. However, further research investigating the effects of these TCEs needs to be conducted in ruminants. Research on the combined

* Corresponding Author: Se-Young Oh, Department of Animal& Poultry Science, University of Guelph, Guelph, Ont. N1G2W1. Email: ugohs530@gmail.com. Tel: (519) 824-4120 ext. 58365.
Niel A. Karrow, Department of Animal& Poultry Science, University of Guelph, Guelph, Ont. N1G2W1. Email: nkarrow@uoguelph.ca. Tel: (519) 824-4120 ext. 53646.

effects of mixed-TCE exposure is especially necessary since this type of exposure typically occurs in nature.

Keywords: *Fusarium* mycotoxins, trichothecenes, ruminants, health

1. INTRODUCTION

Trichothecens (TCEs) are a large group of mycotoxins that are produced by fungi of the *Fusarium* genus. TCEs are typically found in cereals, grains and silages, and are produced by *Fusarium* spp. primarily during plant growth in the field [1]. Furthermore, recent climate changes have been shown to favour *Fusarium* infections of cereals [2]. TCEs are a global concern as they are found in cereals such as barley, corn, oats and wheat that are commonly consumed by livestock and humans [3, 4].

TCEs share a common core structure consisting of an olfenic group, an epoxide group and varying numbers of hydroxyl and acetyl groups. Depending on their functional groups, TCEs can be further classified into one of four groups: A, B, C, and D, where groups A and B are considered to be the most toxic [4, 5]; the TCEs reviewed in this paper belong to groups A or B. In regards to structure, the group A TCEs are distinguished from the group B TCEs by the lack of a carbonyl group on C-8 (Figure 1) [4]. The group A TCEs include T-2 toxin and, the group B TCEs include diacetoxyscirpenol (DAS), deoxynivalenol (DON), nivalenol (NIV), fusarenon-X, and neosolaniol (NEO).

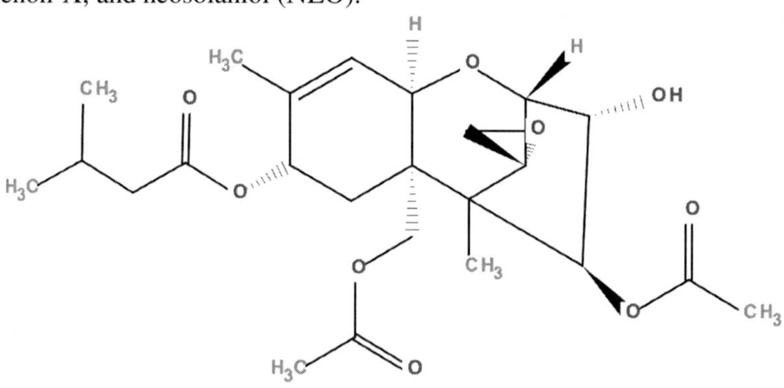

Group A: T-2 Toxin

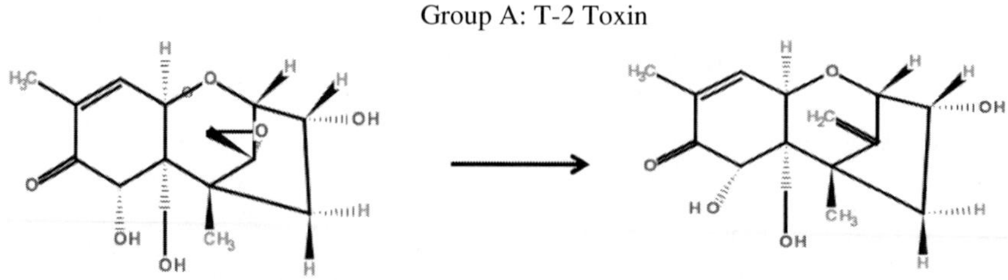

DON DOM-1

Group B: Deoxynivalenol (DON)

Figure 1. Chemical structural differences between groups A & B TCEs and the metabolism of DON to DOM-1 in animals.

Table 1. The effects of TCEs on livestock species

Toxin	Species & Breed	Toxin Source	Concentration (mg/kg feed)	Effects	References
T-2	Holstein calves	Pure (oral exposure)	0.3	- decreased feed consumption	[39]
T-2	Holstein cows	Contaminated corn	0.6	- anorexia - rough hair coats - diarrhea	[40]
T-2	Holstein-friesian cows	Contaminated feed	0.6-1.3	- decreased milk production - absence of visible estrous	[41]
T-2	Ewes	Pure	0.3-0.9	- disrupt ovarian function - prevent ovulation	[42]
T-2	Holstein-friesian x Hungarian red spotted heifers	Pure	9.0	- disrupt ovarian function - prevent ovulation	[42]
T-2	Beef-cross calves	Pure (capsule fed)	0.5 (mg/kg body weight)	- decrease in serum IgG and IgM	[43]
T-2	Beef-cross calves	Pure (capsule fed)	0.5 (mg/kg body weight)	- increase B cell number and proliferative response to mitogens	[44]
T-2	Friesian cattle	Pure	0.1-2.0 ng/ml	- immunosuppressive effect: inhibit proliferation of peripheral mononuclear cells	[45]
T-2	Piglets	Pure	0.5-3.0	- significant reduction in feed intake	[36]
T-2	Pigs	Pure	16.0	- complete feed refusal	[46]
T-2	Piglets	Pure	0.5-3.0	- decrease in leukocyte and T lymphocyte counts - decrease in antibody formation to horse globulin - reduced follicle size in spleen	[36]
T-2	Piglets	Pure	0.54-2.10	- decreased body weight - decreased antibody production - decreased lymphocyte proliferation	[49]
T-2	Sows	Pure	24.0	- sick, dying piglets	[47]
T-2	Sows	Pure (injection)	0.2-0.4 mg/kg body weight	- abortion	[46]
T-2	Broiler Chickens	Contaminated maize	320.5 μg/kg	- decreased body weight gain and feed consumption - decreased nutrient retention - decreased meat quality	[53]
T-2	Broiler Chickens	Pure (added to diet)	4.5-13.5 mg/kg	- decreased body weight gain and feed consumption - increased rate of DNA damage - increased serum IgA	[55]

Table 1. (Continued)

Toxin	Species & Breed	Toxin Source	Concentration (mg/kg feed)	Effects	References
DON	Gilts	Pure	1.0	- decrease in oocyte quality and ovarian dysfunction	[102]
DON	Holstein cattle	Contaminated rolled wheat and ground ear corn	≥6.0	- decreased milk fat content	[66]
DON	German Holstein	Contaminated wheat	7.15-8.05	- increased ammonia concentrations in the rumen - decreased rumen microbial protein synthesis	[7]
DON	Holstein cattle	Contaminated wheat	3.5	- elevated blood urea - reduced neutrophil phagocytosis - increaseIgG response towards OVA	[68]
DON	Holstein steers	Pure	100 pg/ml-10 μg/ml	- impaired functioning of bovine neutrophils	[69]
DON	Pigs	Pure toxin and contaminated corn	0.75-3.0	- decreased feed consumption and weight gain	[64, 73]
DON	Pigs	Pure	0.5-10.0 μM	- intestinal cell membrane damage	[72]
DON	Pigs	Contaminated	0.5-3.1	- intestinal villous and crypt cell apoptosis	[13]
DON	Pigs	Contaminated	2.2-2.5	- increased IgA - decrease lymphocyte proliferation - decreased expression of TGF-β and IFN-γ in the lymph nodes	[17]
DON	Broiler chickens	Naturally contaminated	19.1, 21.4 mg/kg	- decreased body weight and feed consumption during the grower phase (day 21-42)	[74]
DON	Poultry	Artificial contamination	10-20 mg/kg	- decreased feed consumption and weight gain	[75, 76]
DON	Poultry	Artificial contamination	1.7-12.2 mg/kg	- decreased weight gain during 3 weeks of exposure	[81]
DON	Hens	Artificial contamination	18 mg/kg	- decreased egg quality and production	[82]
DON	Hens	Artificial contamination	1.5, 2.0 mg/kg	- decreased egg weight and decreased egg albumen concentration	[83]
DON	Young turkeys	Pure	50 μg/ml	- decreased monocyte phagocytosis	[84]
DON	Chickens	Natural contamination	3.8 μg/g	- increase in total serum IgA and IgG concentrations in coccidia-challenged birds - decreased CD4+ and CD8+ cell populations	[85]
NIV	Cattle	Pure	2 ppm	- ruminal fluid de-epoxidated approximately 80% of NIV	[24]

Toxin	Species & Breed	Toxin Source	Concentration (mg/kg feed)	Effects	References
NIV	Pigs	Contaminated	2.5	- increase in time to consume feed - increase in serum IgA - lesions and thickening of the mucosa in the small intestine	[24]
NIV	Pigs	Contaminated	5.0	- decrease in number of splenic cells - lesions and thickening of the mucosa in the small intestine	[24]
NIV	Pigs	Pure	5mM	- decreased lymphocyte proliferation	[86]
NIV	Chickens	Contaminated	6-12 mg/kg	- decrease in feed consumption and weight gain	[88]
NIV	White leghorn hens	Contaminated	1-5 mg/kg	- decreased feed intake over 50 day period	[89]
Toxin	Species & Breed	Toxin Source	Concentration (mg/kg feed)	Effects	References
NIV + FUS-X	Broiler chicks	Pure	10 μg/ml	- decreased intestinal glucose absorption	[90]
DAS	Lambs	Contaminated	0-5.0	- decreased feed consumption - weight loss - diarrhea	[92]
DAS	Pigs	Pure	0.3-0.5 mg/kg body weight	- lesions and severe hemorrhaging in the small intestine - necrosis of cells in the intestine	[46]
DAS	Pigs	Contaminated	2.5-10	- complete feed refusal and weight loss - decreased total iron binding capacity of the blood	[76]
DAS	Broiler chickens	Pure	12.5, 25 μg/ml	- decreased in number of viable macrophages - decreased phagocytic ability in macrophages	[97]
DAS	Broiler chickens	Pure (treated feed)	5-20 mg/kg feed	- decreased feed consumption and body weight in dose dependent manner - oral lesions	[94]
DAS	Broiler chicken breeders	Pure (treated feed)	5-20 mg/kg feed	- decreased egg production when fed 20mg/kg DAS for 2 weeks. - levels of 10 mg/kg or lower had no detrimental effect on egg production	[96]
DAS	Broiler chicks	Pure (treated feed)	1-2 ppm	- decreased feed intake, body weight - oral lesions - addition of Mycofix counteracted adverse effects of 1 ppm DAS diet on feed intake and	[95]

TCEs can cause a wide range of toxic effects in animals including feed refusal leading to weight loss, diarrhea, vomiting, immunomodulation, and skin lesions [3, 6].

The severity of toxicity of TCEs to animals consuming contaminated feed is dependent on livestock species, metabolism of the toxin, its mode of action, and the animals' physiological defense mechanisms. The mechanisms by which ruminants handle mycotoxins appear to differ from monogastrics, and it has been suggested that due to the rumen's ability to degrade certain mycotoxins, ruminants may be less likely to be adversely affected by mycotoxin contaminated feed [7, 8]. However, studies have shown that the biotransformation of TCEs does not necessarily mean a decrease in toxicity, since some metabolites are actually more toxic than their parent toxin [6]. The effects of each TCE on ruminants and monogastrics are summarized in Table 1. It is possible that TCEs and their metabolites can partition into animal meat and milk products, increasing the potential risk of exposure to humans [9, 10]. For these reasons, a comprehensive understanding of how TCEs are metabolized by different species is required to provide a better understanding of their toxic effects and how exposure to these toxins can be better controlled.

The objective of this paper is to review the effects of the following TCEs: T-2, DON, DAS, FUS-X, NIV and NEO on ruminant health and production, and to identify gaps in our knowledge that can be reduced by follow up research. The effects of these TCEs will also be reviewed for pigs and chickens, and in some cases humans, to compare the different effects of these toxins across ruminant and monogastric species. This work will provide guidance for further study on these TCEs.

2. TCE Mode of Action

TCEs are known for their ability to inhibit protein, RNA and DNA synthesis. This process occurs when TCEs interfere with the active site of peptidyltransferase in ribosomes, thus inhibiting mRNA translation to protein [11]. The precise mechanisms by which TCEs inhibit RNA and DNA synthesis have not been determined, but are suggested to be a secondary effect of inhibited protein synthesis [12]. Studies have also shown that TCEs induce apoptosis both *in vivo* and *in vitro* [13, 14]. TCEs induce apoptosis through mitochondria-mediated and mitochondria–independent pathways [14–16]. In the mitochondria-mediated pathway, TCEs induce apoptosis via increased expression of pro-apoptotic proteins such as: p53 and Bax, which cause a reduction in mitochondrial membrane potential. Along with these effects, TCE treatment causes opening of permeability transition pore complex, cytochrome C release to the cytoplasm, and caspase-3 activation; together, these events promote apoptosis. TCEs can also promote apoptosis through mitochondria-independent pathways, where apoptosis occurs due to caspase-3 activation via the extrinsic pathway. In this case, caspase-3 activation is mediated by the activation of Fas-Fas ligand or TNF receptors on the cell surface [15].

A number of studies have also shown that TCEs can modulate immune responses of animals and humans. TCEs can be either immunosuppressive or immunostimulatory depending on the dose, frequency and duration of exposure, and tissue or cell type being treated. Low-level TCE exposure promotes increased production and secretion of several inflammatory cytokines along with elevated serum immunoglobulin levels [17, 18]. In contrast, high dose TCE exposure has an immunosuppressive effect. The immunosuppressive effects of TCEs are characterized by decreased immunoglobulin secretion, reduced lymphocyte proliferation, and apoptosis of various lymphoid tissues [18, 19].

3. Metabolism of Trichothecenes

TCEs undergo a variety of different biotransformations in animals. Common reactions that take place across all species include hydrolysis, hydroxylation and de-epoxidation of side groups. In monogastrics, de-epoxidation occurs in the small intestine [20, 21]. In ruminants, microorganisms in the rumen are largely responsible for reduction of the epoxide ring [21]. Metabolism of T-2 can occur in the liver, digestive tract, and in the rumen of ruminants. DON, however, is more commonly metabolized by microorganisms in the hindgut or rumen, as opposed to the liver [22, 23]. Ruminants and swine both have the ability to excrete de-epoxidated NIV in the feces [24]. Poultry can rapidly metabolize FX to NIV in the liver and kidney, but they appear to be incapable of metabolizing NIV to less potent de-epoxy NIV [25]. Similarly, chickens also lack the ability to de-epoxidate DAS, whereas, in ruminants and swine this is achieved in the digestive tract and liver [26]. Understanding how TCEs are metabolised by different species may help to explain species-specific effects that are commonly seen after their consumption, and this may have important human health implications for the consumption of different animal products [10].

Most TCEs and their metabolites are excreted via feces and urine, however, there is evidence that trace amounts of TCE residues can be deposited in animal tissues and products [27–31]. For example, ^{3}H radio-labeled T-2 was detected in blood, muscle, skin, and heart tissues 12 hours post intubation of the ^{3}H radio-labeled T-2 toxin, but was highest in bile and liver tissues collected from 6 week-old broiler chicks that were fed 0.5 mg/kg of T-2 for 48 hours [27]. A later study conducted by Visconti et al. (1985) reported high concentrations of T-2 residues in the chicken liver, but not in heart and kidney tissues, and only trace amounts were found in the lungs 17 hours post IP injection with T-2 [31]. Residual T-2 toxin ranging between 2.5 and 8.7 ppb was also detected in cows' milk that was collected 24 hours after three daily oral administrations of T-2 (180 mg/capsule), and the amount was reduced below 1 ppb after 36 hours [32].

Small amounts of DON residues have also been detected in animal tissues. For instance, DON was readily distributed plasma and other tissues, with the majority being found in the small intestine, kidney, proventriculus, liver, spleen, brain, heart, gizzard, fat, muscle and oviduct tissues collected from white leghorn hens fed 2.2 mg/bird/day [30]. Residual DON concentration in these tissues reached a maximum 3-6 hours post feeding, and accounted only 1.33% of the total dose, which was equivalent to 30 μg DON/hen. By 24 hours, the residues were reduced to 0.29%, and most DON was cleared by 96 hours, with only 0.0043% of the total dose remaining in cutaneous fat and kidney tissues. A few studies have reported that extremely low concentrations of DON can be transmitted to eggs and milk. Approximately 0.087% of DON for example, was transmitted into eggs that were laid within 24 hours of DON administration [33], while less than 0.1% of DON (26 ng/ml) was recovered from milk collected over a 10-day period from Holstein and Ayrshire cows fed 66 mg/kg of DON for 5 days [28]. DON residue in the eggs was detected in yolk, albumen, as well as egg shells, and continuous feeding of 2.2 mg/kg of DON for 10 days rapidly increased DON residues by 10-fold in the yolk from 22.1 to 225.2 dpm/g, while only a 3-fold increase was detected in albumen (20.5 to 57.6 dpm/g); for the egg yolk, only about 10% of the residue was identified as unmetabolized DON [33].

Other studies have reported on the toxicokinetics of DAS, FUS-X and NIV residues in animal tissues [34, 35]. Less than 10 ng/g of DAS was detected in liver, kidney, skeletal muscle, mesenteric lymph nodes and spleen tissues isolated from pigs fed up to 1 mg/kg DAS, as well as heifers fed 0.5 mg/kg of DAS [34]. Trace amounts of FUS-X and NIV residues have also been found in liver, kidney and spleen tissues from piglets sampled 24 hours after 1 mg/kg FUS-X administration [35]. The toxicokinetics of these TCE residues in animal products such as eggs and milk are not available at this point in time. Finally, it should be emphasized that, in spite of the above evidences, the presence of TCEs and its derivatives in animal tissues and products presents no appreciable hazards to consumers.

4. Toxicity of Trichothecenes

4.1. T-2

T-2 toxin is one of the most potent of the TCEs and it is produced by *F. acuminatum, F. equiseti, F. poae* and *F. sporotrichoides*. The highest concentrations of T-2 toxin tend to be found in oats and barley [8] and in some countries in maize [36]; wheat is typically less contaminated than these grains [37]. Following consumption, T-2 is rapidly de-acetylated to hydroxy-T-2. A study conducted by Königs et al. (2009) however, indicated that the conversion of T-2 to HT-2 does not seriously affect toxicity [38], thus many studies regard the effects of both these TCEs to be equivalent.

4.1.1. Ruminants

Ruminants, including cattle are adversely affected by exposure to T-2 toxin. Clinical signs commonly displayed in cattle include feed refusal, weight loss, and diarrhoea [8]. Decreased feed consumption was also reported in a study using beef cattle that were orally exposed to 0.3 mg/kg T-2 toxin in artificially contaminated feed [39]. Another study, where dairy calves were fed 0.6 mg/kg T-2 toxin in naturally contaminated feed, showed more severe clinical symptoms such as anorexia, rough hair coats and diarrhoea [40]. In another study where dairy cattle were fed a diet contaminated with 0.6-1.3 mg/kg T-2 toxin, decreased milk production was observed as well as the absence of visible estrous [41]. A later study noted that T-2 had an effect on ovarian activity in ewes and heifers, preventing follicle maturation, which postponed ovulation [42].

Many studies have demonstrated the immunosuppressive effects of T-2 toxin in various species, but strong evidence of an immunomodulatory effect in cattle is lacking. A study conducted by Mann et al. (1983) exposed beef cattle to T-2 toxin for 28 days at 0.5 mg/kg body weight and noted a significant decrease in serum IgG and IgM concentrations in comparison to control animals [43]. The study also noted a significant decrease in albumin concentration as well, which may reflect a general reduction in protein synthesis. A later study by Mann et al. (1984) fed Holstein calves 0.5 mg of T-2/kg body weight for 28 days to determine the effects of T-2 on lymphocyte and neutrophil function [44]. Their data indicated that between days 7 and 19 of exposure, B cell numbers and phytohemagglutinin (PHA)-induced B cell proliferation appeared slightly enhanced, whereas neutrophil function was not significantly affected. This elevation in B cell number and proliferative response to PHA is commonly seen when animals are exposed to T-2, highlighting the toxins ability to enhance the some aspects of immune function [45, 36]. The lack of effect on neutrophil function could be attributed to the concentration of toxin administered in this study. A concentration of 0.5 mg/kg body weight is considered subclinical, and therefore may not have had a significant effect on other immune cells. A more recent *in vitro* study using bovine peripheral blood mononuclear cells (PBMC) demonstrated that T-2 inhibited mitogen-induced proliferation of lymphocytes [45]. This study also examined the effects of ochratoxin A (OTA), T-2 toxin, DON, and fumonisin B1 (FB1) on lymphocyte proliferation induced by the following mitogens: concanavalin A (Con A), PHA and pokeweed mitogen (PWM) as a measure of non-specific cellular immune function. In comparison to DON, OTA and FB1, T-2 toxin was a more potent inhibitor of PBMC proliferation. For instance, concentrations of T-2 that inhibited Con A-, PHA-, and PWM-stimulated PBMC proliferation by 50% (IC50) were: 0.30, 0.40, and 0.18 ng/ml, respectively, whereas IC50s for DON, OTA and FB1 ranged from 0.07-35 μg/ml [45]. These findings demonstrate that T-2 exposure can affect immune function; however, no *in vivo* studies using a pathogen challenge model have been conducted to determine if these changes in immune function affect disease resistance in cattle.

4.1.2. Swine

T-2 toxin has been demonstrated to reduce the feed intake with consequent depression in growth rate in growing pigs at as low dietary concentrations of 0.5 mg/kg [36]. In other experiments, higher dietary concentrations were proven effective e.g., 1.324-2.102 mg/kg (Meissonier et al., 2008). Complete feed refusal was observed in an earlier study where pigs were given an incredibly high dose of 16 mg/kg of T-2 toxin in their feed [46].

T-2 toxin also impacts reproductive performance in pigs. A study by Ványi et al. (1991) for example, looked at the impact of T-2 exposure to pregnant sows fed a concentration of 24 mg/kg T-2 toxin in feed during the last six weeks of pregnancy, and they reported that piglets born from these sows displayed diarrhoea, wasting, coma, and death soon after birth [47]. The piglets were necropsied, and liver and kidney degeneration was observed, as well as enteritis, necrosis of lymphoid tissues, and atrophy of the thymic cortex. Further examination revealed T-2 toxin and its metabolites HT-2, T-2 triol and neosolaniol in the stomach contents of the dead piglets and the sows' milk. In contrast, litters from sows fed 6 mg/kg of T-2 toxin or less appeared normal. In another study, abortion occurred after IV injection of 0.2-0.4 mg/kg body weight of T-2 toxin [46]. However, it should be noted that the exposure levels of T-2 toxin used in these studies were much higher than what is currently detected in livestock feed [48].

In comparison to cattle, the immunosuppressive effect of T-2 toxin in pigs has been well documented. By feeding 7 week old pigs with diets containing 0.5, 1.0, 2.0, or 3.0 mg/kg of purified T-2 toxin, Rafai et al. (1995) reported, for example, significant decreases in total leuokocyte and T lymphocyte counts [36]. The pigs in this study were also immunized with horse globulin to determine the ability of the immune system to develop an antigen-specific antibody response during T-2 toxin exposure; the study found a dose-dependent decrease in antibody production. The study also reported reduced follicle size within the spleen, suggesting a toxic effect of T-2 toxin on this lymphoid organ. In contrast to these findings, a similar study by Meissonnier et al. (2008) reported normal histological morphology of the spleen and Peyer's patches in the ileum of pigs [49]. However, findings from this study that were in agreement with those of Rafai et al. include: decreased body weight, reduced anti-ovalbumin antibody production, and decreased lymphocyte proliferation when pigs were fed diets containing ≥2.0 mg/kg of purified T-2 toxin [49, 50].

4.1.3. Poultry

Several studies have revealed that T-2 toxin significantly reduces feed consumption, body weight gain, and feed efficiency when poultry are exposed to 0.3-16 mg/kg T-2 toxin [51–55]. Altered feather quality is another clinical sign of T-2 mycotoxicosis in poultry. Wyatt et al. (1975) fed broiler chickens 2-16 mg/kg T-2 from hatching until 3 weeks of age and altered feathering was observed in birds fed ≥4 mg/kg T-2 [56]. Oral lesions are another common outcome of T-2 exposure when chickens are exposed to >1 mg/kg T-2 [57, 58].

T-2 toxin can also affect poultry egg production and quality. For example, when hens were exposed to 1-20 mg/kg T-2, a dose-dependent decrease in egg production was observed across all concentrations [59]. Changes in eggshell quality were also detected, but only at a high toxin level of 20 mg/kg [59]. Lower egg and shell weight, and decreased hatchability have also been associated with T-2 consumption at concentrations ranging from 2-10 mg/kg [57, 60].

T-2 toxin can also modulate poultry immune responses. As seen with other species, T-2 can have a stimulatory and suppressive effect on the poultry immune response. For instance, an inhibitory response was seen in an *in vitro* study where lymphocytes from White Leghorn chickens were exposed to T-2 at concentrations of 0.01-10 ng/ml [61]. In this study, lymphocyte proliferation in response to Con-A and PHA stimulation was significantly attenuated at concentrations ≥1 ng/ml, and 10 ng/ml completely inhibited lymphocyte proliferation. A study by Yohannes et al. (2012) fed 2 mg/kg T-2 to broiler chicks for six weeks that were infected or not infected with infectious bronchitis virus (IBV) to evaluate immunopathological effects of T-2 exposure [62]. Similar to the findings by Jaradat et al. (2006), T-2 exposure caused a decrease in Con A-induced lymphocyte proliferation in the uninfected and IBV-infected broilers [61]. The number of $CD4^+$ and $CD8^+$ lymphocytes was significantly reduced in the T-2, IBV, and T-2 + IBV treatment groups suggesting that T-2 has a negative impact T cell populations. In addition to affecting cellular immune function, Yohannes et al. (2012) also demonstrated that T-2, and T-2 + IBV suppressed the antibody response to vaccination with Newcastle disease F strain vaccine, and T-2 suppressed the antibody response to IBV challenge [62]. These data, combined with documented lesions in various lymphoid tissues including the Bursa Fabricus, thymus, spleen and other secondary lymphoid tissues, led Yohannes et al. (2012) to conclude that T-2 has multi-immunosuppressive effects on poultry that can compromise their resistance to IBV infection.

4.2. Neosolaniol

Neosolaniol (NEO) is a metabolite of T-2 toxin that is found after T-2 loses its C-8 isovaleryl group. NEO can be detected in food or feed samples, and can form in vivo after consumption of T-2-contaminated feed [38]. This transformation results in a significant reduction in toxicity [4, 38]. There are no available data concerning the effects of NEO in ruminants, swine, or poultry, however, some rodent and human data are available.

4.2.1. Humans

The majority of the studies examining the effects of NEO involved the use of human or mice cell lines or primary cell cultures, however, these data are also very limited. One recent study by Königs et al. (2009) demonstrated that NEO was cytotoxic to human lung fibroblasts (NHLF) and renal tubular epithelial cells (RPTEC) [38]. In this study, cells were exposed to different concentrations of T-2 and its metabolites HT-2 and NEO. With increasing concentrations of these toxins, a decline in cell viability was observed. Furthermore, RPTEC were able to transform T-2 and HT-2 into NEO during exposure, and the loss of the isovaleryl group during biotransformation dramatically reduced cytotoxicity, which is in agreement with other studies [63, 64]. The Königs et al. (2009) study also examined the apoptotic effects of T-2 and its metabolites. Incubation with NEO significantly increased caspase-3 activity, which is a marker for the induction of apoptosis, and fibroblasts seemed to be more sensitive to caspase-3 activation than RPTEC [38]. Given that NEO is found in animal feed, it would be beneficial to determine whether toxic effects occur in farm animal species, including ruminants. Further research is needed to determine the concentration range at which these toxic effects might occur.

4.3. Deoxynivalenol (DON)

DON and mono-acetylated 3- and 15-acetyl forms of DON are considered the most common occurring TCEs. DON is produced by *F. graminearum* and *F. culmorum*, and is typically found at high levels in corn [48]. It has a lower toxicity than T-2 toxin, but is far more prevalent in cereals and at much higher concentrations [3, 6]. It can also be found as a masked mycotoxin called DON-3 glucoside that can be converted into DON by lactic acid bacteria [65].

4.3.1. Ruminants

The effects of DON in ruminants are less apparent than other TCEs. Several studies have noted that DON is converted to DOM-1 by rumen microbes (Figure 1), which is considerably less toxic [10]. However, since studies have shown that this transformation occurs slowly, DON may potentially affect the rumen microflora or epithelium of the gastrointestinal tract during exposure [21].

Low concentrations of DON do not seem to cause any apparent clinical signs in ruminants. In a study where lactating dairy cows were fed 0-12mg/kg DON in concentrate for example, no decrease in feed intake, body weight or milk production was observed [66]. The study did show, however, that when cows were fed more than 6 mg/kg DON, milk fat content was significantly lower, but the authors suggested that this could have been due to a nutrient imbalance in the feed. These results are in agreement with a similar study conducted by Ingalls et al. (1996), who found that consumption of 6-24 mg/kg DON for 3 weeks had no effect on feed intake, body weight or milk production, and milk fat content was not affected by consumption of the contaminated diet [67].

A study by Dänicke et al. (2005) looked at the effects of consumption of feed artificially contaminated with DON on the microbial environment of the rumen, and its ability to alter the fermentation capacity of the rumen; the study also assessed DON's ability to alter dry matter flow through the duodenum of cows [7]. Interestingly, this study found that rumen ammonia concentration was significantly increased 1.5 to 5 hours after feeding a diet contaminated with 7-8 mg/kg DON. The increase in ammonia concentration should lead to increased utilization by rumen microbes for enhanced net protein synthesis. However, this did not occur, as indicated by the decreased flow of microbial protein at the duodenum. This decreased microbial protein synthesis could be due to an altered gut microbial profile, that could be caused by DON. The study did make note, however, that while DON was the main contaminant found in the feed, there were a number of other *Fusarium* mycotoxins present in the contaminated wheat. Also, artificial contamination of the wheat with *Fusarium* mycotoxins was found to have altered the physical and chemical properties of the grain. The authors suggested that this alteration of the grain might have contributed to the outcome of the experiments. Analysis of mycotoxin flow through the duodenum of cows from this study revealed different concentrations of DON and de-epoxy DON compared to control animals, which suggests that either rumen microbes play an important role in degrading DON to a less toxic compound, or that DON itself is readily absorbed within the digestive tract. In agreement with this study, findings by Korosteleva et al. (2009) show elevated blood urea when cows were fed a naturally contaminated DON diet. An elevation in blood urea suggests that protein utilization by rumen microflora and efficiency of microbial protein synthesis are both reduced by DON [68]. However, in contrast to the Dänicke et al. study, no effect of diet on rumen ammonia concentration was observed. These authors attributed these differences to the total mixed ration (TMR) diet that they used in their study, which may have reduced the toxicity of the contaminated ingredients by diluting the concentration of mycotoxins.

The study by Korosteleva et al. (2009) also showed that a TMR naturally contaminated with DON affected immune function [68]. For example, cows on this diet exhibited reduced neutrophil phagocytosis. Neutrophils play an important role as effector cells during the innate immune response. Another study looked at the effects of DON on the production of bovine neutrophil antimicrobial reactive oxygen species (ROS) [69]. In this study, bovine neutrophils were incubated with 100 pg/ml to 10 μg/ml of DON, and then stimulated with phorbol myristate acetate (PMA) to induce ROS production. DON concentrations >5 pg/ml were shown to significantly decrease ROS production. A similar study by Takayama et al. (2005) also evaluated ROS production in PMA-stimulated bovine and porcine neutrophils exposed to 0.001-10 μg/ml of DON, and they showed a dose-dependent decrease in ROS production [70]. While it is unclear where the inhibitory effect of DON occurs in the cascade of events that lead to decreased ROS production, these findings do show that DON has a suppressive effect on ruminant neutrophil function.

In regards to specific immunity, cattle from the Korosteleva et al. (2009) study were immunized with OVA and an enhanced OVA-specific IgG response was detected in cows consuming the TMR naturally contaminated with DON [68]. This finding is in agreement with other studies involving different species that have shown that low concentrations of TCEs enhance specific immunity, whereas high concentrations lead to immunosuppression.

4.3.2. Swine

Many porcine studies have been carried out to study the toxic effects of DON because pigs are highly susceptible to DON mycotoxicosis [71]. The toxic effects of DON include reduced weight gain and feed efficiency, diarrhoea, vomiting and gastrointestinal inflammation, and necrosis of several different tissues [72]. In feeding trials where DON was naturally or artificially mixed into the diet, DON has been shown to decrease feed consumption and weight gain at doses from 0.75-3.0 mg DON/kg feed [64, 73].

Significant morphological changes have been noted in the intestine of pigs fed a DON-contaminated diet. For example, a study assessing the in vivo effects of DON on the small intestine of pigs showed that low concentrations of DON (0.5-3.1 mg/kg) cause villous and crypt cell apoptosis in the small intestine [13]. An *in vitro* study was also carried out to determine if DON altered intestinal epithelial cell function [72], and these authors showed that even the lowest doses (0.5-2.5 μM) of DON reduced cell viability and increased the release of lactate dehydrogenase into the culture medium, which is an indication of cell membrane damage.

Porcine studies have further shown DON's ability to modulate immune function. A study by Pinton et al. (2008) for example, showed that ingestion of feed naturally contaminated with DON affected the immune response to OVA antigen [17]. Pigs were fed a naturally contaminated diet containing 2.2-2.5 mg/kg DON for 9 weeks, and during this period an increase in total IgA and OVA-specific IgA and IgG was noted in OVA vaccinated pigs that consumed the contaminated diet. This study also evaluated the effect of DON on mitogen-and antigen-induced lymphocyte proliferation. While DON had no effect on mitogen-stimulated lymphocyte proliferation, it significantly increased antigen-specific lymphocyte proliferation after 21 days of exposure, whereas at day 35, lymphocyte proliferation was significantly decreased. This study also measured gene expression levels of cytokines TGF-β, IFN-γ, IL-4 and IL-6 in the spleen, ileum and lymph nodes. In the lymph node, mRNA expression levels for both TGF-β and IFN-γ were found to be significantly lower in animals fed the DON-contaminated diet. The cytokines TGF-β and IFN-γ play an important role in modulating T-cell differentiation, proliferation, and function. Decreased expression of these cytokines may contribute to increased susceptibility of disease in animals exposed to DON.

4.3.3. Poultry

Many studies have been published on the effects of DON on poultry. Feed refusal, reduced weight gain and feed consumption have been shown in several studies. These effects are typically seen when poultry are exposed to high concentrations of DON. For instance, a study by Swamy et al. (2004) fed grain naturally contaminated with 19.1 and 21.4 mg/kg DON to broiler chickens and these authors observed a decrease in body weight and feed consumption during the grower phase (day 21-42) [74]. However, feed intake and weight gain of these birds recovered during the finisher phase. In other studies, where poultry were fed 10-20 mg/kg DON, a significant decrease in feed consumption and weight gain was observed [75, 76]. Lower concentrations of DON, for example between 0.1-5.0 mg/kg, do not appear to affect weight gain and feed consumption [77–80]. Although, findings from a study by Yunus et al. (2012) showed that both low and high concentrations (1.7-12.2 mg/kg) of DON added to feed caused a decrease in weight gain during the first 3 weeks of exposure [81].

An early study suggested that only high concentrations of DON (18 mg/kg) can also significantly affect egg quality and production [82]. However, a study by Lee et al. (2012) observed decreased egg weight in hens fed 1.5 and 2.0 mg/kg DON, as well as decreased egg albumen concentration [83]. After DON was removed from the diet, hens went through a 4-week recovery phase after which all egg quality measurements returned to levels equal to that of eggs produced by the control birds. Data from this experiment indicate that DON exposure has the potential to affect egg quality, even at concentrations as low as 1.5-2.0 mg/kg.

DON also has an immunosuppressive effect in poultry. An *in vitro* study by Kidd et al. (1995) for example, exposed monocytes from young turkeys to 50 and 200 μg/ml DON and evaluated the ability of DON to depress the phagocytic ability [84]. This study found that 50 μg/ml DON decreased monocyte phagocytosis of opsonized sheep RBC (sRBC), and 200 μg/ml decreased the phagocytosis of unopsonized sRBC. This study also showed that DON dose-dependently reduced macrophage cell viability. Another study by Girgis et al. (2008) investigated the effects of feeding grains naturally contaminated with 3.8 μg/g DON on various immune parameters of chickens during an enteric coccidial infection [85]. Their results showed an increase in total serum IgA and IgG concentrations in coccidia-challenged birds that had been fed the DON-contaminated diet. The percentage of CD4+ and CD8+ cell populations also decreased in birds fed the contaminated diet; the authors of the study suggested this could be due to T cell migration from the peripheral blood to the intestine, where coccidia replication occurs. Findings from this study suggest that DON can modulate immune responses in birds.

4.4. Fusarenon-X &Nivalenol

FUS-X and NIV belong to the B group of TCEs, and FUS-X is distinguished from NIV by an acetyl group at the C4 position (Figure 2). FUS-X is rapidly deacetylated to NIV in vivo, which is why their effects are commonly assessed in conjunction with each other [5]. FUS-X and NIV are produced by *F. graminearum, F. culmorum, F. poae,* and *F. cerealis* and tend to co-occur with DON [6].

4.4.1. Ruminants

There is a major lack of data on the effects of FUS-X and NIV in ruminants, however, one study was found regarding the ability of rumen microbes to transform NIV. This study by Hedman et al. (1997) involved incubating NIV with ruminal fluid obtained from fistulated cows [24]. Their results showed that the ruminal fluid was able to metabolize approximately 80% of NIV at a concentration of 2 ppm, through de-epoxidation. While findings from this in vitro study are useful in determining the ability of the rumen to metabolize NIV, it may not accurately represent what happens in the actual rumen of the cow.

4.4.2. Swine

Feeding NIV to pigs has not been associated with a decrease in feed intake or body weight, which typically occurs when feeding TCEs to pigs. One early study, where young pigs were fed 2.5 or 5mg/kg NIV for 3 weeks, noted an increase the in time to consume feed contaminated with NIV, however, this had no effect on quantity consumed or body weight [24].

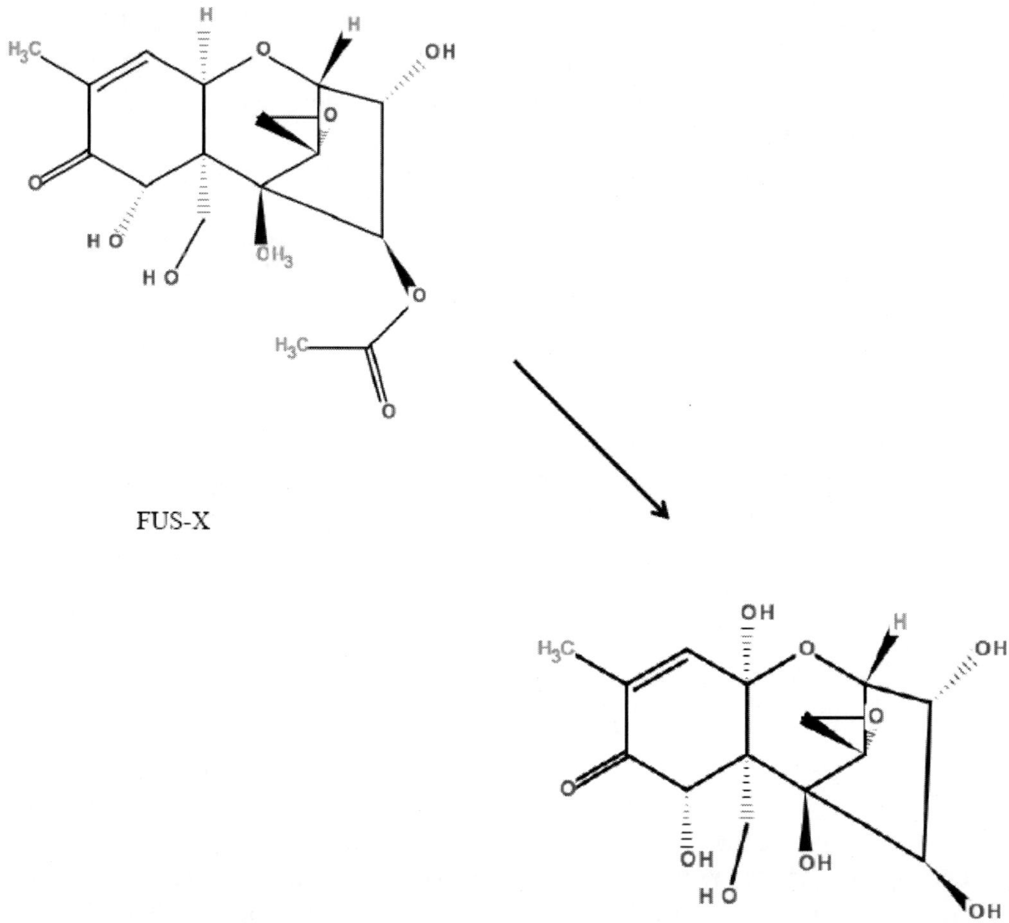

FUS-X

NIV

Figure 2. Deacetylation of FUS-X to NIV in animals.

The study by Hedman et al. (1997) also showed morphological changes in the small intestine and kidney of pigs exposed to NIV. Pigs that received a contaminated diet containing 2.5 or 5 mg/kg of NIV exhibited lesions and thickening of the mucosa in the small intestine, suggesting that NIV can impair proper functioning of the gastrointestinal tract in pigs.

NIV appears to have an effect on porcine immunity as well. A study by Taranu et al. (2010) for example, found that in vitro exposure of pig lymphocytes to NIV inhibited the PMA-stimulated cell proliferation [86]. This study also reported cell-specific differences in sensitivity to NIV. For instance, porcine splenocytes were more sensitive to NIV than human lymphocytes.

4.4.3. Poultry

Information on the toxic effects of NIV to poultry is also limited. Studies by Pettersson et al. (1995) and Hedman et al. (1995) showed a dose-dependent decrease in feed consumption and weight gain when chickens were fed 6-12 mg/kg NIV [87, 88]. Birds from these studies that were fed less than 5 mg/kg NIV, were not affected by the diet. Similar findings from a study by Garaleviciene et al. (2002) showed that 1-5 mg/kg NIV fed for 50 days to White leghorn hens resulted in decreased feed intake, but no effect on body weight, egg production or egg quality was observed [89].

An in vitro study by Awad et al. (2008) using epithelial cells of the jejunum of male broiler chicks was conducted to evaluate the effects of the B-TCEs DON, NIV and FUS-X on intestinal glucose absorption [90]. Intestinal transport was determined by changes in the short-circuit current (Isc), proportional to ion trans-membrane flux. When cells were exposed to D-glucose, an increase in Isc was observed. Once these cells were exposed to 10 μg/ml of each of the B-TCEs, a decrease in Isc occurred. These findings suggest that NIV and FUS-X cause decreased absorption of glucose due to the inhibitory effect of B-TCEs on D-glucose/sodium dependent transport.

4.5. Diacetoxyscirpenol

Diacetoxyscirpenol (DAS) belongs to the group A TCEs. It is a highly toxic mycotoxin produced by *F. sambucinum* that is typically found at very low concentrations (0.4, 2.7 μg/kg) in cereals [6, 91].

4.5.1. Ruminants

Data on the effects of DAS on ruminant health and production are sparse. One study, by Harvey et al. (1995) exposed lambs to 0-5 mg/kg of DAS in their feed [92]. They observed a decrease in feed consumption and significant weight loss when lambs were fed the higher concentrations. Lambs also exhibited diarrhoea for the first two weeks of the experiment at the higher concentrations. This study also examined immunologic and hematologic parameters, however, no change in white blood cell count, neutrophil count and lymphocyte response was observed.

4.5.2. Swine

Similar to ruminants, there are limited data available regarding the effects of DAS in swine. One study conducted by Weaver et al., (1978) administered purified DAS via intravenous (IV) injection to pigs [46]. Several pigs that were administered DAS displayed lesions and severe haemorrhaging and edema in the small intestine and colon. Additionally, the mucosal crypts of the small intestine and mucosa from the colon of these pigs contained necrotic cells. The findings of this study may not be physiologically relevant, as exposure to DAS via IV injection is likely to cause more severe effects than exposure via consumption of contaminated grains. Furthermore, lesions observed in this study were not consistently reproducible. Another study by Harvey et al. (1991) showed that DAS caused weight loss and complete feed refusal for pigs fed 2.5, 7.5, and 10mg/kg DAS [76]. In contrast to the study by Weaver et al. (1978), pigs in this experiment did not develop any small intestinal lesions.

Interestingly, DAS induced a decrease in the total iron-binding capacity of the blood, which also occurs during T-2 exposure [46]. This suggests that DAS as well as T-2 may both interfere with iron absorption or transport [93].

4.4.3. Poultry

In comparison to ruminants and swine, the effects of DAS have been well studied in poultry. Similar to other mycotoxins, DAS reduces feed intake in poultry exposed to the toxin. A study by Brake et al. (2000) for example, exposed broiler breeder chickens to 10-20 mg/kg of DAS for three weeks [94]. Chickens exposed to the highest concentration showed the greatest decline in feed consumption. The study also found a dose-dependent effect on the occurrence and severity of oral lesions caused by DAS consumption. Another study by Diaz et al. (2002) reported similar effects when broiler chicks were fed aDAS-contaminated diet. In this study, day old chicks received 1-2 mg/kg DAS for 21 days [95]. The most significant decline in feed intake and weight gain occurred when chicks were exposed to 2 mg/kg DAS.

DAS was also found to affect broiler breeder egg production [96]. In this study, chickens were exposed to 10 or 20 mg/kg DAS for three weeks. No significant effect on egg production was observed at 10 mg/kg DAS, however, egg production was reduced when chickens were fed 20 mg/kg DAS within two weeks. Since the adverse effects on egg production were no longer seen after the DAS treatment ended, these findings suggest that short-term consumption of pure DAS at levels that may naturally occur in feed have a minimal effect on egg production in broiler parent flocks.

DAS may also be immunotoxic to poultry. An in vitro study by Qureshi et al. (1998) tested the effects of DAS on peritoneal macrophage viability, phagocytic ability and Fc-receptor expression at concentrations of 12.5 or 25 μg/ml 1 DAS [97]. Treatment with DAS resulted in decreased cell viability at the highest DAS concentration. Also, when exposed to DAS, a decrease in the percentage of macrophages phagocytising RBC was observed. sRBC were also used to assess macrophage Fc-receptor expression, which was also reduced by DAS. These findings indicate that DAS suppresses macrophage phagocytic ability in vitro, which may compromise the local and overall immune response in chickens.

5. INTERACTION BETWEEN *FUSARIUM* TOXINS

Most TCE studies have investigated the effects of one particular toxin or measured individual toxins in naturally contaminated feeds. However, the presence of only one toxin in naturally contaminated feed does not typically occur. For instance, Eckard et al. (2011) recently found varying concentrations of DON, Ac-DON, NIV, HT-2, and T-2 in 19 different silage maize samples [98]. Therefore, the different effects seen in experiments whereby animals are fed naturally contaminated feed versus pure toxin feed could be explained by the interactions occurring with other mycotoxins found in the diet [99].

Some studies have reported combined effects of TCEs both in vitro and in vivo. A study involving feeding DON and the *Fusarium* mycotoxin fumonisin (FB) to piglets demonstrated additive, synergistic and antagonistic effects on morphological and immunological changes in the small intestine [13]. When piglets were fed a combined diet containing 3 mg/kg DON and 6 mg/kg FB for example, additive and synergistic effects were observed in regards to the expression of important immunoregulatory cytokine genes such as TNF-α, IL-1β, IFN-γ, IL-6 and IL-10 in the small intestine. An antagonistic effect was also noted whereby FB was able to prevent DON-induced intestinal expression of the cytokine IL-6, which is responsible for initiating IgA class switching, among other things. These findings suggest that these combined mycotoxins may alter intestinal barrier function, which may further predispose animals to microbial invasion [13]. Another study investigated the combined effects of DON and zearalenone (ZEN) on metabolism and immunity in dairy cows [14]. In this study, wheat, corn, hay and corn silage that was naturally contaminated with DON and ZEN (0.2 mg/kg each) was fed to 6 dairy cows for 63 days. While the contaminated diet had no effect on dry matter intake, body weight, or milk production and composition, neutrophil phagocytosis was reduced in cows fed the contaminated diet. In this study, cows were also immunized with OVA, and an increase in primary antibody IgG response to OVA was observed in animals fed the contaminated diet. Another study by Thuvander et al. (1999) showed the combined effects of T-2, DAS, NIV and DON on human lymphocyte function [18]. The results of this study demonstrated mainly additive or antagonistic interactions among mycotoxins. In regards to inhibiting lymphocyte proliferation for example, T-2 or DAS in combination with DON had antagonistic interactions. In contrast, NIV in combination with T-2 and DAS showed an additive inhibitory effect on lymphocyte proliferation. The antagonistic findings may be due to less potent toxins competing with more potent toxins for common binding sites. Furthermore, the additive interactions found in this study suggest that certain toxins work via similar mechanisms [18].

CONCLUSION

Little information about the toxicity of TCEs in ruminants is available. There are some data available concerning the toxic effects of T-2 and DON in ruminants, however, we are unaware of any published studies on the effects of NEO, and very few studies have been published on the effects of FUS-X, NIV, and DAS in ruminants. The available studies have indicated potential effects on feed intake, weight gain, and reproductive performance. Studies have also shown subclinical effects on different haematological and immunological parameters.

Compared to ruminants and other farm animals, pigs are clearly the most sensitive to TCEs. Pigs display clinical signs of exposure including reduced feed intake, weight loss, feed refusal, emesis, and diarrhoea. There were no data concerning the effects of NEO in any farm animal species. The studies assessing the effects of NEO typically involved human or mice cell cultures or cell lines, and cytotoxic effects on NHLF and RPTEC have been reported. These data indicated the potential for TCEs to impair proper functioning tissues, and possibly organs, in different livestock species.

There is not enough information available to conduct a scientific evaluation of the risks associated with TCEs exposure to ruminants. However, there is a significant amount of evidence indicating the occurrence of high concentrations of TCEs in ruminant feed [98, 100]. Several studies have suggested that ruminants are tolerant or resistant to these toxins, but this does not appear to be the case. It is true that some TCEs can be degraded in the rumen, but this detoxifying activity is reduced in the case of rumen dysfunction, as is seen with exposure to DON [7]. Furthermore, there are little data concerning the effects of TCE exposure to calves. In comparison to adult cattle, calves are likely more susceptible to the adverse effects of TCEs due to their underdeveloped immune system and rumen. Veal calves, in particular, are likely especially at risk to the effects of TCE exposure since they are typically fed a corn-based diet at a young age. For these reasons, further research focusing on the effects of TCE exposure to calves is warranted. Regardless of the species, the mechanisms by which multi-TCE exposure affects livestock species have still not been established. Furthermore, most studies have focussed on the effects of one toxin in a certain species. While the findings of these studies are valuable, the reality is that animals are not exposed to one particular TCE at a time [98, 101]. Current research has shown that TCEs have the ability to interact with each other to potentially enhance or inhibit their known effects when combined together. Therefore, further research should investigate the combined effects of TCEs in ruminants and other livestock species. The findings from such research may help to better understand the mechanisms of action for various TCEs and assess risk of exposure to livestock in the context of health and production. It will also be important to conduct both in vitro and in vivo studies with ruminants. In vitro studies may contribute to a better understanding of mechanisms of action and organ and target tissues of these toxins. In vivo studies on the other hand, will help to establish individual differences in uptake, distribution, metabolism and excretion of these mycotoxins. These studies will also help to show how the toxin is affecting the animal at a clinical level.

ACKNOWLEDGMENTS

Funding for this review was provided through Alltech Inc., Nicholasville, Kentucky.

Conflict of Interest

The authors declare that there are no conflicts of interest

REFERENCES

[1] Hussein HS, Brasel JM: Toxicity, metabolism, and impact of mycotoxins on humans and animals. *Toxicology* 2001, 167:101–134.

[2] Parikka P, Hakala K, Tiilikkala K: Expected shifts in Fusarium species' composition on cereal grain in Northern Europe due to climatic change. *Food Addit Contam Part Chem Anal Control Expo Risk Assess* 2012, 29:1543–1555.

[3] Eriksen GS, Pettersson H: Toxicological evaluation of trichothecenes in animal feed. *Anim Feed Sci Technol* 2004, 114:205–239.

[4] Wu Q, Huang L, Liu Z, Yao M, Wang Y, Dai M, Yuan Z: A comparison of hepatic in vitro metabolism of T-2 toxin in rats, pigs, chickens, and carp. *Xenobiotica Fate Foreign Compd Biol Syst* 2011, 41:863–873.

[5] Shank RA, Foroud NA, Hazendonk P, Eudes F, Blackwell BA: Current and Future Experimental Strategies for Structural Analysis of Trichothecene Mycotoxins—A Prospectus. *Toxins* 2011, 3:1518–1553.

[6] Wu Q, Dohnal V, Huang L, Kuca K, Yuan Z: Metabolic pathways of TCEs. *Drug Metabolism Reviews* 2010, 42:250–67.

[7] Dänicke S, Matthäus K, Lebzien P, Valenta H, Stemme K, Ueberschär K-H, Razzazi-Fazeli E, Böhm J, Flachowsky G: Effects of Fusarium toxin-contaminated wheat grain on nutrient turnover, microbial protein synthesis and metabolism of deoxynivalenol and zearalenone in the rumen of dairy cows. *J Anim Physiol Anim Nutr* 2005, 89:303–315.

[8] Mostrom MS, Jacobsen BJ: Ruminant mycotoxicosis. *Vet Clin North Am Food Anim Pract* 2011, 27:315–344, viii.

[9] Beasley VR, Lambert RJ: The apparently minimal hazard posed to human consumers of products from animals fed trichothecene-contaminated grains. *Vet Hum Toxicol* 1990, 32 Suppl:27–39.

[10] Jouany J, Yiannikouris A, Bertin G: Risk assessment of mycotoxins in ruminants and ruminant products. *Nutri and foraging ecol of sheep and goats* 2009, 85:205–225.

[11] Rocha O, Ansari K, Doohan FM: Effects of trichothecene mycotoxins on eukaryotic cells: a review. *Food Addit Contam* 2005, 22:369–378.

[12] Ueno Y: The toxicology of mycotoxins. *Crit Rev Toxicol* 1985, 14:99–132.

[13] Bracarense A-PFL, Lucioli J, Grenier B, Drociunas Pacheco G, Moll W-D, Schatzmayr G, Oswald IP: Chronic ingestion of deoxynivalenol and fumonisin, alone or in interaction, induces morphological and immunological changes in the intestine of piglets. *Br J Nutr* 2012, 107:1776–1786.

[14] Pestka JJ: Deoxynivalenol: Toxicity, mechanisms and animal health risks. *Anim Feed Sci Technol* 2007, 137:283–298.

[15] Arunachalam C, Doohan FM: Trichothecene toxicity in eukaryotes: cellular and molecular mechanisms in plants and animals. *Toxicol Lett* 2013, 217:149–158.

[16] Moon Y, Yang H, Lee SH: Modulation of early growth response gene 1 and interleukin-8 expression by ribotoxin deoxynivalenol (vomitoxin) via ERK1/2 in human epithelial intestine 407 cells. *Biochem Biophys Res Commun* 2007, 362:256–262.

[17] Pinton P, Accensi F, Beauchamp E, Cossalter A-M, Callu P, Grosjean F, Oswald IP: Ingestion of deoxynivalenol (DON) contaminated feed alters the pig vaccinal immune responses. *Toxicol Lett* 2008, 177:215–222.

[18] Thuvander A, Wikman C, Gadhasson I: In vitro exposure of human lymphocytes to trichothecenes: individual variation in sensitivity and effects of combined exposure on lymphocyte function. *Food Chem Toxicol Int J Publ Br Ind Biol Res Assoc* 1999, 37:639–648.

[19] Uzarski RL, Islam Z, Pestka JJ: Potentiation of trichothecene-induced leukocyte cytotoxicity and apoptosis by TNF-alpha and Fas activation. *Chem Biol Interact* 2003, 146:105–119.

[20] Eriksen GS, Pettersson H, Johnsen K, Lindberg JE: Transformation of trichothecenes in ileal digesta and faeces from pigs. *Arch Für Tierernähr* 2002, 56:263–274.

[21] Swanson SP, Nicoletti J, Rood HD Jr, Buck WB, Cote LM, Yoshizawa T: Metabolism of three trichothecene mycotoxins, T-2 toxin, diacetoxyscirpenol and deoxynivalenol, by bovine rumen microorganisms. *J Chromatogr* 1987, 414:335–342.

[22] Dänicke S, Valenta H, Döll S: On the toxicokinetics and the metabolism of deoxynivalenol (DON) in the pig. *Arch Anim Nutr* 2004, 58:169–180.
[23] Kiessling KH, Pettersson H, Sandholm K, Olsen M: Metabolism of aflatoxin, ochratoxin, zearalenone, and three trichothecenes by intact rumen fluid, rumen protozoa, and rumen bacteria. *Appl Environ Microbiol* 1984, 47:1070–1073.
[24] Hedman R, Pettersson H, Lindberg JE: Absorption and metabolism of nivalenol in pigs. *Arch Für Tierernähr* 1997, 50:13–24.
[25] Hedman R, Pettersson H: Transformation of nivalenol by gastrointestinal microbes. *Arch Für Tierernähr* 1997, 50:321–329.
[26] Swanson SP, Helaszek C, Buck WB, Rood HD Jr, Haschek WM: The role of intestinal microflora in the metabolism of trichothecene mycotoxins. *Food Chem Toxicol Int J Publ Br Ind Biol Res Assoc* 1988, 26:823–829.
[27] Chi M: Excretion and tissue distribution of radioactivity from tritium-labeled T-2 toxin in chicks*1, *2. *Toxicol Appl Pharmacol* 1978, 45:391–402.
[28] Côté L-M, Dahlem AM, Yoshizawa T, Swanson SP, Buck WB: Excretion of Deoxynivalenol and Its Metabolite in Milk, Urine, and Feces of Lactating Dairy Cows. *J Dairy Sci* 1986, 69:2416–2423.
[29] Matsumoto H, Ito T, Ueno Y: Toxicological approaches to the metabolities of fusaria. XII. Fate and distribution of T-2 toxin in mice. *Jpn J Exp Med* 1978, 48:393–399.
[30] Prelusky D: Tissue distribution and excretion of radioactivity following administration of 14C-labeled deoxynivalenol to White Leghorn hens*1. *Fundam Appl Toxicol* 1986, 7:635–645.
[31] Viscont A, Mirocha CJ: Identification of Various T-2 Toxin Metabolites in Chicken Excreta and Tissues. *Appl Environ Microbiol* 1985, 49:1246–50.
[32] Yoshizawa T, Mirocha CJ, Behrens JC, Swanson SP: Metabolic fate of T-2 toxin in a lactating cow. *Food Cosmet Toxicol* 1981, 19:31–39.
[33] Prelusky DB, Trenholm HL, Hamilton RMG, Miller JD: Transmission of [14C]deoxynivalenol to eggs following oral administration to laying hens. *J Agric Food Chem* 1987, 35:182–186.
[34] Coppock RW, Swanson SP, Gelberg HB, Buck WB: Tissue residues of diacetoxy-scirpenol in pigs and calves after intravenous dosing. *Am J Vet Res* 1988, 49:1997–1999.
[35] Saengtienchai T, Poapolathep S, Isariyodom S, Ikenaka Y, Ishizuka M, Poapolathep A: Toxicokinetics and tissue depletion of Fusarenon-X and its metabolite nivalenol in piglets. *Food Chem Toxicol Int J Publ Br Ind Biol Res Assoc* 2014, 66:307–312.
[36] Rafai P, Tuboly S, Bata A, Tilly P, Ványi A, Papp Z, Jakab L, Túry E: Effect of various levels of T-2 toxin in the immune system of growing pigs. *Vet Rec* 1995, 136:511–514.
[37] Pettersson H: Toxicity and risks with T-2 and HT-2 toxins in cereals. *Plant Breed Seed Sci* 2011, 64:65–76.
[38] Königs M, Mulac D, Schwerdt G, Gekle M, Humpf H-U: Metabolism and cytotoxic effects of T-2 toxin and its metabolites on human cells in primary culture. *Toxicology* 2009, 258:106–115.
[39] Osweiler GD, Hook BS, Mann DD: Effects of T-2 toxin in cattle. In *Proceedings United States Animal Health Association, 85th Annual Meeting*. St. Louis (MO); 1981:214–31.

[40] Hsu IC, Smalley EB, Strong FM, Ribelin WE: Identification of T-2 toxin in moldy corn associated with a lethal toxicosis in dairy cattle. *Appl Microbiol* 1972, 24:684–690.

[41] Kegl T, Venyi A: T-2 fusariotoxicosis in cattle stock. *Magyar Allatorvosock Lapja* 1991, 46:467–71.

[42] Huszenicza G, Fekete S, Szigeti G, Kulcsár M, Fébel H, Kellems RO, Nagy P, Cseh S, Veresegyházy T, Hullár I: Ovarian consequences of low dose peroral Fusarium (T-2) toxin in a ewe and heifer model. *Theriogenology* 2000, 53:1631–1639.

[43] Mann DD, Buening GM, Hook B, Osweiler GD: Effects of T-2 mycotoxin on bovine serum proteins. *Am J Vet Res* 1983, 44:1757–1759.

[44] Mann DD, Buening GM, Osweiler GD, Hook BS: Effect of subclinical levels of T-2 toxin on the bovine cellular immune system. *Can J Comp Med Rev Can Médecine Comparée* 1984, 48:308–312.

[45] Charoenpornsook K, Fitzpatrick JL, Smith JE: The effects of four mycotoxins on the mitogen stimulated proliferation of bovine peripheral blood mononuclear cells in vitro. *Mycopathologia* 1998, 143:105–111.

[46] Weaver GA, Kurtz HJ, Mirocha CJ, Bates FY, Behrens JC: Acute toxicity of the mycotoxin diacetoxyscirpenol in swine. *Can Vet J* 1978, 19:267–71.

[47] Ványi A, Glávits R, Gajdács E, Sándor G, Kovács F: Changes induced in newborn piglets by the trichothecene toxin T-2. *Acta Vet Hung* 1991, 39:29–37.

[48] Driehuis F, Spanjer MC, Scholten JM, te Giffel MC: Occurrence of mycotoxins in feedstuffs of dairy cows and estimation of total dietary intakes. *J Dairy Sci* 2008, 91:4261–4271.

[49] Meissonnier GM, Laffitte J, Raymond I, Benoit E, Cossalter A-M, Pinton P, Bertin G, Oswald IP, Galtier P: Subclinical doses of T-2 toxin impair acquired immune response and liver cytochrome P450 in pigs. *Toxicology* 2008, 247:46–54.

[50] Rafai P, Bata A, Ványi A, Papp Z, Brydl E, Jakab L, Tuboly S, Túry E: Effect of various levels of T-2 toxin on the clinical status, performance and metabolism of growing pigs. *Vet Rec* 1995, 136:485–489.

[51] Diaz GJ, Squires EJ, Julian RJ, Boermans HJ: Individual and combined effects of T-2 toxin and DAS in laying hens. *Br Poult Sci* 1994, 35:393–405.

[52] Kubena LF, Smith EE, Gentles A, Harvey RB, Edrington TS, Phillips TD, Rottinghaus GE: Individual and combined toxicity of T-2 toxin and cyclopiazonic acid in broiler chicks. *Poult Sci* 1994, 73:1390–1397.

[53] Liu YL, Meng GQ, Wang HR, Zhu HL, Hou YQ, Wang WJ, Ding BY: Effect of three mycotoxin adsorbents on growth performance, nutrient retention and meat quality in broilers fed on mould-contaminated feed. *Br Poult Sci* 2011, 52:255–263.

[54] Raju MV, Devegowda G: Influence of esterified-glucomannan on performance and organ morphology, serum biochemistry and haematology in broilers exposed to individual and combined mycotoxicosis (aflatoxin, ochratoxin and T-2 toxin). *Br Poult Sci* 2000, 41:640–650.

[55] Rezar V, Frankic T, Narat M, Levart A, Salobir J: Dose-dependent effects of T-2 toxin on performance, lipid peroxidation, and genotoxicity in broiler chickens. *Poult Sci* 2007, 86:1155–1160.

[56] Wyatt RD, Hamilton PB, Burmeister HR: Altered feathering of chicks caused by T-2 toxin. *Poult Sci* 1975, 54:1042–1045.

[57] Chi MS, Mirocha CJ, Kurtz HF, Weaver G, Bates F, Shimoda W: Effects of T-2 toxin on reproductive performance and health of laying hens. *Poult Sci* 1977, 56:628–637.

[58] Kubena LF, Harvey RB, Huff WE, Corrier DE, Philips TD, Rottinghaus GE: Influence of ochratoxin A and T-2 toxin singly and in combination on broiler chickens. *Poult Sci* 1989, 68:867–872.

[59] Wyatt RD, Doerr JA, Hamilton PB, Burmeister HR: Egg production, shell thickness, and other physiological parameters of laying hens affected by T-2 toxin. *Appl Microbiol* 1975, 29:641–645.

[60] Tobias S, Rajíc I, Ványi A: Effect of T-2 toxin on egg production and hatchability in laying hens. *Acta Vet Hung* 1992, 40:47–54.

[61] Jaradat ZW, Viià B, Marquardt RR: Adverse effects of T-2 toxin on chicken lymphocytes blastogenesis and its protection with Vitamin E. *Toxicology* 2006, 225:90–96.

[62] Yohannes T, Sharma AK, Singh SD, Goswami TK: Immunopathological effects of experimental T-2 mycotoxocosis in broiler chicken co-infected with infectious bronchitis virus (IBV). *Vet Immunol Immunopathol* 2012, 146:245–253.

[63] Gutleb AC, Morrison E, Murk AJ: Cytotoxicity assays for mycotoxins produced by Fusarium strains: a review. *Environ Toxicol Pharmacol* 2002, 11:309–320.

[64] Rotter BA, Thompson BK, Lessard M, Trenholm HL, Tryphonas H: Influence of low-level exposure to Fusarium mycotoxins on selected immunological and hematological parameters in young swine. *Fundam Appl Toxicol Off J Soc Toxicol* 1994, 23:117–124.

[65] Berthiller F, Crews C, Dall'Asta C, Saeger SD, Haesaert G, Karlovsky P, Oswald IP, Seefelder W, Speijers G, Stroka J: Masked mycotoxins: a review. *Mol Nutr Food Res* 2013, 57:165–186.

[66] Charmley E, Trenholm HL, Thompson BK, Vudathala D, Nicholson JW, Prelusky DB, Charmley LL: Influence of level of deoxynivalenol in the diet of dairy cows on feed intake, milk production, and its composition. *J Dairy Sci* 1993, 76:3580–3587.

[67] Ingalls JR: Influence of deoxynivalenol on feed consumption by dairy cows. *Anim Feed Sci Technol* 1996, 60:297–300.

[68] Korosteleva SN, Smith TK, Boermans HJ: Effects of feed naturally contaminated with Fusarium mycotoxins on metabolism and immunity of dairy cows. *J Dairy Sci* 2009, 92:1585–1593.

[69] Wada K, Nabae A, Higuchi H, Nagahata H: Effects of mycotoxins on chemiluminescent response and cytokine mRNA expression of bovine neutrophils. *J Vet Med Sci Jpn Soc Vet Sci* 2010, 72:1507–1511.

[70] Takayama H, Shimada N, Mikami O, Murata H: Suppressive effect of deoxynivalenol, a Fusarium mycotoxin, on bovine and porcine neutrophil chemiluminescence: an in vitro study. *J Vet Med Sci Jpn Soc Vet Sci* 2005, 67:531–533.

[71] Pestka JJ, Smolinski AT: Deoxynivalenol: toxicology and potential effects on humans. *J Toxicol Environ Health B Crit Rev* 2005, 8:39–69.

[72] Awad WA, Aschenbach JR, Zentek J: Cytotoxicity and metabolic stress induced by deoxynivalenol in the porcine intestinal IPEC-J2 cell line. *J Anim Physiol Anim Nutr* 2012, 96:709–716.

[73] Prelusky DB, Gerdes RG, Underhill KL, Rotter BA, Jui PY, Trenholm HL: Effects of low-level dietary deoxynivalenol on haematological and clinical parameters of the pig. *Nat Toxins* 1994, 2:97–104.

[74] Swamy HVLN, Smith TK, Karrow NA, Boermans HJ: Effects of feeding blends of grains naturally contaminated with Fusarium mycotoxins on growth and immunological parameters of broiler chickens. *Poult Sci* 2004, 83:533–543.

[75] Ghareeb K, Awad WA, Böhm J: Ameliorative effect of a microbial feed additive on infectious bronchitis virus antibody titer and stress index in broiler chicks fed deoxynivalenol. *Poult Sci* 2012, 91:800–807.

[76] Harvey RB, Kubena LF, Elissalde MH, Corrier DE, Huff WE, Rottinghaus GE, Clement BA: Cocontamination of swine diets by aflatoxin and diacetoxyscirpenol. *J Vet Diagn Investig Off Publ Am Assoc Vet Lab Diagn Inc* 1991, 3:155–160.

[77] Awad WA, Böhm J, Razzazi-Fazeli E, Zentek J: Effects of feeding deoxynivalenol contaminated wheat on growth performance, organ weights and histological parameters of the intestine of broiler chickens. *J Anim Physiol Anim Nutr* 2006, 90:32–37.

[78] Bergsjø B, Kaldhusdal M: No association found between the ascites syndrome in broilers and feeding of oats contaminated with deoxynivalenol up to thirty-five days of age. *Poult Sci* 1994, 73:1758–1762.

[79] Hamilton RM, Thompson BK, Trenholm HL, Fiser PS, Greenhalgh R: Effects of feeding white Leghorn hens diets that contain deoxynivalenol (vomitoxin)-contaminated wheat. *Poult Sci* 1985, 64:1840–1852.

[80] Hulan HW, Proudfoot FG: Effects of feeding vomitoxin contaminated wheat on the performance of broiler chickens. *Poult Sci* 1982, 61:1653–1659.

[81] Yunus AW, Ghareeb K, Twaruzek M, Grajewski J, Böhm J: Deoxynivalenol as a contaminant of broiler feed: effects on bird performance and response to common vaccines. *Poult Sci* 2012, 91:844–851.

[82] Kubena LF, Harvey RB, Phillips TD, Holman GM, Creger CR: Effects of feeding mature White Leghorn hens diets that contain deoxynivalenol (vomitoxin). *Poult Sci* 1987, 66:55–58.

[83] Lee JT, Jessen KA, Beltran R, Starkl V, Schatzmayr G, Borutova R, Caldwell DJ: Effects of mycotoxin-contaminated diets and deactivating compound in laying hens: 2. effects on white shell egg quality and characteristics. *Poult Sci* 2012, 91:2096–2104.

[84] Kidd MT, Hagler WM Jr, Qureshi MA: Trichothecene mycotoxins depress the mononuclear-phagocytic system of young turkeys. *Immunopharmacol Immunotoxicol* 1995, 17:385–398.

[85] Girgis GN, Sharif S, Barta JR, Boermans HJ, Smith TK: Immunomodulatory effects of feed-borne Fusarium mycotoxins in chickens infected with coccidia. *Exp Biol Med Maywood NJ* 2008, 233:1411–1420.

[86] Taranu I, Marina DE, Burlacu R, Pinton P, Damian V, Oswald IP: Comparative aspects of in vitro proliferation of human and porcine lymphocytes exposed to mycotoxins. *Arch Anim Nutr* 2010, 64:383–393.

[87] Hedman R, Pettersson H, Engström B, Elwinger K, Fossum O: Effects of feeding nivalenol-contaminated diets to male broiler chickens. *Poult Sci* 1995, 74:620–625.

[88] Pettersson H, Hedman R, Engström B, Elwinger K, Fossum O: Nivalenol in Swedish cereals--occurrence, production and toxicity towards chickens. *Food Addit Contam* 1995, 12:373–376.

[89] Garaleviciene D, Pettersson H, Elwinger K: Effects on health and blood plasma parameters of laying hens by pure nivalenol in the diet. *J Anim Physiol Anim Nutr* 2002, 86:389–398.

[90] Awad WA, Razzazi-Fazeli E, Böhm J, Zentek J: Effects of B-trichothecenes on luminal glucose transport across the isolated jejunal epithelium of broiler chickens. *J Anim Physiol Anim Nutr* 2008, 92:225–230.

[91] Streit E, Schwab C, Sulyok M, Naehrer K, Krska R, Schatzmayr G: Multi-mycotoxin screening reveals the occurrence of 139 different secondary metabolites in feed and feed ingredients. *Toxins* 2013, 5:504–523.

[92] Harvey RB, Edrington TS, Kubena LF, Elissalde MH, Corrier DE, Rottinghaus GE: Effect of aflatoxin and diacetoxyscirpenol in ewe lambs. *Bull Environ Contam Toxicol* 1995, 54:325–330.

[93] Sato S: Comparative biochemical changes associated with the TCE toxicoses. *Current Topics in Veterinary Medicine and Animal Science* 1986, 33:113–23.

[94] Brake J, Hamilton PB, Kittrell RS: Effects of the trichothecene mycotoxin diacetoxyscirpenol on feed consumption, body weight, and oral lesions of broiler breeders. *Poult Sci* 2000, 79:856–863.

[95] Diaz GJ: Evaluation of the efficacy of a feed additive to ameliorate the toxic effects of 4,15-diacetoxiscirpenol in growing chicks. *Poult Sci* 2002, 81:1492–1495.

[96] Brake J, Hamilton PB, Kittrell RS: Effects of the tricothecene mycotoxin diacetoxyscirpenol on egg production of broiler breeders. *Poult Sci* 2002, 81:1807–1810.

[97] Qureshi MA, Brundage MA, Hamilton PB: 4 beta, 15-Diacetoxyscirpenol induces cytotoxicity and alterations in phagocytic and Fc-receptor expression functions in chicken macrophages in vitro. *Immunopharmacol Immunotoxicol* 1998, 20:541–553.

[98] Eckard S, Wettstein FE, Forrer H-R, Vogelgsang S: Incidence of Fusarium species and mycotoxins in silage maize. *Toxins* 2011, 3:949–967.

[99] Grenier B, Loureiro-Bracarense A-P, Lucioli J, Pacheco GD, Cossalter A-M, Moll W-D, Schatzmayr G, Oswald IP: Individual and combined effects of subclinical doses of deoxynivalenol and fumonisins in piglets. *Mol Nutr Food Res* 2011, 55:761–771.

[100] Van Asselt ED, Azambuja W, Moretti A, Kastelein P, De Rijk TC, Stratakou I, Van Der Fels-Klerx HJ: A Dutch field survey on fungal infection and mycotoxin concentrations in maize. *Food Addit Contam Part Chem Anal Control Expo Risk Assess* 2012, 29:1556–1565.

[101] Stratton GW, Robinson AR, Smith HC, Kittilsen L, Barbour M: Levels of five mycotoxins in grains harvested in Atlantic Canada as measured by high performance liquid chromatography. *Arch Environ Contam Toxicol* 1993, 24:399–409.

[102] Alm H, Brüssow K-P, Torner H, Vanselow J, Tomek W, Dänicke S, Tiemann U: Influence of Fusarium-toxin contaminated feed on initial quality and meiotic competence of gilt oocytes. *Reprod Toxicol Elmsford N* 2006, 22:44–50.

In: Mycotoxins
Editor: Cesar Rios

ISBN: 978-1-63483-544-2

Chapter 3

Occurrence of Mycotoxins in Goat's Milk

Edlayne Gonçalez *** and *Joana D'arc Felicio***
Centro de Pesquisa e Desenvolvimento de Sanidade Animal,
Instituto Biologico, São Paulo, Brazil

Abstract

Mycotoxins are toxic secondary metabolites produced by certain fungi in agricultural commodities. Among mycotoxigenic fungi, *Aspergillus, Penicillium* and *Fusarium* are the most important genera. The contamination of feed with mycotoxins is a worldwide problem in animal production and the direct consequences are the reducing of feed intake, production, reproduction and weight gain. The complex diet of ruminants consisting of forages and concentrates can be a source of diverse mixtures of mycotoxins. The mycotoxins can be transferred to animal products mainly in milk. Once milk and milk products are a good source of many nutrients such as proteins and calcium, mycotoxin residues in milk can pose a hazard to human health. The cow milk is the most consumed; however, goat milk could substitute it because there is a percentage of the population that has allergies and gastro-intestinal disorders with cow milk.

Researchers demonstrate the beneficial effects of goat milk, but there is few studies about the incidence of mycotoxins in goat milk. Several countries have established regulations for mycotoxins in feeds, milk and its products due the risk for consumers and the adverse effects on animal health. This chapter describes the occurrence of mycotoxins in goat milk and its dairy products.

* Corresponding author: Edlayne Gonçalez. Centro de Pesquisa e Desenvolvimento de Sanidade Animal, Instituto Biologico, Av. Conselheiro Rodrigues Alves, 1252, CEP 04014-002, São Paulo, Brazil. Telephone: +55-11-50871753, fax: +55-11-50871754, e-mail address: goncalez@biologico.sp.gov.br.

INTRODUCTION

Consumption of dairy products and goat milk in particular is associated with human health benefits [1]. Goat milk and their products have threefold significance in human nutrition: (1) feeding more starving and malnourished people in the developing world than cow milk; (2) treating people afflicted with cow milk allergies and gastro-intestinal disorders, which is a significant segment in many populations of developed countries and (3) fills the gastronomic needs of connoisseur consumers, which is a growing market share in many developed countries [2]. Besides, goat serves as a source of high quality protein for rural families mainly because of their small size and affordable nature [3].

The chemical characteristics of goat milk manufacture a wide variety of products, including fluid beverage products (low fat, fortified, or flavored) and UHT milk (ultra high temperature), cheese, buttermilk, yogurt, ice cream, sweets and candies. In addition, other specialty products to hair, skin care and cosmetic products are made from goat milk [4]. Although dairy goat products are generally in specific markets (dietetic milks, fresh or ripened cheeses), their profitability and competitive advantage depends on their relative price and special organization of the goat production systems (seasonal production, size of herds, goat milk productivity, goat milk characteristics). An informal survey of the situation in each country showed that probably less than 5% of the total goat milk is traded [5].

Milk production of goats is likely to be much greater than in official statistics, because of the large amounts of unreported home consumption, especially in developing countries [2]. Most of the livestock is in Asia (59.5%), followed by Africa (34%), the Americas (4.3%), Europe (1.8%) and Oceania (0.4%), which proves that the highest share is in the developing countries where goat milk is a basic food especially for rural population [6].

Ruminants can be exposed to abroad range of mycotoxins that occur in concentrate and forage, such as silage and hay [7]. When animals consume contaminated feed, the toxins can be excreted before or after of the biotransformation [8]. The carryover of mycotoxins in tissues and milk is a public health problem. Human exposure to mycotoxins occurs directly through the intake of contaminated agricultural products (cereals, corn, fruits, etc.) or indirectly through the consumption of products of animal origin (milk, eggs, etc.) prepared or obtained from animals that were fed with contaminated material [9]. Children are the largest consumers of milk, as it is one of the principal foods during their first years of life. Thus, it is very important that milk is free of toxic compounds that can be harmful for humans, but especially for children which are more susceptible to the action of toxic compounds. Moreover, the presence or absence of mycotoxins in milk could be used for indirect analysis of animal health [10]. Ensuring of safe and healthy feeds or foods free from any dangerous contamination levels of various mycotoxins and protecting the humans or animals from various food-born diseases is an important challenge and mission of many food safety specialists all over the world [11].

GOAT HANDLING FEED

Goats are present on all continents [5]. They were domesticated in 6-7 BC, as evidenced by archaeological remains collected in Western Asia. It has role in the evolvement of human civilization around the world. It is particularly important in the tropics and subtropics where it is used as a major source of meat, milk, skin and manure in many traditional societies. Goats have become important livestock in arid and semi-arid regions of the world because of their characteristic of versatility in harvesting forage and their ability to survive adverse foraging conditions [12].

The most numerous goat livestock is in Asia (India, China, Pakistan, Bangladesh, Iran), Africa (Nigeria, Ethiopia, Somalia, Sudan, South Africa), Europe (Greece, Spain, France), North America (Mexico, US and Canada), Central America (Haiti, Jamaica and Cuba), South America (Brazil, Argentina, Venezuela, Peru, Bolivia and Columbia) [13].

There are many types of goats raised under quite varied production conditions throughout the world [14]. The two major systems of small ruminant farming are pasture and indoor systems. Between these two farming systems there is a wide scope of mixed systems such as summer pasture/winter indoors or alternatively indoors/outdoors subject to climatic differences. People consider more often pasture systems as more extensive than indoor systems [1]. Indoor farming systems (Figure 1) require higher investments than pasture systems. The factors of production are intensified, particularly the feeding systems. Some farmers have chosen maize silage enabling to produce the same biomass of high nutritive value on less acreage than from hay or grass silage. In North European countries, ewes and goats are let outdoors only after starting of vegetation growth.

In South European countries, animals are at grazing even if the available vegetation is poor in drought periods or in arid areas. Accordingly, the level of intensification is subject to variations and depends on the available biomass, and its nutritive value varies according to seasons as well as to complementary feed supplies (forages and concentrates) [15].

Both the dietary concentrate level and nature of specific concentrate and forage feedstuffs impact level of milk production and characteristics of milk and milk products [14].

Photo by Edlayne Gonçalez.

Figure 1. Indoor farming systems of dairy goat in Ibiúna city, Brazil.

MYCOTOXIN'S OCCURRENCE

The complex diet of ruminants consisting of forages, concentrates and silages, can be a source of diverse mixture of mycotoxins that contaminate individual feed components [16]. Both, feed grains (e.g., corn, wheat, barley, cottonseed) and forages support fungal growth that leads to mycotoxins production [17].

Mycotoxins are toxic secondary metabolites produced by certain fungi in agricultural commodities, which are susceptible to mold infestation. The cereals are usually invaded by fungi in the field and after harvest and this circumstance increases significantly the potential for multi-toxin contamination. The production of mycotoxins is often unavoidable and depends on target environmental conditions during the plant growth and the subsequent food storage. Temperature stress is an important cause of fungi growth on crops in the field, and high moisture content (water activity) and temperature are associated with the growth of fungi in stored grain [18].

Moretti et al., [19] emphasize the importance of knowledge of biodiversity of toxigenic fungi to better understand factors that contribute to mycotoxins production, assessment of risks posed by mycotoxigenic fungi, and reduction of mycotoxins contamination in feed and food crops. The mycotoxigenic fungi involved in the food and feed chain belong mainly to the *Aspergillus*, *Penicillium* and *Fusarium* genera. Silva et al., [20] isolated from dairy goat feed and forage the fungi belong to the genera: *Penicillium*, *Mucor*, *Fusarium*, *Aspergillus*, *Cladosporium*, *Rhizopus*, *Trichoderma*, *Colletotrichum*, *Eurotium* and *Paecilomyces*. The authors also isolated *A. flavus* producer of the aflatoxins and cyclopiazonic acid (CPA).

The contamination of animal feed with mycotoxins is a worldwide problem in animal production. The major classes of mycotoxins of concern in feedstuffs include aflatoxins, trichothecenes, zearalenone, fumonisins, ochratoxin and ergot alkaloids [21].

In general, ruminants tend to be less susceptible to mycotoxins compared with monogastrics [22], because rumen microflora can degrade and inactivate mycotoxins, as a result, they are among the least susceptible animal species [23]. Rumen flora is supposed to be a defense against mycotoxins due to the fact that some of them: ochratoxin A (OTA), deoxynivalenol (DON), aflatoxin B1 and zearalenone (ZEA) are metabolized to less toxic compounds. However, other mycotoxins, such as patulin (PAT) or fumonisins, pass the rumen barrier unchanged [23].

Another point of concern is that some mycotoxins have a high carry-over rate from feed to milk, possibly contributing to mycotoxin intake in human populations [23]. Aflatoxin B1 (AFB1) and B2 (AFB2) ingested by ruminants are metabolized at hepatic level and the hydroxylated metabolites M1 (AFM1) and M2 (AFM2) are excreted through the milk, faeces and urine [24].

AFM1 is classified as a possible carcinogen for human (International Agency for Research on Cancer) [25], and it is a toxic agent that can contaminate milk and consequently any dairy product, when lactating animals are fed with feed contaminated with AFB1 [26]. One of the major xenobiotic contaminants of milk is AFM1 that comes from AFB1 [27].

Aflatoxin M1, the major metabolite of aflatoxin B1 in goat's milk [24], was excreted from less than 1 to 3% of the ingested dose of AFB1 [27]. Levels of aflatoxin at 95 mg/kg of feed offered to weanling goats had no effects on body weight gain and did not show any noticeable signs of toxic effects [28].

Battacone et al., [29] administered 0.8 mg of AFB1 orally in five dairy goats in lactation. The peak values for AFM1 concentration in milk was observed in milk collected during the milking at 3 and 6 h. After the peak, the AFM1 in milk disappeared with a trend that fitted well a monoexponential decreasing function, and the toxin was not detected after 84 h. Only about 0.17% of the amount of AFB1 administered was detected as AFM1 in milk, and about 50% of this was excreted in the first liter of milk yielded after AFB1 intake.

The occurrence of two mycotoxins, AFB1 and ZEA, in samples of feedstuff for dairy cows (963), ewes (42), and goats (131) produced in Portugal was related by Almeida et al., [30]. AFB1 was found in 15 samples of cow feed (1.6%), 3 samples of ewe feed (2.3%) and in 2 samples of goat feed (4.8%). All samples contained AFB1 at levels below the European Union maximum level (5 g/kg). Nearly half (45%) of the samples were contaminated with ZEA, but its levels were relatively low, at 5–136.9 μg/kg. Feed for dairy goats were found more frequently contaminated with AFB1 (4.8%) followed by ewes (2.3%) and cows (1.6%) [30]. The AFM1 content was also studied in 41 hard goat cheese samples from 12 cheese factories in Sardinia (Italy). AFM1 was detected in 36 bulk-tank goat milk samples (17.3%) at concentrations of between 5 and 40 ng/L and it was also detected in 4 (9.8%) out of 41 samples of ripened goat cheese at levels of between 79.5 ng/kg and 389 ng/kg [31].

AFM1 was analyzed in milk samples from buffalo, cow, goat and camel species collected in Ismailia, Egypt. Most milk (80%, 74%, 66% and 52% of the camel, goat, cow and buffalo milks, respectively) was below the European Union maximum of AFM1. AFM1 levels were influenced by feeding practices, wherein goats are fed a combination of prepared ration as a supplement to pasture grazing [32]. In the region of Serbia 80% of the goat milk samples had AFM1 [33].

In Iran, a total of 110 samples from bovine (60), ovine (42), and caprine (48) bulk milk samples were analyzed, and the incidence rates of AFM1 were 66.7, 31.0, and 35.4%, respectively. The concentration of AFM1 in 37.5% of AFM1 positive bovine milk samples and 5.9% of AFM1-positive caprine milk samples were higher than 50 ng/L. The low incidence of AFM1-positive ovine and caprine milk samples in this study could be at least partially due to the fact that sheep and goat herds are fed on stored grains for only 3–4 months and are raised on range lands with free grazing range plants for major part of the year [34].

During November 2007 to December 2008, 311 samples of raw milk from cow, buffalo, camel, sheep, and goat were collected in the Ahvaz (southwest Iran). AFM1 was found in 42.1% of the samples by average concentration of 43.3 ± 43.8 ng/kg. The incidence rates of AFM1 in raw cow, buffalo, camel, sheep, and goat milks were, 78.7%, 38.7%, 12.5%, 37.3%, and 27.1%, respectively [35].

A total of 137 milk specimens from Turkish (39 from goats, 53 from cows, and 45 from UHT market milks) were analyzed. AFM1 positivity was detected in 14 (35.8%) goat milk samples, in 46 (86.7%) cow milk samples and 33 (73.3%) UHT milk samples [36].

In Pakistan were analyzed 169 milk samples from five species made up of 55, 40, 30, 24, and 20 samples of buffalos, cows, goats, sheep, and camels respectively. The percentage of AFM1 contamination in buffalo milk, cow milk, goat milk, and sheep milk has been found to be 34.5%, 37.5%, 20%, and 16.7%, respectively. AFM1 contamination has not been detected in the camel milk [37].

Samples (36) of pasteurized, UHT treated and goat milk powder traded in the city of Campinas, Brazil was analyzed for aflatoxin M1 (AFM1). Results showed 25 (69.4%) positive samples for AFM1 at levels of 0.011–0.161 g/L of milk, which were below the tolerance limit of 0.500 g/L as adopted for AFM1 in milk by Brazilian regulations. Mean levels of AFM1 in pasteurized, UHT and goat milk powder were 0.072 ± 0.048, 0.058 ± 0.044 and 0.056 ± 0.031 g/L, respectively [38]. Another study evaluate the presence of AFB1, AFB2, AFG1, AFG2 and AFM1 and CPA in feed of goats in lactation and their residues in milk at 11 properties located in the southwest of Sao Paulo state, Brazil.

Of the 128 feed samples analyzed, 2.34% were contaminated with aflatoxin at levels: 3.65 to 36.93 μg/kg for AFB1, 0.35 to 2.46 μg/kg AFB2, traces to 46.06 μg/kg for AFG1 and 0.61 to 1.65 μg/kg AFG2. Only one sample showed concentrations of 87.1 mg/kg, above that allowed by Brazilian law of 50 μg/g. For the CPA, 3.12% of feed samples were positive at concentrations ranging from traces to 1.90 μg/Kg. AFM1 and CPA were not detected in the 120 milk samples analyzed [39]. Batatinha et al., [40] analyzed aflatoxins in goat feed samples from Bahia state, Brazil, and the presence of these toxins were not detected. AFM1 in goat milk was detected, in Brazil, by Barrios et al., [41] and the results showed 45% of the samples contained in AFM1 levels range from 20-200 mg/kg. The occurrence of aflatoxins M1 and M2 in 39 goat milk samples manufactured in three Brazilian states, consisting of 15 UHT milk samples and 24 frozen pasteurized milk samples were analyzed. None of the analyzed milk samples was contaminated with aflatoxin M2, and aflatoxin M1 was detected in 7 (18%) samples in a range of from 20 to 98 ng/L [42].

Samples of ewes and goats raw milk (234) from organic and conventional farms in Greece were analyzed and 191 (81.6%) samples AFM1 was not detected, while 4 (1.7%) were found above the EU maximum tolerable limit. There was no conventional samples found over the maximum limit for AFM1 (0/117), while 4/117 (3.4%) organic samples exceeded 50 ng/kg. The complex of the associated risk factors in AFM1 contamination found (winter season, feed storage practices and feeding pea), indicates that these should be associated when assessing AFM1 contamination risk in milk and that constant monitoring and increased farmer's awareness is needed [26].

Another study in Greece, samples found above the maximum level were 1.79%, while AFM1 was detected at 73.3% and 66.7% for raw sheep and goat milk, respectively [43].

Study on bovine milk has demonstrated that AFM1 binds to milk proteins, and in particular to CN, and that it is concentrated in the curd during manufacture of cheese [44], and, can be for these reasons, AFM1 concentrations are 3 to 4 fold higher in cheese than in milk [45]. Information on the binding of AFM1 to proteins in ovine and caprine milk is important because part of them is processed into cheese [46, 47].

In a study from Italy, AFM1 was detected in ~10% of samples of ripened goat cheese at levels of between 79.5 and 389 ng/kg. All contaminated samples were obtained from the same cheese manufacturer [31]. In another study from the city of Kilis in Turkey, AFM1 was detected in ~85% of the samples, of which 6% were contaminated by AFM1 at hazardous levels for human health [48]. Several ochratoxigenic mycobiota can grow in grains and forages used for animal feeding and produce OTA and other mycotoxins. Thus, farm animals may be exposed to the toxic effects of OTA together with those of other toxins. Similarly to cattle, ochratoxicosis has been rarely reported in small ruminant species, such as sheep and goats [49]. Ruminants are more resistant to ochratoxin, compared with monogastrics, because rumen protozoal metabolism quickly converts OTA to the less toxic OTA. The elimination half-life in ruminants is short, about 17.3 hours, compared with 100 hours in pigs [49]. However, rapid changes in rations and an increased percentage of concentrates can decrease ruminant metabolism of OTA to less toxic compounds.

Regulatory concerns for OTA in ruminants are chronic exposure and potential for accumulation in tissues or milk residues [22].

OTA was studied in goats given a single oral dose of 3H-OTA (0.5 mg/kg). More than 90% of the radioactivity was found to be excreted in 7 days and the majority (53%) was found in faeces. Thirty-eight percent, 6.0% and 2.26% of the activity was found in urine, milk and serum, respectively. Less than 0.03% of free OTA was found in milk during the 7 day period [50]. A total of 85 milk samples from cows, sheep and goats were collected from the same dairy farms and the AFM1 and OTA were not found [51]. Similar results, as the AFM1 content concerns, have been reported by Saccà et al., [52] for raw milk samples from 17 dairy goat farms from North-eastern Italy.

Samples (39 bovine, 15 goats and 9 sheep) coming from organic farms were analyzed for OTA. Only 3 out of 63 samples resulted positive for OTA, with amounts comprised between 0.07 and 0.11 ppb [53]. Low frequency of OTA occurrence and concentration reported in the above cited studies suggest that, even if milk can be a possible source of this toxin for human diet, it does not represent a particular risk for adult consumers. However, OTA presence in milk may represent a danger for some categories such as children, who consume large amounts of milk [49].

CONCLUSION

Due to mycotoxins toxicity, and in order to ensure human and animal health and these toxins are not allowed to be present above maximum permitted levels in milk and milk products. So, permanent monitoring is an essential tool for the prevention of the mycotoxin contamination in goat milk. The procedures that prevent the formation of mycotoxins or fungal contamination represent a good way to avoid the exposure to animal feed and foods for human consumption. Agricultural good practices have a fundamental role in the prevention of the biosynthesis of the mycotoxins in crops, and a systematic monitoring of feeds and raw materials to control the exposure of the animals are important ways to prevent the contamination of goat milk and dairy products.

REFERENCES

[1] Silanikovea, N., Leitnerb, G., Merinc, U., Prosserd, C. G. Recent advances in exploiting goat's milk: Quality, safety and production aspects. *Small Ruminant Research*, 2010, 89, 110–124.

[2] Haenlein, G. F. W. Goat milk in human nutrition. *Small Ruminant Research*, 2004, 51, 155–163.

[3] McMillin, K. W., Webb, E. C., Donkin, E. F., Pinkarton, F. Goat meat production system. In: Mahgoud, O., Kadim, I., Webb, E. (editors). *Goat meat production and quality*. CAB International, 2012, pp. 15-31.

[4] Ribeiro, A. C., Ribeiro, S. D. A. Specialty products made from goat milk. *Small Ruminant Research*, 2010, 89, 225–233.

[5] Dubeuf, J. P., Morand-Fehr, P., Rubino, R. Situation, changes and future of goat industry around the world. *Small Ruminant Research*, 2004, 51, 165–173.

[6] Popescu, A. study regarding the trends in the world and european goat milk production. *Lucrări Ştiinţifice-Seria Zootehnie*, 2013, 59, 127-132.

[7] Chelia, F., Campagnolib, A., Dell'Orto, V. Fungal populations and mycotoxins in silages: From occurrence to analysis. *Animal Feed Science and Technology*, 2013, 183, 1–16.

[8] Finoli, C., Vecchio, A. Occurrence of aflatoxins in feedstuff, sheep milk and dairy products in Western Sicily. *Italian Journal of Animal Science*, 2003, 2, 191-196.

[9] Capriotti, A. L., Caruso, G., Cavaliere, C., Foglia, P., Samperi, R., Lagana, A. Multiclass mycotoxin analysis in food, environmental and biological matrices with chromatography/mass spectrometry. *Mass Spectrometry Reviews*, 2012, 31, 466-503.

[10] Flors-Flores, M. E., Lizarraga, E., Cerain, A. L., Gonzalez-Penas, E. Presence of mycotoxins in animal milk: A review. *Food Control*, 2015, 53, 163-176.

[11] Stoev, S. D. Food safety and increasing hazard of mycotoxin occurrence in foods and feeds. *Critical Reviews in Food Science and Nutrion*, 2013, 53, 887–901.

[12] Mahgoud, O., Kadim, I. T., Lu, C. D. Overview of the global goat meat sector In: Mahgoud, O., Kadim, I., Webb, E. (editors). *Goat meat production and quality*. CAB International, 2012, pp. 1-14.

[13] Meza-Herrera, C. A. Goat production in the world: Main trends in Mexico and the Comarca Lagunera, 2010, in www.uco.es.

[14] Goetsch, A. L., Zeng, S. S., Gipson, T. A. Factors affecting goat milk production and quality. *Small Ruminant Research*, 2011, 10, 155–63.

[15] Morand-Fehr, P., Fedele, V., Decandia, M., Le Frileux, Y. Influence of farming and feeding systems on composition and quality of goat and sheep milk. *Small Ruminant Research*, 2007, 68, 20–34.

[16] Smith, T. K., Korosteleva, S. N. The significance of feed-borne mycotoxins in ruminant nutrition. In: Gonçalez, E., Felicio, J. D., Aquino, S. *Mycotoxicoses in animals economically important*. New York: Nova Science Publishers, Inc., 2012, pp. 35-66.

[17] Osweiler, G. D. Equine mycotoxicosis In: Gonçalez, E., Felicio, J. D., Aquino, S. *Mycotoxicoses in animals economically important*. New York: Nova Science Publishers, Inc., 2012, p. 1-34.

[18] Stoev, S. D. Foodborne mycotoxicoses, risk assessment and underestimated hazard of masked mycotoxins and joint mycotoxin effects or interaction. *Environmental Toxicology and Pharmacology*, 2015, 39, 794-809.

[19] Moretti, A. Molecular biodiversity of mycotoxigenic fungi that threaten food safety. *International Journal Food Microbiology*, 2013, 167, 57-66.

[20] Silva, J. L., Aparecido, C. C., Hansen, D., Pereira, T. A. M., Felicio, J. D., Gonçalez, E. Identification of toxigenic Aspergillus species from diet dairy goat using a polyphasic approach. *Ciência Rural*, 2015. In press.

[21] CAST. Mycotoxins: Risks in plant, animal, and human systems. Task Force Report No. 139. *Council for Agricultural Science and Technology*, Ames, IA, 2003.

[22] Mostrom, M. S., Jacobsen, B. J. *Ruminant Mycotoxicosis*. Veterinary Clinic of North Amarica: Food Animal Practice, 2011, 27, 315–344.

[23] Fink-Gremmels, J. Mycotoxins in cattle feeds and carry-over to dairy milk: a review. *Food Additives and Contaminants. Part A, Chemistry, Analysis, Control, Exposure and Risk Assessment*, 2008, 25, 172-180.
[24] Goto, T., Hsieh, D. P. Fractionation of radioactivity in the milk of goats administered 14C-AXatoxin B1. *Journal of AOAC*, 1985, 68, 456–458.
[25] International Agency for Research on Cancer (IARC). Aflatoxins: Natural occurring aflatoxins (Group 1), aflatoxin M_1 (Group 2B). *Lyon: IARC Scientific Publications*, 1993, 56, pp. 245.
[26] Malissiova, E., Tsakalof, A., Arvanitoyannis, I. S., Katsafliaka, A., Katsioulis, A., Tserkezou, P., Koureas, M., Govaris, A., Hadjichristodoulou, C. Monitoring Aflatoxin M1 levels in ewe's and goat's milk in Thessaly, Greece; potential risk factors under organic and conventional production schemes. *Food Control*, 2013, 34, 241-248.
[27] Applebaum, R. S., Brackett, R. E., Wieseman, D. W., Marth, E. H. Aflatoxin: toxicity to dairy cattle and occurrence in milk and milk products. A review. *Journal Food Protection*, 1982, 45, 752-777.
[28] Gurung, N. K., Rankins, D. L., Shelby, R. A., Goel, S. Effect of fumonisin B1-contaminated feeds on weanling angora goats. *Journal Animal Science*, 1998, 76, 2863–2870.
[29] Battacone, G., Nudda, A., Rassu, S. P. G., Decandia, M., Pulina, G. Excretion pattern of aflatoxin M1 in milk of goats fed a single dose of aflatoxin B1. *Journal of Dairy Science*, 2012, 95, 2656–2661.
[30] Almeida, I. F. M., Guerra, M. M., Martins, H. M. L., Costa, J. M. G., Bernardo, F. M. A. Aflatoxin B1 and zearalenone in dairy feeds in Portugal, 2009–2011. *Mycotoxin Research*, 2013, 29, 131–133.
[31] Virdis, S., Corgiolu, G., Scarano, C., Pilo, A. L., De Santis, E. P. L. Occurrence of Aflatoxin M1 in tank bulk goat milk and ripened goat cheese. *Food Control*, 2008, 19, 44–49.
[32] Motawee, M. M., Bauer, J., McMahon, D. J. Survey of Aflatoxin M1 in Cow, Goat, Buffalo and Camel Milks in Ismailia-Egypt. *Bulletin of Environmental Contamination and Toxicology*, 2009, 83, 766–769.
[33] Kos, J., Lević, J., Duragić, O., Kokić, B., Miladinović, I. Occurrence and estimation of aflatoxin M1 exposure in milk in Serbia. *Food Control*, 2014, 38, 41–46.
[34] Rahimi, E., Mehrdad, A. A Survey of Aflatoxin M1 Contamination in Bulk Milk Samples from Dairy Bovine, Ovine, and Caprine Herds in Iran. *Bulletin of Environmental Contamination and Toxicology*, 2012, 89, 158–160.
[35] Rahimi, E., Bonyadian, M., Rafei, M., Kazemeini, H. R. Occurrence of aflatoxin M1 in raw milk of five dairy species in Ahvaz, Iran. *Food and Chemical Toxicology*, 2010, 48, 129–131.
[36] Delialioğlu, N., Otağ, F., Ocal, N. D., Aslan, G., Emekda, G. Investigation of aflatoxin M1 levels in raw and market milks in Mersin Province, Turkey. *Mikrobiyoloji Bulteni*, 2010, 44, 87-91.
[37] Hussain, I., Anwar, J., Asi, M. R., Munawar, M. A., Kashif, M. Aflatoxin M1 contamination in milk from five dairy species in Pakistan. *Food Control*, 2010, 21, 122–124.
[38] Oliveira, C. A. F., Ferraz, J. C. O. Occurrence of aflatoxin M1 in pasteurised, UHT milk and milk powder from goat origin. *Food Control*, 2007, 18, 375–378.

[39] Freitas, T. E. Monitoramento de micotoxinas em ração destinada à caprinos em lactação e seus resíduos no leite. *Master Dissertation*, Instituto Biológico, São Paulo, Brazil, 2012.

[40] Batatinha, M. J. M., Botura, M. B., Simas, M. M. S., Mallmann, C. A., Bittencourt, T. B. C. S. T., Sabino, M. Otimização de método analítico para determinação de aflatoxinas em rações de caprinos e ocorrência no estado da Bahia. *Arquivo Brasileiro de Medicina Veterinária e Zootecnia*, 2008, 60, 503-505.

[41] Barrios, M. J., Gualda, M. J., Cabanas, J. M., Medina, L. M., Jordano, R. Occurrence of aflatoxin M_1 in cheeses from de south of Spain. *Journal of Food Protection*, 1996, 59, 898-900.

[42] Iha, M. H., Abe, L. T., Okada, M. M., Fávaro, R. M. D. Occurrence of aflatoxins M1 and M2 in goat milk marketed commercialized in the region of Ribeirão Preto-SP, Brazil. *Revista do Instituto Adolfo Lutz*, 2007, 66, 57-62.

[43] Roussi, V., Govaris, A., Varagouli, A., Botsoglou, N. A. Occurrence of aflatoxin M1 in raw and market milk commercialized in Greece. *Food Additives and Contaminants*, 2002, 19, 863-868.

[44] Lopez, C., Ramos, L., Ramadan, S., Bulacio, L., Perez, J. Distribution of aflatoxin M1 in cheese obtained from milk artificially contaminated. *International Journal Food Microbiology*, 2001, 64, 211–215.

[45] Kiermeier, F., Buchner, R. E. On the aflatoxin M1 content of the cheese during ripening and storage. *Z Lebensm Unters Forsch*, 1977, 164, 87–91.

[46] Battacone, G., Nudda, A., Palomba, M., Pascale, M., Nicolussi, P., Pulina, G. Transfer of aflatoxin B1 from feed to milk and from milk to curd and whey in dairy sheep fed artificially contaminated concentrates. *Journal of Dairy Science*, 2005, 88, 3063–3069.

[47] Bognanno, M., Fauci, L. L., Ritieni, A., Tafuri, A., De Lorenzo, A., Micari, P., Renzo, L. D., Ciappellano, S., Sarullo, V., Galvano, F. Survey of the occurrence of aflatoxin M1 in ovine milk by HPLC and its confirmation by MS. *Molecular Nutrition and Food Research*, 2006, 50, 300–305.

[48] Ozdemir, M., Determination of aflatoxin M-1 levels in goat milk consumed in Kilis province. *Ankara Universitesi Veteriner Fakultesi Dergisi*, 2007, 54, 99–103.

[49] Battacone, G., Nudda, A., Pulina, G. Effects of Ochratoxin A on Livestock Production. *Toxins*, 2010, 2, 1796-1824.

[50] Nip, W. K., Chu, F. S. Fate of ochratoxin A in goats. *Journal Environmental Science Health Part B*, 1979, 14, 319-33.

[51] Tsiplakou, E., Anagnostopoulos, C., Liapis, K., Haroutounian, S. A., Zervas, G. Determination of mycotoxins in feedstuffs and ruminant's milk using an easy and simple LC–MS/MS multiresidue method. *Talanta*, 2014, 130, 8-19.

[52] Saccà, E., Boscolo, D., Vallati, A., Ventura, W., Bigaran, F., Piasentier, E. Aflatoxin occurrence in milk and supplied concentrates of goat farms of north-eastern Italy. *Journal of the Science of Food and Agriculture*, 2009, 89, 487–493.

[53] Pattono, D., Gallo, P. F., Civera, T. Detection and quantification of Ochratoxin A in milk produced in organic farms. *Food Chemistry*, 2011, 127, 374–377.

In: Mycotoxins
Editor: Cesar Rios

ISBN: 978-1-63483-544-2

Chapter 4

DETERMINATION OF MYCOTOXINS IN FOOD WITH THE AID OF CHEMOMETRIC TOOLS

Agustina V. Schenone[1,2] ***and Nilda R. Marsili***[1,*]
[1]Cátedra de Química Analítica II,
Facultad de Bioquímica y Ciencias Biológicas,
Universidad Nacional del Litoral,
Ciudad Universitaria, Santa Fe, Argentina
[2]Instituto de Desarrollo Tecnológico para la
Industria Química (INTEC), Consejo
Nacional de Investigaciones Científicas y
Técnicas (CONICET) and Universidad
Nacional del Litoral (UNL), Santa Fe, Argentina

ABSTRACT

This chapter presents an overview of current chemometric tools utilized for the evaluation of several mycotoxins often found in foodstuff, such as cereals, peanuts, pistachio nuts and condiments. Mycotoxins are compounds produced as a result of the secondary metabolism of fungus, such as *Aspergillus*, *Fusarium* and *Penicillium*. Because of their harmful effect in animals and humans, their levels in food are strictly controlled.

Discussions are focused on the analysis of first and second-order data generated through several techniques such as near infrared spectroscopy, fluorescence spectroscopy and high performance liquid chromatography with diode array detector. This review manuscript describes the algorithms frequently used to model this type of data, including Partial Least Squares (PLS) regression, Parallel Factor Analysis (PARAFAC) and Multivariate Curve Resolution Alternating Least Squares (MCR-ALS). These multivariate calibration methods allow the quantification of multiple analytes in complex food samples with the application of minimum pre-treatment techniques. Another important chemometric tool commented in this review is the Response Surface Method

* Corresponding Author: E-mail: nmarsili@fbcb.unl.edu.ar.

(RSM), commonly applied in the optimization of experimental procedures and chemical measurements.

This chapter discusses the different strategies implemented for the analysis of a wide spectrum of mycotoxins, as well as their advantages and disadvantages.

INTRODUCTION

Mycotoxins are compounds produced as a result of the secondary metabolism of fungus, such as *Aspergillus*, *Fusarium* and *Penicillium*. These toxins are often found in corn, rice and other cereals, peanuts, and other food, as a consequence of contamination before harvest, during storage or prior to processing the material [1]. In addition, they can also contaminate cow milk and consequently, dairy products such as cheeses and yoghurt [2].

Within mycotoxins we can mention aflatoxins (produced by *Aspergillus flavus* and *A. parasiticus*), ochratoxins (produced by *Aspergillus ochraceus* and *Penicillium verrucosum*), fumonisins (produced by *Fusarium verticillioides* and *F. proliferatum*), deoxynivalenol (produced by *Fusarium graminearum* and *F. culmorum*), patulin (mostly produced by *Penicillium expansum*), zearalenone (produced by *F. graminearum* or *F. culmorum*) and T-2 toxin (produced principally by *F. sporotrichioides*). Table 1 presents the most common mycotoxins found in foodstuff as well as their toxic properties.

Several legislations and regulations presented by the Food Agricultural Organization (FAO), the Codex Alimentarius, the Food and Drug Administration (FDA) and the European Union (EU), establish the maximum mycotoxin level in foodstuff and raw materials because of their harmful effect in animals and humans. In the case of aflatoxins, the B1 toxin is the most important from the toxicological point of view and because of its frequency. The EU accepts 15 $\mu g\ Kg^{-1}$ of total aflatoxins in peanut without processing, and less than 4 $\mu g\ Kg^{-1}$ of the same mycotoxin in peanuts ready to human consumption. Ochratoxin A is the major mycotoxin of its group and a limit of 5 $\mu g\ Kg^{-1}$ and 3 $\mu g\ Kg^{-1}$ are accepted for unprocessed and processed cereals, respectively. Patulin can be found principally in fruit products, such as apple juice. The EU sets 10 $mg\ Kg^{-1}$ as the limit of this toxin in baby food. Accordingly, their identification and quantification are essential [3-6].

A wide range of analytical techniques for the analysis of mycotoxins is available in the literature. Separation methods, such as high performance liquid chromatography (HPLC) coupled with fluorescence detection or with mass spectrometry detector [7-11], gas chromatography-mass spectrometry (GC-MS) [12-14] and capillary electrophoresis (CE) [15, 16], are the most frequently used quantitative methods in research and routine analyses of mycotoxins. Various clean-up methods have been reported for different food matrices prior to HPLC analysis. Among them, multifunctional column [17], immunoaffinity column (IAC) [18, 19] and solid phase extraction (SPE) [20-22] have been applied for different food samples. However, these techniques are known to use large quantities of contaminant solvents and long analysis time. Recently, enzyme-linked immunosorbent assay (ELISA) [23-25] became very popular due to the easy application of the commercially available kits and its high sensitivity. Nevertheless, it presents some disadvantages, such as high cost and that it cannot resolve the presence of matrix effect in the sample. In any case, it is important that the analytical method maintains adequate selectivity, sensitivity, accuracy and precision, as well as other qualities in order to properly quantify mycotoxins in complex matrices.

Table 1. Mycotoxins and their toxic properties

Mycotoxin	Toxic Properties	Molecular Structure
Aflatoxin B1	- Hepatotoxic, liver damage in animals - Immunosuppressive - Carcinogenic, teratogenic and mutagenic [26]	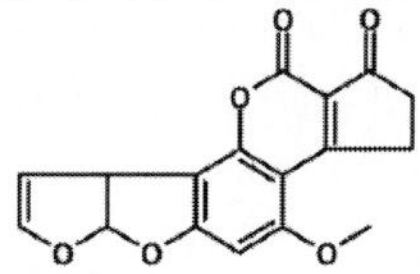
Ocrhatoxin A	- Damage in kidney and liver [27]	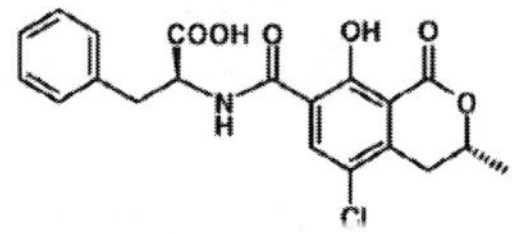
Fumomisin B1	- Leukoencephalomalacia in horses [28] - Tumors in the liver and kidney - Esophageal tumors [29, 30]	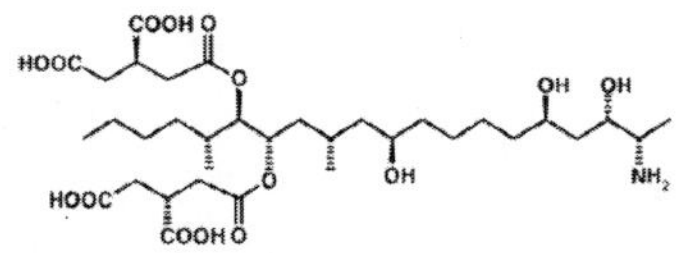
Deoxynivalenol	- Vomiting syndrome - Immunosuppressive - Kidney problems in animals [31]	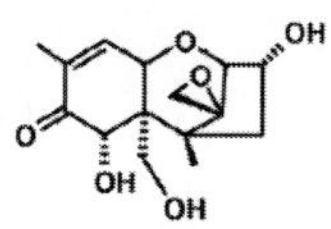
Patulin	- Immunological, neurological and gastrointestinal outcomes [32]	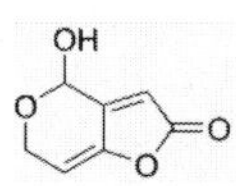
Zearalenone	- Estrogenic effects [33]	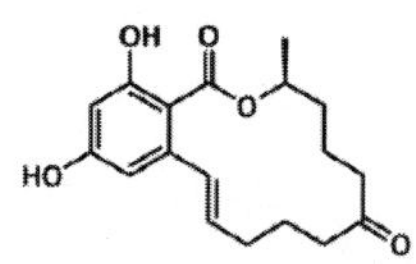
T-2 toxin	- Inhibits protein synthesis - Decreases antibody levels and humoral factors. [34, 35]	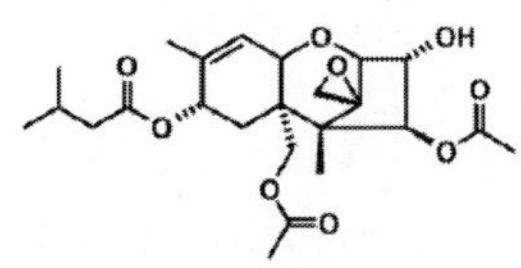

The multivariate calibration methods have impacted significantly the development of analytical methods, especially in the determination of one or more target analytes in samples with complex matrices. The aim of these approaches is to develop simple, selective and sensitive techniques avoiding sample preprocessing and clean-up methods.

CHEMOMETRIC METHODS

Chemometrics is a sub-discipline of analytical chemistry and it is divided in three fields [36]: a) design and/or selection of optimal experimental procedures, b) calibration, validation and significance of analytical measurements and c) proportion of maximum relevant chemical information by analyzing chemical data.

Multivariate calibration is the set of chemometric methods which objective is the quantification of one or more components of an unknown sample with multiple components [37]. Another field in which chemometric is applied is the optimization of experimental procedures and the chemical measurements. Since several factors generally influence a particular system, the application of an optimization technique is usually required to achieve a good method performance.

Multivariate Calibration Methods

Classification of Instrumental Data

Multivariate calibration is closely related to the different types of analytical instruments, since the calibration method must be chosen according to the obtained type of data. Instrumental signal can be classified in [38, 39]:

- Zero order data: they are obtained by instruments that generate an only data point per sample and the calibration set consists in a vector. This type of data is modeled through univariate calibration. The atomic absorption photometer and the ion selective electrodes are examples of this kind of instruments.
- First order data: each sample generates a data vector and the calibration set forms a matrix (two-way arrangement). This type of data is analyzed by multivariate calibration methods of first order. Instruments that provide this type of signal are, for example, spectrophotometers, chromatographs and spectrofluorometers that produce absorption spectra, fluorescence emission spectra and chromatograms, respectively.
- Second order data: in this case, a data matrix is generated per sample and the set of calibration consists in a three-way array. The matrices can be produced by a single instrument, such as excitation-emission fluorescence matrix or by monitoring chemical reaction kinetics with a spectrophotometer. Furthermore, second order data can be generated by hyphenated techniques, including tandem mass spectrometry (MS/MS) and gas chromatography coupled to mass spectrometry (GC/MS).
- Third order data: an example of this type of data is the excitation-emission fluorescence spectroscopy resolved in time when reaction kinetics or fluorescence decay is monitored. For each sample, a three-way array is obtained and the calibration set is a four-way data array.

Chemometric Tools: Determination of Mycotoxins

In all the cases, the aim of the calibration step is to determine a model between the data (e.g., a series of spectra) and the concentrations of standard samples. There are several methods depending on the data order and on how much information is known about the system under study.

The quality of the found model can be assessed on an independent "test" set, which consists of a number of samples with different concentrations respect to the calibration samples.

First Order Algorithms

Several multivariate calibration methods use first order responses. Instead of calibrating a single variable (e.g., a peak height from a chromatogram) to another single variable (e.g., a concentration), several variables (e.g., spectroscopic absorbencies at several wavelengths) are calibrated to one or more variables. The first developed methods were based on direct calibration: Ordinary Least Square (OLS) [36], Classical Least Square Regression (CLS) [40-42], as well as inverse calibration: Inverse Least Square (ILS) [43, 44]. The Principal Component Regression (PCR) [45] method represents one of the first attempts to bring together the main advantages of CLS, OLS and ILS, but using decomposition concepts and spectral factors. The Partial Least Square (PLS) regression improved the results obtained by PCR [46], becoming a routine laboratory tool.

These algorithms have the limitation that each species that present a response signal must be included in the calibration set. First order methods have the ability of detecting the presence of interferents in unknown samples. In that case, the sample is identified as outlier because the spectral residues are abnormally large compared to the instrumental noise. This property is known as the first order advantage [38]. Accordingly, the outlier sample may not be properly analyzed and/or the prediction result will be wrong.

Since PLS is probably the most widespread multivariate first order method, this algorithm will be briefly explained.

PLS

PLS approach is done by decomposition of both, the concentration vector **y** (or a matrix **Y**) and signal matrix **X** of the mixture standard solutions in calibration set, into latent variables. The algorithm generally work in three steps: 1) the mutual latent variables for **X** and **y** are calculated based on maximum covariance; 2) a diagonal regression matrix is created for all PLS components; and 3) the concentrations of unknown samples (test set) are predicted using the diagonal regression matrix and a new **X** matrix (signal from test samples).

Usually, the cross validation method proposed by Haaland and Thomas [37] is used for the selection of the optimum number of factors. The calibration step is done leaving out one sample at a time. With the created model, the concentration of the sample that was left aside is predicted. The process is then repeated leaving out another calibration sample and so on. With the predicted concentrations (c_{pred}) and the actual concentrations (c_{act}), the error is calculated through the statistical prediction error sum of squares:

$$PRESS = \sum_{1}^{m}(c_{act} - c_{pred})^2 \qquad \text{Eq. 1}$$

Then, the following F ratio is computed: $F_i = PRESS_i/PRESS_{lowest}$ in which $PRESS_i$ is each calculated error and $PRESS_{lowest}$ is the lowest obtained error, for $I = 1,2,\ldots,I/2 + 1$. The optimum number of factors is suggested to correspond to the first F_i which probability is less than 75%.

The quality of PLS calibration can be evaluated by the root mean square error of prediction:

$$RMSEP = [\sum_{i=1}^{n} \frac{(c_{act} - c_{pred})^2}{n}]^{1/2} \qquad \text{Eq. 2}$$

and the relative error of prediction:

$$REP(\%) = \frac{100}{\bar{c}}[\frac{1}{n}\sum_{1}^{n}(c_{act} - c_{pred})^2]^{1/2} \qquad \text{Eq. 3}$$

in which $\bar{c}$ is the average concentration in the n calibration mixtures.

Literature Examples of First Order Methods

Pettersson and Aberg developed a spectroscopic method for the determination of the mycotoxin deoxynivalenol (DON) in wheat samples [47]. Dilution series of wheat field infected with *Fusarium culmorum* was used in the calibration trials. They modeled near infrared spectra with PCA and PLS obtaining good statistical parameters (slope = 0.949, a correlation coefficient of 0.984 and a standard error of 381 μg DON per kg). Three wavelength regions were tested and the reduced wavelength range (670–1100 nm) had the best results. In this case, the number of factors was 11.

The PLS method was also applied to model synchronous fluorescence spectra in order to determine aflatoxin B1 (AFB1) in pistachio samples [48]. AFB1 was obtained after a few extraction steps with methanol/water, ether and hexane. Finally, the solvents were evaporated and a solution of β-cyclodextrine was added and dissolved in 10% (v/v) methanol/water solution. It was demonstrated that cyclodextrines enhance the molecular fluorescence of AFB1. Calibration and test samples were prepared by spiking blank pistachio samples with concentrations ranging from 0 to 15 ppb. Spiked as well as naturally contaminated samples were used to validate the results. After obtaining the optimum $\Delta\lambda$, fluorescence spectra were recorded maintaining a constant difference $\Delta\lambda = 71$ nm, in the emission range 300–430 nm. These spectra were used as analytical signal to make the calibration models. After cross validation, the factor number selected as optimum was 3. The values of RMSEP and REP were 0.458 and 6.081%, respectively.

Aflatoxin B1 was also evaluated by Tripathi and Mishra [49]. In this case, the sample was red chili powder and the investigation was done by using Fourier transform near-infrared (FT-NIR) spectroscopy in diffuse reflectance mode. Aflatoxin free chili powder samples were spiked with known amount of standard AFB1 in methanol ranging from 15 to 500 μg kg^{-1} and used for calibration model building based on PLS regression algorithm. The training set consisted of 40 samples, while 25 samples were kept as test set (from 16.63 to 495.85 μg kg^{-1}). AFB1 was extracted from chilies with dichloromethane acidified with hydrochloric acid, cleaned up with lead acetate solution, again extracted with dichloromethane and purified by using silica gel column chromatography. In this research, three data preprocessing techniques were analyzed: straight line subtraction (SLS), constant offset elimination (COE) and minimum maximum normalization (MMN). The wavelength regions of 6900.3–4998.8 cm^{-1} and 4902.3–3999.8 cm^{-1} were selected as optimal on the basis of RMSEP. The best results were obtained with the application of SLS and 5 PLS factors, giving RMSEP = 0.65 and R^2 = 98% (R^2 is the correlation coefficient of the regression line between predicted and actual values). Furthermore, naturally contaminated and market samples were analyzed with the developed method and the results were compared with those obtained with HPLC and thin layer chromatography (TLC) methods. There were no statistically significant differences in the means of FT-NIR method and reference chemical methods.

Sugiyama and co-workers developed a method for predicting total aflatoxin (aflatoxins B1, B2, G1, and G2) in nutmeg extract using fluorescence spectra and PLS regression [50]. Nutmeg extract was artificially spiked with an aflatoxin reagent and prepared from uncontaminated nutmeg powder. For the extraction step, 90% acetonitrile solution was used. After filtration, the nugmet extract was spiked with aflatoxin at 5% volume of the extract. Twenty-one samples with seven aflatoxin concentrations were prepared and divided into two groups: 11 were used as the calibration data set to develop a calibration model, and the remaining 10 were used as the test set to validate the calibration model. Then, the obtained data was preprocessed to remove signals not related to fluorescence. Three latent variables were used to develop the PLS model by cross validation. R^2 was 0.993 and the standard error of calibration was 0.2 μg L^{-1} for the calibration data set. Moreover, significant correlations were observed between the actual and predicted values for the validation data set, with R^2 of 0.773 and standard error of prediction of 1.0 μg L^{-1}.

As can be observed, all the mentioned methods use uncontaminated samples in the calibration step. This is because each component that presents a response signal must be included in the calibration set, as mentioned before. Notice that several factors must be included in this step and it is usually higher than the number of the analytes under study.

Another important point to mention is the fact that none of the previous cited methods reported limits of detection or limits of quantification. These are important parameters that allow the comparison between different methods and with the maximum values accepted by the international food control agencies.

Second Order Algorithms

Among the algorithms used to model second order data, the most important are: Parallel Factor Analysis (PARAFAC) [51], Generalized Rank Annihilation Method (GRAM) [52], Direct Trilienar Decomposition (DTLD) [53], Multivariate Curve Resolution Alternating Least Squares (MCR-ALS) [54], Bilinear Least Squares (BLLS) [55, 56], Alternating Trilinear Decomposition (ATLD) [57] and the Multiway Partial Least Square (N-PLS) [58] combined with Residual Bilinearization (RBL) [59].

Second order calibration methods have a special property that allows the analysis of two way data even in the presence of uncalibrated components, which is known as second order advantage [38]. Accordingly, these algorithms can predict the concentration of an analyte in a mixture with interferences. This fact overcomes the disadvantages of lower order analytical methods, i.e. the need to be selective (in the case of univariate calibration) and the availability of calibration samples containing potential interfering species (in the case of first order models). Furthermore, the analysis of several analytes is possible. This last characteristic can also be achieved with first order methods, but a higher number of calibration samples are necessary and they must have the same composition of the unknown sample. On the other hand, second order algorithms only require a small number of standard samples.

The description of the most used algorithms can be found elsewhere [51, 54, 59, 60]. In the literature, there are few publications in this field, and most of them describe quantification methods based on PARAFAC. Another used multivariate method is MCR-ALS. Therefore, brief descriptions of these algorithms are given below.

PARAFAC

The PARAFAC model decomposes three way data ($\underline{\mathbf{X}}$) in their individual compounds (analytes and interferences) according to the individual contributions of each of them. This means that, spectral overlap problems can be solved and that the profiles of both variation modes (e.g., excitation and emission spectra) can be estimated for each compound in particular. The data decomposition is performed in triads or trilinear components, yielding one scores vector and two loadings vectors for each component. In PARAFAC, a generic element x_{ijk} of the matrix $\underline{\mathbf{X}}$ is defined as:

$$x_{ijk} = \sum_{n=1}^{N} a_{in} b_{jn} c_{kn} + e_{ijk} \quad \text{Eq. 4}$$

in which N is the total number of chemical components, e_{ijk} is the residual error, a_{ni}, b_{nj}, c_{nk} are the elements of vectors $\mathbf{a}_n$, $\mathbf{b}_n$ and $\mathbf{c}_n$ which correspond to relative concentration, known as scores $[(I + n_{unk})\times 1]$ and both loadings for each N component $(J \times 1)$ and $(K \times 1)$, respectively. In chemistry, the columns of the matrices **B** and **C** can be assigned with a physical meaning, e.g., excitation and emission spectra. I is the number of standards and n_{unk} is the number of unknown samples. The adequate number of components is usually selected by use of the Core Consistency [61].

MCR-ALS

Generally, a column-wise augmented matrix **D** is created by joining together the matrices of the unknown sample and the calibration data. The bilinear decomposition of **D** is then performed according to the expression:

$$\mathbf{D} = \mathbf{C}\,\mathbf{S}^{\mathrm{T}} + \mathbf{E} \quad \text{Eq. 5}$$

in which the rows of **D** contain, in the case of excitation-emission data, the emission spectra measured for different samples at several excitation wavelengths, the columns of **C** contain the excitation spectra of the intervening species, the columns of **S** their related emission spectra, and **E** is a matrix of residuals not fitted by the model.

Decomposition of **D** is achieved by iterative least-squares minimization of the Frobenius norm of **E**, under suitable constraining conditions during the ALS procedure. MCR-ALS requires initialization with system parameters as close as possible to the final results. The most employed method is the SIMPLISMA (simple to use interactive self-modeling mixture analysis) methodology [62]. After MCR-ALS decomposition of **D**, the information contained in **C** can be used for quantitative predictions. If the area under each profile is considered proportional to each component concentration, then a univariate model can be built and used to make the predictions on test and real samples.

Literature Examples of Second Order Methods

Parallel Factor Analysis was used to model second order data obtained with high performance liquid chromatography with diode-array detection (DAD). In this case, aflatoxins B1, B2, G1 and G2 were simultaneously determined in pistachio nuts in the presence of matrix interferences [63]. Sample preparation was based on solvent extraction with a methanol/water/hexane solution and the hexane portion was then diluted with a methanol/water solution. A pre-concentration and clean-up step with a general and non-polar sorbent C18 was used and represented a low-cost technique relative to standard method of aflatoxin analysis (immune affinity column assay). A calibration set of five standard solutions (concentrations between 0.05 and 0.75 mg L^{-1}) of each analyte was prepared in methanol. Furthermore, nine validation samples (containing a blank test) were prepared by spiking different amounts of the four analytes into blank pistachio samples. Finally, two naturally contaminated Iran pistachio samples were analyzed with the developed method. Photometric detection using DAD detector was done between 320 and 420 nm (step of 1 nm and integration period of 0.4 s per spectrum) for all the samples.

Appropriate pre-processing steps were applied to correct background signals and the effect of retention time shifts. For calibration step, chromatographic data obtained from direct injection of standard analytes into HPLC system were standardized into matrix matched standards using piecewise direct standardization (PDS), saving both chemicals and time of analysis. Moreover, since the clean-up step based on a general and non-polar sorbent was not selective enough to separate analytes of interest from interfering matrix constituents, many of matrix interferences were retained on the sorbent and consequently they were eluted to the final sample. Because of this, it was necessary to exploit the second-order advantage through the application of PARAFAC. In this study, non-negativity and unimodality constraints were used in all modes and chromatographic mode, respectively. The estimated numbers of components for all validation samples (after pre-processing steps) were 2, 3, 3 and 3 for Af-B1, Af-B2, Af-G1 and Af-G2, respectively. Acceptable recoveries values were obtained for validation samples (mean values between 73 and 115%), considering the low concentration of spiked samples and despite the complexity of real samples. RMSEP and REP values were also presented and showed the validity of the method. The results of PARAFAC on naturally contaminated pistachio nuts indicated that the four aflatoxins could be successfully determined. Moreover, multivariate analytical figures of merit were calculated. Limits of detection (LOD) vary according to the aflatoxin between 0.2 and 3 μg Kg^{-1}.

The same investigation group proposed another strategy to determine aflatoxins B1, B2, G1 and G2 in pistachio samples without the need to divide the data into four sub-sections and create four different models, which can be a very time-consuming process [64]. Therefore, Vosough and co-workers developed a multi-target and fast HPLC-DAD method, through a new optimized gradient elution program and in a single three-way array without the need for developing specific three-way model for each analyte. In this work, PARAFAC2 was used to model the data, which is less affected by retention time shifts than PARAFAC. In this case, the chromatogram lasted 4.9 min (the previous analysis time was about 13 min). The recovery values obtained for the prediction samples by PARAFAC2 for AFB1, AFB2, AFG1 and AFG2 were 91.3–109.8%, 90.6–108.8%, 99.4–108.8% and 93.0–116.8%, respectively. The maximum values of RMSEP and REP% were 0.18 and 6.7%, respectively for AFB1.

Figures of merit were also evaluated. The LOD and sensitivity (SEN) values for AFB1, AFB2, AFG1 and AFG2 were 2.12, 1.24, 2.26, 2.63 μg kg^{-1} and 1.38, 2.34, 1.41 and 1.15 μg kg^{-1}, respectively. Among the analytes, AFG2 showed the lowest sensitivity. The most important issue was that SEN and LOD values depend strongly on the other sample constituents such as interferences and the other calibrated analytes. While the LOD were not better than the ones obtained with the previous investigation [64], the analysis time was shorter and no transference of calibration method was necessary.

Recently, Luna et al. presented a method for the simultaneous determination of AFB1 and AFG2 in peanuts using PARAFAC [65]. Excitation-emission data were registered. Samples were subjected to extraction with methanol/water and to a clean-up technique which involved an immunoaffinity column containing monoclonal antibody for aflatoxins B1, B2, G1 and G2. Moreover, samples were derivatized with bromine before the measurements. The fluorescence spectra were recorded in the excitation range from 300 to 370 nm (step of 2 nm) and in the emission range from 427 to 550 nm (step of 1 nm) in order to avoid the presence of Rayleigh scattering. In this work, the matrix effect was investigated trough the comparison of the slopes obtained from the calibration curve over the range of 10–50 ng mL^{-1} with the slope of the peanut spiked samples processed as describe before, for both analytes. The *t*-test was used and no statistically significant difference was found. Therefore, the regression linear curve obtained from the pure standards could be used to quantify aflatoxins in spiked peanut sample without interference of matrix. The results of PARAFAC on a set of spiked and naturally contaminated peanuts indicated that the two aflatoxins could be successfully determined. LOD were 0.05 and 0.04 μg kg^{-1} for aflatoxins B2 and G2, respectively. The limits of quantification (LOQ) were 0.16 and 0.12 μg kg^{-1} for aflatoxins B2 and G2, respectively. These values are better than those obtained in previous works cited in this chapter. Moreover, they are two orders smaller than the allowed values of total aflatoxins in peanuts by the European Union. However, a more sophisticated and selective clean-up process was applied.

An inherent problem of many fluorimetric procedures is the absorption of exciting and/or emitted radiation by dissolved species (either fluorescent or not) or by the fluorophore itself, leading to a variation in the signal intensity [66]. This drawback can be corrected by the well known standard addition method (SAM), which compensates matrix effects arising from the specific composition of the sample. Combinations of second order algorithms and SAM have been proposed to solve both inconveniences: presence of unknown components and matrix effect.

Spectrofluorimetric determination of aflatoxin B1 in wheat by standard addition method combined with PARAFAC has been reported [67]. The method was based on the enhanced fluorescence of AFB1 by β-cyclodextrin in 10% (w/w) methanol/water solution and did not require immunoaffinity column, extended cleanup steps and derivatization process. The sample treatment consisted of a few extraction procedures with methanol/water solution and chloroform. Because of the presence of matrix effect caused by wheat matrix, four standard additions were performed for each sample. Excitation-emission matrices were acquired by measuring the fluorescence spectra in the excitation range from 320 to 368 nm (step of 4 nm) and in the emission range from 390 to 490 nm (step of 2 nm). Then, a specific PARAFAC model was built from three-way arrays formed by the 5 excitation-emission matrices (EEM) (sample plus 4 additions). In all cases, three factors and no constraints were used and the models accounted for more than 99.90% of the total data variance. The loadings related to sample mode (scores) were used for calibration through a linear regression ($R^2 > 0.990$ and Standard Deviation < 0.3 over the range of 0–18 μg kg^{-1} of AFB1). The limit of detection of the method was 0.9 μg kg^{-1} for the determination of AFB1 in wheat samples. Finally, AFB1 was analyzed in two naturally contaminated wheat samples and three spiked wheat samples. Furthermore, the results of the two naturally contaminated wheat samples were compared with HPLC method. The values obtained by the proposed method were statistically comparable to that obtained with the chromatographic technique.

So far, most of the methods found in the literature were developed in order to determine aflatoxin, in its different forms. Recently, the determination of aflatoxins and ochratoxin A (OTA) using fluorescence emission-excitation matrices and PARAFAC was investigated [68]. The calibration set as well as five artificial mixtures were prepared in methanol/water. In this case, the real sample was sorghum grains and minimum pretreatment was necessary. The grains were suspended in a methanol/water solution. After sonicating and filtering, the toxins pass into the methyl alcohol solution, which was then used for quantification. The EEM were collected between 250 and 400 nm and from 400 to 500 nm for excitation and emission sensors, respectively. Figure 1 shows the EEM from both mycotoxins standard solutions.

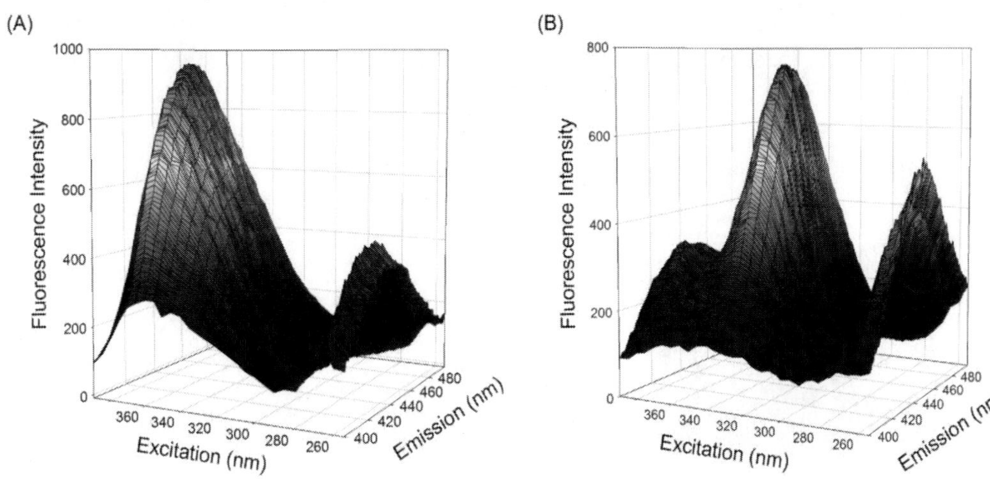

Figure 1. (A) EEM obtained for AFs standard solution. (B) EEM obtained for OTA standard solution.

The second order algorithm N-PLS/RBL was also evaluated. The predicted concentration of the artificial mixtures by PARAFAC showed recoveries between 95.6 and 105.6% and between 95.0 and 105.1% for AFs and OTA, respectively. N-PLS/RBL method showed recoveries between 68.0 and 86.2% for AFs and between 90.0 and 125.8% for OTA. When the former method was applied, it was found that to predict artificial samples concentration, the produced error was approximately 30%. For this reason, in the studied conditions, N-PLS/RBL could not be applied to determine the concentration of the mycotoxins correctly. On the other hand, the results obtained by applying PARAFAC to predict the concentrations of the artificial samples were satisfactory, with an error less than 6% in the validation set. Hence, only this algorithm was used to predict the mycotoxins in real sorghum grains samples. Moreover, a commercial competitive ELISA kit was also used and considered as a reference method, and both results were compared. The recoveries produced by PARAFAC were between 82.9 and 101.7% for AFs and between 95.9 and 100.9% for OTA, with respect to ELISA results. The limits of detection achieved by this method expressed in the solid samples (between 2.0 and 3.5 μg kg^{-1} for AFs and 3.5 and 4.0 μg kg^{-1} for OTA) are consistent with the maximum values established by the FDA and the European Commission in the EU. The developed method was able to quantify both, AFs and OTA, even in the presence of a complex sample matrix. Although the limits of detection were higher than the method presented by Luna et al. [65], it showed to be a relatively inexpensive strategy, of medium complexity, rapid and simple, and also did not require sample preconcentration or a laborious clean-up technique.

A novel technique was developed by Saidi and co-workers, in which antibody coated gold nanoparticles (GNP) were used as fluorescence quenchers for simultaneous determination of aflatoxins (B1, B2) [69]. It was found that GNPs with definite surface plasmon absorption can quench the fluorescence of aflatoxins. The quenching of the fluorescence of EEMs of both aflatoxins, provoked by gold-labeled antibodies, was studied by MCR–ALS. Unlike fluorescence or enzyme-detection systems, gold-labeled antibodies are more stable and easy to use. There are no needs for strict operations as incubation, washing and enzymatic reactions during signal generation, which surely shorten the detection time. Furthermore, nano scale surfaces provided by GNPs could accelerate antibody–antigen reaction sufficiently, which supply an amplified signal. Moreover, with the increasing of the concentration of GNPs, the fluorescence intensity of aflatoxins decreased correspondingly.

The MCR–ALS procedure decomposed the augmented data matrix into two matrices: one matrix containing the excitation spectra as function of the concentration, and the second matrix containing the emission spectra common to all concentrations. Non-negativity constraint and equality constraint in spectral profiles were applied in excitation and emission modes.

Quantitative results obtained from aflatoxins in pistachio nut samples (concentration of AFs between 0.00 and 64.5 μg kg^{-1}) by MCR–ALS are compared to those obtained using the HPLC method. There were no significant differences between the methodology proposed and the standard one, and may be a good alternative to the traditional methods of analysis for aflatoxins.

The use of the chemometric algorithms allowed solving problems such as presence of interferences in the sample matrix and simultaneous analytes determination even in the case of overlapped signals. Besides, several works demonstrated the capability of the use of EEMs for the determination of mycotoxins in complex matrices, such as peanuts and pistachio nuts with simple sample pretreatment.

In general terms, PARAFAC is the preferred algorithm to process this kind of data, but further work should be done in order to explore other second order methods widely used in the literature.

Another interesting consideration when analyzing complex samples is the matrix effect caused by the composition of the sample. In this sense, few works were found which describe the combination of this type of multivariate methods and the standard addition method to determine mycotoxins.

Response Surface Methodology

The Response Surface Methodology (RSM) is a chemometric tool and consists of a group of empirical techniques designed to evaluate the relationship between several experimental factors and the measured response according to one or more selected criteria [70, 71]. For this purpose, a series of mathematical and statistical techniques are applied to create a model which relates experimental data to describe the behavior of the response. RSM study each factor in three or more levels in order to model the curvature of the system. The response surface is represented as a surface in a tridimensional space. To better visualize the shape of a response surface, its contours are plotted on the plane x_1, x_2. Each contour corresponds to a specific height of the response surface.

The response surface methodology is a sequential technique and the aim is to determine optimum operating conditions for a system, or to determine the region of space of the factors on which the operating conditions are satisfied. The main advantage of RSM is that fewer experiments are needed in comparison with the classical analysis, where the response is investigated for each variable, keeping the other variables at a constant level involving the consumption of more time and effort.

The selection between the existing designs depends basically on the number of factors to optimize, the experimental environment and the efficiency and complexity of each of them. The most applied experimental designs are Complete Factorial Design with three levels, Central Composite Design and Box-Behnken. The detailed description of these methodologies can be found elsewhere [70, 71].

Literature Examples of RSM

HPLC conditions were optimized for the quantification of aflatoxins B1, B2, G1 and G2 in contaminated peanuts [72]. For this purpose, aflatoxins were quantified by using high performance liquid chromatography with fluorescent detector. A central composite design (CCD) was used to investigate the effect of three variables, i.e., mobile phase composition with different ratios of acetonitrile/water/methanol (*X*1), temperature (*X*2) and flow rate (*X*3). In this optimization study, the peak areas of the four target aflatoxins were considered as response variables (i.e., B1 (*Y*1), B2 (*Y*2), G1 (*Y*3) and G2 (*Y*4)).

To determine the statistical significance of model terms and fit a regression relationship relating the experimental data to the independent variables, analysis of variance (ANOVA) and regression surface analysis were conducted. The results indicated that the polynomial models were significant ($p < 0.05$) with high coefficient of determination (R^2) ranging from 0.939 to 0.999.

The overall optimum condition was then determined by superimposing all response surface plots. A numerical optimization was also carried out by the response optimizer for determining the exact optimum level of HPLC variables providing the highest response. The highest quantification value (i.e., highest peak area) for target aflatoxins could be obtained under the following HPLC conditions: the mobile phase composition of acetonitrile/water/methanol: 8/54/38, temperature of 24°C and flow rate of 0.4 mL min^{-1}.

The recommended HPLC conditions were practically applied to evaluate the detection value of target aflatoxins extracted from the spiked peanut. The results showed that corresponding response values obtained from the experimental data were close to those values predicted by the model. The authors concluded that the interaction between temperature and flow rate as well as mobile phase and flow rate should be considered as critical parameters for the detection of target aflatoxins.

The same investigation group, mentioned in the previous cited work, developed a new HPLC method to determine ochratoxin A [73]. In this case, the variables under study were: pH (*X*1), temperature (*X*2) and flow rate (*X*3). The objectives of this study were to determine the optimal HPLC conditions that provide the highest recovery of OTA; and to compare the newly optimized condition with two currently available methods used for OTA quantification. Once again, the investigation was based on the central composite design principle and five levels of each factor were chosen. After the ANOVA test, a high coefficient of determination equal to 0.999 was obtained. Furthermore, the pH and the flow rate had a significant ($p < 0.05$) effect on the detection value of OTA. The interaction effects of flow rate with the temperature and pH were also significant ($p < 0.05$) in the final model. Thus, the interaction effects of all of the variables should be considered as critical parameters for the detection of OTA. The numerical optimization showed that the combination set of the mobile phase consisting of 5 mM sodium acetate/acetonitrile/methanol (40:30:30) (pH adjusted at 2.36 with acetic acid), low oven temperature (24°C) and relatively low flow rate (0.4 mL min^{-1}) resulted in the highest detection value of OTA. The recommended optimal HPLC conditions, applied to evaluate the detection value of OTA extracted from spiked peanuts, resulted in a significant ($p < 0.05$) increase (at least 2.2-fold) in the detection value of OTA when compared to the two commonly used HPLC conditions.

In this case, some figures of merit were calculated. The LOD and LOQ were 0.02 and 0.06 μg Kg^{-1}, respectively.

RSM was also applied to optimize the extraction of ochratoxin A in bread samples by the QuEChERS methodology [74]. This extraction technique is simple and rapid compared to the majority of the reported methods which generally involve extraction with mixtures of polar solvents followed by a clean-up procedure that usually employs solid-phase extraction. A factorial design was used to examine the effect on extraction efficiency of the QuEChERS conditions (*X*1: extraction time, *X*2: extraction solvent volume and *X*3: sample mass). Extracts were analyzed by liquid chromatography with fluorescence detection. The OTA recovery after QuEChERS extraction from bread spiked samples was taken as the dependent variable or response of the design experiments. The analysis of variance was used for determination of the significant factors affecting OTA recovery. The model reached statistical significance ($p < 0.05$) and the lack of fit was not significant ($p > 0.05$) as desired. The model presented a quadratic correlation coefficient of 0.8847, fitting the statistical model quite well.

The mathematical model was used to generate the response surface plots and to calculate the conditions for maximum OTA recovery from bread samples. The optimum conditions were: extraction time = 3.0 min, extraction solvent volume = 15.0 mL and sample mass = 5.0 g. Moreover, the detection and quantification limits were calculated and resulted in 0.02 and 0.05 μg Kg^{-1} wet weight, respectively, in the bread samples.

The method was applied to different types of bread highly consumed in Portugal and none of the samples exceeded the legal European limit of 3 μg Kg^{-1} wet weight for cereal based products. However, maize-based breads seem to present higher OTA levels.

The QuEChERS sample preparation method was also optimized for the analysis of 58 mycotoxins in dairy products by the application of RSM [75]. Ultrahigh-performance liquid chromatography and electrospray ionization quadrupole Orbitrap high-resolution mass spectrometry (UHPLC/ESIQ-Orbitrap) was used for the separation and detection of all the analytes. RSM was employed to investigate the variations in recovery rates with respect to the extraction solvent volume, the amounts of sodium acetate, bondesil primary secondary amine (PSA), and C_{18} sorbent. The optimal composition of the 4 variables was determined by using a central composite design approach. In the case of zearalenone, the response surfaces generated suggested that the best extraction conditions were extraction volume 10 mL, sodium acetate 1.45 g, PSA 108 mg, C_{18} 405 mg. The extraction recoveries were in a range of 86.6–113.7%, with coefficient of variation <6.2%. The 58 compounds could be detected in the 0.001–100 μg kg^{-1} concentration range, with correlation coefficient >0.99. The limits of detection for the analytes were in the range 0.001–0.92 μg kg^{-1}. The repeatability was lower than 6.4%. This method was successfully applied on screening of mycotoxins in commercial dairy product samples. Once the proposed method was optimized and validated, it was applied to investigate the occurrence of the fifty-eight mycotoxins in a total of thirty commercial milks, milk beverages and yogurt products. Compared with traditional methods, the sensitivity was enhanced, and the accuracy was improved by more than 5 times, leading to a powerful method for screening mycotoxins in foods.

The above cited methods exploited the advantages of RSM, so it was possible to detect the existence of interaction effects between the different factors in the response. Moreover, it proved to be a more efficient strategy than the classical approach when examining the influence of a factor keeping constant other factors.

CONCLUSION

An overview of the application of chemometric tools for the determination of mycotoxins was assessed in this chapter. In general, real samples were quite complex and contain matrix interferences. However, rapid and simple extraction techniques were used. The cited examples, which included first and second order data, showed the ability of the different algorithms to model these types of data. In addition, the benefit of exploiting the second order advantage was demonstrated.

It was also shown that careful optimization of the corresponding operational parameters is extremely important in order to simplify the analytical work as well as to diminish the analysis time.

Although some publications can be found in the scientific literature, this topic needs to be consolidated since highly satisfactory results were obtained in recent studies. In general, the contribution of chemometrics, especially multivariate calibration methods (e.g., PLS, PARAFAC and MCR-ALS), seems to be crucial to extend the applicability of multicomponent determinations to solve more complex analytical problems.

ACKNOWLEDGMENTS

The authors are grateful to Universidad Nacional del Litoral. A. V. Schenone thanks CONICET (Consejo Nacional de Investigaciones Científicas y Técnicas) for her fellowship.

REFERENCES

[1] Richard J.L. (2007) *International Journal of Food Microbiology*, 119, 3–10.

[2] Prelusky, D.B., Scott, P.M., Trenholm, H.I., Lawrence, G.A. (1990) *Journal of Environmental Science and Health, Part* B 25, 25, 87–103.

[3] European Comission (2010) Regulation (UE) Nº 212/2010, which modifies the Regulation (CE) Nº 669/2009.

[4] Food Agricultural Organization (FAO) (2004) Worldwide regulation for mycotoxins in food and feed in 2003. FAO Food and nutrition paper 81. http://www.fao.org/ docrep/007/y5499s/y5499e00.htm

[5] Codex Alimentarius (2003a) Code of Practice for the Prevention and Reduction of Patulin Contamination in Apple Juice and Apple Juice Ingredients in Other Beverages. CA/RCP50-2003.
www.codexalimentarius.org/input/download/standards/405/CXP_050e.pdf

[6] Codex Alimentarius (2003b) Code of Practice for the Prevention and Reduction of Mycotoxin Contamination in Cereals. CAC/RCP 51-2003. www.codexali mentarius. org/input/download/standards/406/CXP_051e.pdf

[7] Berthiller F., Sulyok M., Krska R., Schuhmacher R. (2007) *InternationalJournal of Food Microbiology*, 119, 33–37.

[8] Nguyen M.T., Tozlovanu M., Tran T.L., Leszkowicz A.P. (2007), *Food Chemistry*, 105, 42–47.

[9] Konga W., Weia R., Logrieco A.F., Wei J., Wen J., Xiao X., Yang M. (2014) *Food Chemistry*, 146, 320–326.

[10] García-Moraleja A., Font G., Mañes J., Ferrer E. (2015) *Food ResearchInternational*, 72, 247–255.

[11] Arroyo-Manzanares N., Huertas-Pérez J.F., Gámiz-Gracia L., García-Campaña A.M. (2015) *Food Chemistry*, 177, 274–279.

[12] Onji Y., Aoki Y., Tani N, Umebayashi K., Kitada Y., Dohi Y. (1998) *Journal of Chromatography A*, 815, 59–65.

[13] Nielsen K.F., Thrane U. (2001) *Journal of Chromatography A* 929, 75–87.
[14] Rodríguez-Carrasco Y., Moltó J.C., Berrada H., Mañes J. (2014) *FoodChemistry,* 146, 212–219.
[15] Peña R., Alcaraz M.C., Arce L., Ríos A., Valcárcel M. (2002) *Journal of Chromatography A*, 967, 303–14.
[16] González-Peñas E., Leache C., López de Cerain A., Lizarraga E. (2006) *Food Chemistry*, 97, 349–354.
[17] Trucksess M.W., Stack M.E., Nesheim S., Albert R.H., Romer T.R. (1994) *Journal of the Association of Official Analytical Chemists*, 77, 1512–1521.
[18] Stroka J., Anklam E., Jorissen U., Gilbert J. (2000) *Journal of theAssociation of Official Analytical Chemists*, 83, 320–340.
[19] MacDonald S.J., Anderson S., Brereton P., Wood R. (2003) *Journal of AOAC International*, 86, 1164–1171.
[20] Hernández M.A., Valme G.M., Duran E., Guillen D., Barroso C.G. (2006) *Analytica Chimica Acta*, 566, 117–121.
[21] Katerere D.R, Stockenström S., Shephard G.S. (2008) *Food Control,* 19, 389–392.
[22] Mariño-Repizo L., Kero F., Vandell V., Senior A., Sanz-Ferramola M.I., Cerutti S., Raba J. (2015) *Food Chemistry*, 172, 663–668.
[23] Lee N.A., Wang S., Allan R.D., Kennedy I.R. (2004) *Journal ofAgricultural and Food Chemistry*, 52, 2746–2755.
[24] Li P., Zhang Q., Zhang W., Zhang J., Chen X., Jiang J., Xie L., Zhang D. (2009) *Food Chemistry*, 115, 313–317.
[25] Wang Y-K, Wang Y-C, Wang H., Ji W-H, Sun J-H, Yan Y-X. (2014) *Food Control*, 40, 41–45.
[26] Miller, D.M., Wilson, D.M. (1994) In: Eaton, D.L., Groopman, J.D. (Eds.), The Toxicology of Aflatoxins. Academic Press, Inc., New York, pp. 347–364.
[27] Pfohl-Leszkowicz, A., Petkova-Bocharova, T., Chernozemsky, I.N., Castegnaro, M. (2002) *Food Additives and Contaminants* 19, 282–302.
[28] Marasas,W.F.O., Kellerman, T.S., Gelderblom, W.C.A., Coetzer, J.A.W., Thiel, P.G., van der Lugt, J.J. (1988) *Onedersteport Journal of Veterinary Research*, 55, 197–203.
[29] Marasas,W.F.O. (1993) South African Medical Journal, 83, 382–383.
[30] Marasas,W.F.O. (1996) In: Jackson, L.S., DeVries, J.W., Bullerman, L.B. (Eds.), Fumonisins in Food. Plenum Press, New York, pp. 1–17.
[31] Pestka, J.J., Bondy, G.S. (1994) In: Miller, J.D., Trenholm, H.L. (Eds.), Mycotoxins in Grain-Compounds other than Aflatoxin. Eagan Press, St. Paul, pp. 339–358.
[32] Puel O., Galtier P., Oswald I.P. (2010) *Toxins* 2010, 2, 613–631.
[33] Prelusky D.B., Rotter B.A., Rotter R.G. (1994) In: Miller, J.D., Trenholm, H.L. (Eds.), Mycotoxins in Grain-Compounds other than Aflatoxin. Eagan Press, St. Paul, pp. 359–403.
[34] Niyo K.A., Richard J.L., Tiffany L.H. (1988) *American Journal ofVeterinary Research*, 49, 1766–1773.
[35] Richard J.L. (1991) In: Bray, G.A., Ryan, D.H. (Eds.), Mycotoxins, Cancer, and Health, Pennington Center Nutrition Series. Louisiana State University Press, Baton Rouge, Louisiana, pp. 197–220.

[36] Massart D.L., Vandeginste B.G. M., Buydens L.M.C., De Jong, S., Lewi, P.J., Smeyers-Verbeke J. (1997) Handbook of Chemometrics and Qualimetrics. Elsevier. Amsterdam.
[37] Haaland D.M., Thomas E.V. (1988) *Analytical Chemistry*, 60, 1193–1202.
[38] Booksh K.S., Kowalski, B.R. (1994) *Analytical Chemistry*, 66, 782A–791A.
[39] Escandar G.M., Faber N.K.M., Goicoechea H.C., Muñoz de la Peña A., Olivieri A.C., Poppi R.J. (2007) *TrAC Trends in Analytical Chemistry*, 26, 752–765.
[40] Haaland D.M., Han L., Niemczyk T.M. (1999) *Applied Spectroscopy,* 53, 390–395.
[41] Galeano Diaz T., Guiberteau A., Ortiz Burguillos J.M., Salinas F. (1997) *Analyst,* 122, 513–517.
[42] Brown S.D. (1991) *Chemometrics and Intelligent Laboratory Systems*, 10, 87–105.
[43] Faber N.M. (2000) *Chemometrics and Intelligent Laboratory Systems*, 50, 107–114.
[44] Thomas E.V. (1994) *Analytical Chemistry*, 66, 795A–804 A.
[45] Naes T., Martens H. (1998) *Journal of Chemometrics*, 2, 155–167.
[46] Martens H., Naes T. (1989). Multivariate Calibration. John Wiley & Sons Ltd., Chichester, England.
[47] Pettersson H., Aberg L. (2003) *Food Control*, 14, 229–232.
[48] Aghamohammadi M., Hashemi J. (2007) *Analytica Chimica Acta,* 582, 288–294.
[49] Tripathi S., Mishra H.N (2009) *Food Control*, 20, 840-846.
[50] Sugiyama J., Fujita K., Tsuta M., Shibata M., Kokawa M., Onda H. (2013) *Food Science and Technology Research*, 19, 539–545.
[51] Bro R. (1997) *Chemometrics and Intelligent Laboratory Systems*, 38, 149–171.
[52] Sanchez E., Kowalski B.R. (1986) *Analytical Chemistry*, 58, 496–499.
[53] Sanchez E., Kowalski B.R. (1990) *Journal of Chemometrics*, 4, 29–45.
[54] de Juan A., Casassas E., Tauler R. (2000) Soft modeling of analytical data. Meyers, R.A. Encyclopedia of Analytical Chemistry. Wiley, Chichester, UK, p. 9800–9837.
[55] Linder M., Sundberg R. (1998) *Chemometrics and IntelligentLaboratory Systems*, 42, 159–178.
[56] Linder M., Sundberg R. (2002). *Journal of Chemometrics*, 16, 12–27.
[57] Wu H.L., Shibukawa M., Oguma K. (1998) *Journal of Chemometrics*, 12, 1–26.
[58] Bro R. (1996) *Journal of Chemometrics*, 10, 47–61.
[59] Olivieri A.C. (2005) *Journal of Chemometrics*, 19, 253–265.
[60] de Juan A., Tauler R. (2001) *Journal of Chemometrics*, 15, 749–772.
[61] Bro R., Kiers H.A.L. (2003) *Journal of Chemometrics*, 17, 274–286.
[62] Windig W., Guilment J. (1991) *Analytical Chemistry*, 63, 1425–1432.
[63] Vosougha M., Bayata M., Salemib A. (2010) *Analytica Chimica Acta,* 663, 11–18.
[64] Vosougha M., Salemib A. (2011) *Food Chemistry*, 127, 827–833.
[65] Luna A.S., Luiz R.A., Lima I.C., Março P.H., Valderrama P., Boqué R., Ferré J. (2013) *Analytica Chimica Acta*, 17, 9–14.
[66] Lakowicz J.R. (1983) Principles of Fluorescence Spectroscopy, Plenum Press, New York.
[67] Hashemi J., Asadi Kram G., Alizadeh N. (2008) *Talanta,* 75, 1075–1081.
[68] Rodríguez M.C, Sánchez G.H, Sobrero M.S, Schenone A.V., Marsili N.R. (2013) *Microchemical Journal*, 110, 480–484.
[69] Saidi A., Mirzaei M. (2015) *International Journal of Nano Dimension*, 6, 31–38.

[70] Montgomery D.C. (2005) Design and Analysis of Experiments, John Wiley and Sons, New York, USA.

[71] Myers R., Montgomery D. (2002). Response Surface Methodology: Process and Product Optimization Using Designed Experiments. Wiley, Estados Unidos.

[72] Afsah-Hejri L., Jinap S., Arzandeh S., Mirhosseini H. (2011) *Food Control*, 22, 381–388.

[73] Afsah-Hejri L., Jinap S., Mirhosseini H. (2012) *Food Control*, 23, 113–119.

[74] Paíga P., Morais S., Oliva-Teles T., Correia M., Delerue-Matos C., Duarte S.C., Pena A., Matos Lino C. (2012) *Food Chemistry*, 135, 2522–2528.

[75] Jia W., Chu X., Ling Y., Huang J., Chang J. (2014) *Journal of Chromatography A*, 1345, 107–114.

In: Mycotoxins
Editor: Cesar Rios

ISBN: 978-1-63483-544-2

Chapter 5

MYCOTOXINS IN CEREALS AND CEREAL PRODUCTS: OCCURRENCE, TOXICITY AND PREVENTION

Agnieszka Waśkiewicz* and Piotr Goliński
Department of Chemistry, Poznań University of Life Sciences,
Wojska Polskiego, Poznań, Poland

ABSTRACT

Mycotoxins as a metabolic product of fungi (moulds) may be formed in many crops at various stages, starting from the plant growth phase through harvest time and up to crop storage. The most common mycotoxins are related to three species: *Aspergillus*, *Penicillium* and *Fusarium*. The first two are generally recognised as contaminants of food during drying and later storage, while *Fusarium* species are pathogens damaging plant and producing mycotoxins during vegetation or just after harvest. Mycotoxin content is an important indicator of cereal grain hygiene status and their quality. Due to the broad spectrum of their toxic properties and mode of action, followed – in general – by high resistance to temperature, the presence of mycotoxins in food and feeds generates a serious health hazard both for humans and animals. The consequence is of prime social concern. Since mycotoxicoses – poisonings and diseases caused by fungal toxins – are worldwide and agency problems are observed, international organisations emphasise the pressing need of continuous quality assessment for plant and animal food in terms of toxic fungal metabolites. Symptoms of mycotoxicoses depend on the type and concentration of consumed mycotoxins, time of exposure as well as the age, health conditions and sex of the organism with significant influence of diet.

Keywords: *Fusarium*, *Aspergillus*, *Penicillium*, mycotoxins, pathogens, fumonisins, zearalenone, deoxynivalenol, ochratoxins, aflatoxins, cereals, cereal products, toxicity

* Corresponding author: Agnieszka Waśkiewicz, Department of Chemistry, Poznań University of Life Sciences, Wojska Polskiego 75, 60-625 Poznań, Poland, tel +48 61 848 78 41, fax +48 61 848 78 24, e-mail: agat@up.poznan.pl.

1. Introduction

Microscopic fungi and their metabolites are not only a contemporary issue, since for centuries the problem of both food and feed poisoning has been observed and described. Although the beginnings of the history of mycotoxins are ancient, dim and distant in time, the association of poisoning properties and disease symptoms with particular secondary metabolites of microscopic fungi is relatively recent. The term mycotoxin appeared for the first time in the scientific literature in 1955 (Forgacs and Carll 1955), while wide intensive studies on the subject started in 1960, the year of the turkey X disease in eastern and southern England, where at least 100 thousand turkeys died after consumption of feed containing Brazilian peanut meal contaminated with toxins (aflatoxins) (Blount 1961, Forgacs 1962). To date, detection and determination of the structures of approximately 400 mycotoxins along with their derivatives, as a consequence yielding several thousand scientific publications, including around a dozen books, has been achieved.

The term mycotoxin originates from the Greek word *mycos* for fungi combined with the Latin *toxicum* meaning poison. Mycotoxins as metabolic products of fungi (moulds) may be formed in many crops and at various stages, from the plant growth phase through harvest time up to crop storage. Mycotoxin content is an important indicator of cereal grain hygiene status and quality. Economic losses caused by the presence of toxigenic fungi and their metabolites in different cereals are estimated at approximately 1 billion dollars annually in the USA alone. It also needs to be stressed here that in countries with advanced standards of agronomy the risk of food contamination with mycotoxins is significantly reduced. Due to the broad spectrum of their toxic properties and mode of action, followed – in general – by high resistance to temperature, the presence of mycotoxins in food and feeds generates a serious health hazard both for humans and animals, which in consequence is of prime social concern.

Since mycotoxicoses – poisonings and diseases caused by fungal toxins - are a worldwide problem, international organizations emphasise the pressing need of continuous assessment of quality for plant and animal origin food in terms of the presence of toxic fungal metabolites.

2. Classification of Microscopic Fungi and Their Brief Characterisation

Fungi are organisms commonly present worldwide in all kind of matrices, and they develop wherever they find sufficient nutrients in the substrate. As heterotrophic organisms (lacking chlorophyll) definitely most often they are obligate saprotrophs with an important role in organic matter decomposition, while others are parasites of animals and plants, the latter being much more frequent. Species parasitising plants include both obligate parasites and opportunistic parasites and saprotrophs. The vegetative organ of most fungi, called the mycelium, is composed of single, filiform hyphae (Marcinkowska 2003).

The taxonomy of fungi has been changing over the years. For a long time they were considered to be plants and classified as such. Based on cytochrome C analysis, fungi were recognised as a phylogenetically separate kingdom, and as a result they were distinguished as a separate taxonomic unit, assigned the kingdom of Mycota (Fungi) within the domain Eukaryota (Marcinkowska 2003).

The most common toxigenic fungi include numerous species of the genera *Penicillium*, *Aspergillus* and *Fusarium* (Goliński et al., 2009), the most representative being *Penicillium* and *Aspergillus* (Hedayati et al., 2007, Zain 2011).

The latter fungi, as saprophytes, develop on cereals after harvest, during storage, and they are responsible for biosynthesis of such mycotoxins as ochratoxins, citrinin, penicillic acid, etc. (Wawrzyniak and Waśkiewicz 2014). *Fusarium* due their polyphagous character may develop on crops during their vegetation (growth and development), and after harvest – during storage – forming the most common mycotoxins, e.g., zearalenone, deoxynivalenol, nivalenol, fumonisins, moniliformin, etc. (Goliński et al., 2009, 2010, Waśkiewicz et al., 2008, 2010a, b, Wiśniewska et al., 2014). Increasing contamination of agricultural products with fungi of the genus *Fusarium* may be closely related to monoculture (no crop rotation) and no or minimal tillage systems (Edwards 2004, Blandino et al., 2009).

Table 1. Major *Fusarium* spp. and their mycotoxin profiles

Fusarium spp.	Major mycotoxins
F. verticillioides	Fumonisins
F. proliferatum	Fumonisins, moniliformin, beauvericin
F. oxysporum	Moniliformin
F. graminearum	Zearalenone, deoxynivalenol with derivatives, nivalenol
F. culmorum	Zearalenone, deoxynivalenol with derivatives, nivalenol
F. avenaceum	Moniliformin, enniatins
F. poae	Nivalenol, enniatins, beauvericin
F. tricinctum	Moniliformin
F. equiseti	Zearalenone, nivalenol

Such farming practices lead in the agronomic environment to accumulation of inoculum of *Fusarium* fungi and development on postharvest residues and debris. Ear blight has been classified as the most important disease of wheat in many regions (Waśkiewicz et al., 2008, Zhang et al., 2008, Goliński et al., 2009), which encouraged worldwide studies on the structure of the most toxic fungal metabolites of the genus *Fusarium* followed by implementation of standards and official regulations concerning permitted (allowed), safe levels of mycotoxins in agricultural products. Table 1 presents major fungal species of the genus *Fusarium* and their metabolic profile. Generally the development of fungi in agricultural products followed by biosynthesis of their mycotoxins is accompanied by the simultaneous development of non-toxigenic fungi, with synergistic as well as antagonistic effects, which often reduces or even inhibits development of pathogens and biosynthesis of mycotoxins. Fungi of the genus *Trichoderma* are a well-recognized example of confirmed antagonistic action against toxin-forming fungi (Rojo et al., 2007, Matarese et al., 2012).

3. Mycotoxins – Their Function and Significance

Pathogenic fungi, because of their common occurrence, frequently degrade considerable amounts of organic matter in plants. Moreover, the harmful character of fungi is related to the spoilage of stored agricultural products and food, as well as the formation (biosynthesis) of a broad spectrum of metabolites toxic to humans, animals and plants (Bennett and Klich 2003).

It should be stressed that moulds do not develop in animal organisms, but proliferate on plants. These fungi are adapted to colonisation and development on substrates with a broad range of moisture and nutrient contents (Waśkiewicz et al., 2010a, Wawrzyniak and Waśkiewicz 2014). Their growth leads to biosynthesis of secondary metabolites, i.e., mycotoxins, and their release to the substrate, which does not affect development and growth of fungi (Goliński et al., 2009, Waśkiewicz et al., 2010a).

At present approximately 400 mycotoxins together with their derivatives have been identified, and systematised in approximately 25 structural types (Bennet and Klich 2003). They are compounds of diverse chemical structure, mostly aromatic (sometimes aliphatic) hydrocarbons, typically characterised by low molecular weight, which determines their resistance to environmental factors as well as a lack of or weak immunogenic properties (Cavaliere et al., 2005). The above metabolites form stable complexes with DNA, disturbing transcription of information from DNA to mRNA, which in consequence disturbs protein synthesis. They also influence the genetic information, leading to disorders in fetal development (Zain 2011), and exhibit carcinogenic, mutagenic, teratogenic, nephropathic, neurotoxic and hepatogenic properties (Khoury and Atoui 2010), while some of them are well known as strong oestrogenic factors, causing e.g., false oestrus (Gajęcka 2012, Gajęcka et al., 2013). These compounds are chemically very stable, and are often resistant to hydrothermic technological processes (Raters and Matissek 2008, Kabak 2009).

3.1. Factors Determining Biosynthesis of Mycotoxins

The physical presence of a toxigenic fungus is a *sine qua non* condition, but it is – because of the requirement of optimal biosynthesis conditions – not always sufficient for the formation of mycotoxins, so not all samples containing fungal pathogens are detected to contain their toxic metabolites.

Cereal grain contamination with mycotoxins depends on weather conditions and represents a serious agronomic problem in many countries. It is estimated that every year about 25% of world food production is deteriorated because of the presence of the above metabolites (Reddy et al., 2010), while their types and concentrations to a significant degree depend on fungi distribution and climatic conditions (zone) (Zain 2011).

Table 2 presents several factors influencing the development of fungi in the field, during harvest and storage, temperature and water activity (a_w) being two major ones (Bouras et al., 2009, Waśkiewicz et al., 2010a, Wawrzyniak and Waśkiewicz 2014).

Table 2 lists factors directly influencing pathogenicity of fungi and susceptibility of plants to infection caused by microorganisms. A significant role is also played by the substrate where the fungus develops, including macro- and micronutrients (particularly zinc, magnesium, cobalt), as well as synergistic and antagonistic interactions with accompanying microorganisms (Moretti et al., 2010).

Environmental conditions not only influence the development of the fungus and its survival rate, but also significantly change biosynthesis of its secondary metabolites. Change in one of the parameters may result in significant changes in the biosynthesis products, in terms of both their quality and quantity.

Table 2. Factors influencing fungal development before, during and after harvest of crops

Type of factor	Vegetation period	Harvest	Storage
physical			
Moisture content	+	+	+
Drying rate	+	+	+
Repeated wetting	+	+	+
Relative humidity	+	+	+
Temperature	+	+	+
Mechanical damage	+	+	+
Milling and grain mixing	-	+	+
Material self-heating	-	-	+
chemical			
CO_2	-	-	+
O_2	-	-	+
Substrate	+	-	+
Mineral fertilisation	+	-	+
Plant protection preparation	+	-	+
biological			
Stress (e.g., temperature)	+	-	+
Pests	+	+	+
Fungal infestation	+	+	+
Cultivars	+	-	+
Fungal strain	+	-	+
Number of spores	+	+	+
Ecological niches	+	+	+

Inappropriate storage conditions, high moisture content (above 15%) in combination with optimal temperature (approx. 25°C) as well as inadequate grain cleaning to remove glumes, dust and screenings, obviously affect rapid development of fungi (Bouras et al., 2009, Waśkiewicz et al., 2010a, Wawrzyniak and Waśkiewicz 2014). Literature data indicate enhanced biosynthesis of mycotoxins after application of plant protection preparations, with a similar effect described for certain pesticides.

3.2. Sources and Routes of Exposure to Fungal Secondary Metabolites Toxic to Humans and Animals

Mycotoxins occur in a broad range of materials and agricultural products under different conditions. The most important source of mycotoxins is food, plant and animal origin foodstuffs naturally contaminated with the metabolites (Bennett and Klich 2003, Goliński et al., 2009, Reddy et al., 2010).

Many crops grown worldwide may be substrates for fungal growth and mycotoxin formation (Bennett and Klich 2003, Goliński et al., 2010, Reddy et al., 2010, Wiśniewska et al., 2014). Agricultural products may be contaminated from plant vegetation in the field, through harvest, up to processing, storage and transport of the final product (Waśkiewicz et al., 2013, Yang et al., 2014). At each of these stages the species composition may vary and the product may become contaminated with different mycotoxins. It is worth underlining that individual fungal species may produce one or more mycotoxins, while individual mycotoxins may be produced by different fungal species, e.g., aflatoxins, produced by several fungal species, have diverse structures and thus, depending on the species, affect animal organisms in a variety of ways (Ehrlich et al., 2007).

Mycotoxins penetrate to animal organisms together with contaminated feeds or their components. From the alimentary tract they may penetrate to blood, systemic fluids, tissues and organs, and in consequence to animal origin food, i.e., meat, milk and eggs. Mycotoxins may penetrate to the human organism with food, by inhalation and through skin, with their toxicity being approximately 40 times higher when compared to food/feed consumption (Kelman et al., 2004).

3.3. Toxicity Properties

Mycotoxins – after consumption of contaminated food and feeds – may cause diseases called mycotoxicoses in humans and animals (Marczuk et al., 2012, White 2012). These diseases include acute and chronic poisoning, caused by consumption of toxins formed by fungi, and they are not contagious diseases. Symptoms of mycotoxicoses depend on type and concentration of consumed mycotoxins, time of exposure as well as the age, health condition and sex of humans and/or animals, with a significant influence of composition of the diets (Richard 2007).

Young, fast-growing organisms respond to mycotoxins most strongly. These substances penetrate to blood and tissues, with the rate of their transport and resorption in the organism being varied and dependent on the mycotoxin dose, while the disappearance of metabolites from the organism is usually slow (Waśkiewicz et al., 2014).

Moreover, the severity of mycotoxin poisoning may be related to such conditions as vitamin deficiency, exhaustion, alcohol abuse or contagious diseases. On the other hand, mycotoxicoses may increase sensitivity to microbial diseases and lead to malnutrition (Bryden 2007).

It is worth underlining that clinical symptoms of mycotoxicoses are difficult to diagnose due to the quite common synergistic action of mycotoxins (Speijers and Speijers 2004). Since symptoms of toxin exposure start to appear after a certain concentration is reached, potential accumulation of the metabolite has to be considered. Very often an atypical clinical picture is a result of mixed poisoning and interactions between mycotoxins (Creppy et al., 2004, Tritscher and Page 2004). To link a particular mycotoxin with a respective disease found in humans or animals, the following parameters need to be considered: the frequency of occurrence of a given mycotoxin in food, the degree of exposure of humans to its action, as well as the appearance of characteristic, stable disease symptoms in examined animals and humans.

Consumption of products highly contaminated with toxins may lead to different types of poisoning, while long-term exposure to low concentrations of mycotoxins increases the cancer incidence rate (Zain 2011). Lungs, the alimentary tract, the placenta and the mammary glands are most sensitive to the action of mycotoxins. Mycotoxins may also influence the cardiovascular system, lead to dysfunction of the liver and kidneys, and cause allergies, sialorrhoea and damage to the mucosa (Schütze et al., 2010). In the metabolic pathways they may disturb protein synthesis and proper functioning of the nervous system, as well as causing pathological changes starting with skin irritation and necroses with significant reduction of immunity (Campbell et al., 2004). Toxic metabolites in feeds cause diseases in farm animals, such as liver damage and hepatic cancer in the case of aflatoxins, kidney damage due to ochratoxin A, hyperoestrogenic effects with reduced fertility when zearalenone was present in feed for pigs, and loss of appetite, reduction of body gain and feed effectiveness together with vomiting and possibly alimentary aleukia caused by consumption of deoxynivalenol (vomitoxin) (Abdulkadar et al., 2004, Hope and Hope 2012).

Recent scientific reports indicate that mycotoxins need to be ascribed an indirect effect on the development of civilisation-related diseases, such as diabetes, arthritis or atherosclerosis, while according to epidemiological data from Asian countries they may also cause an increasing incidence of many diseases such as gout, hyperlipidaemia and atherosclerosis (Hope and Hope 2012).

Mycoses – in contrast to mycotoxicoses – are other fungal diseases caused by the development of a fungus on the surface or inside the human or animal organisms. Frequent cases of dermatomycosis or aspergillosis in young growing organisms (particularly poultry) fed friable feed contaminated with fungi or the use of old straw overgrown with fungi as bedding are good examples of such diseases (Sajid et al., 2006).

Although the harmful effect of mycotoxins on human and animal organisms has been very well recognised, permanent monitoring of the toxic effects of these substances is required, particularly as the profile of mycotoxins characteristic of a given climatic zone undergoes constant changes due to the cosmopolitan nature of fungi producing them.

3.4. Classification of Mycotoxins Present in Cereals

The most common mycotoxins are related to three genera: *Aspergillus*, *Penicillium* and *Fusarium* (Goliński et al., 2009). The first two are generally recognised as contaminants of food during drying and later storage, while *Fusarium* species are pathogens damaging plants and producing mycotoxins during vegetation or just after harvest (Waśkiewicz et al., 2008, Goliński et al., 2010, Wiśniewska et al., 2014).

Economically and toxicologically significant mycotoxins belong to five groups: aflatoxins, ochratoxins, trichothecenes (mainly deoxynivalenol), zearalenone and fumonisins (Goliński et al., 2009, Zain 2011).

The first two groups of metabolites include toxins formed by storage fungi (*Aspergillus* and *Penicillium*), i.e., just after harvest, during inappropriate storage of cereal grain, oil crop seeds and their processed products (Amadi and Adeniyi 2009). The other three groups of mycotoxins are formed by *Fusarium* fungi. Infection of ears during cereal flowering, under high humidity, with relatively low temperature, causes strong development of mycelium followed by poorly formed, light shrivelled and discoloured kernels, containing pathogen mycelium deposits, responsible for biosynthesis of large amounts of mycotoxins, instead of portions of the endosperm and the aleuronic layer (Waśkiewicz et al., 2008, Goliński et al., 2010, Wiśniewska et al., 2014).

3.4.1. Aflatoxins

The structure of aflatoxins was described in 1963, shortly after the mass-scale outbreak of turkey X disease in New England (Goldblatt 1969). This group of compounds is biosynthesised mainly by the species *Aspergillus flavus* and *A. parasiticus* commonly found in soil and in stored agricultural produce (Yu et al., 2005, Reddy et al., 2009, Zrari 2013). *A. parasiticus* often contaminates oilseeds and produces B_1, B_2, G_1 and G_2 aflatoxins, while *A. flavus* can usually be found on cereals and it produces only B_1 and B_2 aflatoxins (Iqbal et al., 2006, Romagnoli et al., 2007, Ramesh et al., 2013). Other, less common species include *A. niger*, *A. nomius*, *A. pseudotamari* and *A. bombycids* (Mushtaq et al., 2012). Fungi produce aflatoxins in the presence of higher moisture levels, higher temperature and adequate substratum, with biosynthesis being the most effective when humidity is above 13% and temperature between 24 and 37°C (Jakic-Dimic et al., 2009), which generates the most serious mycotoxin problems in countries with subtropical and tropical climates (Toteja et al., 2006, Giray et al., 2007, Aydin et al., 2008).

Chemically, aflatoxins belong to the bifuranocoumarin group. Among the over 20 aflatoxin analogues currently identified, only four are naturally found in cereal cultures (aflatoxins B_1, B_2, G_1, G_2) (Lutfullah and Hussain 2012), with aflatoxin B_1 being predominant. Toxicity of the metabolites is determined by the presence of the lactone ring forming with the neighbouring benzene ring with the skeleton identical to that in coumarin (Figure 1).

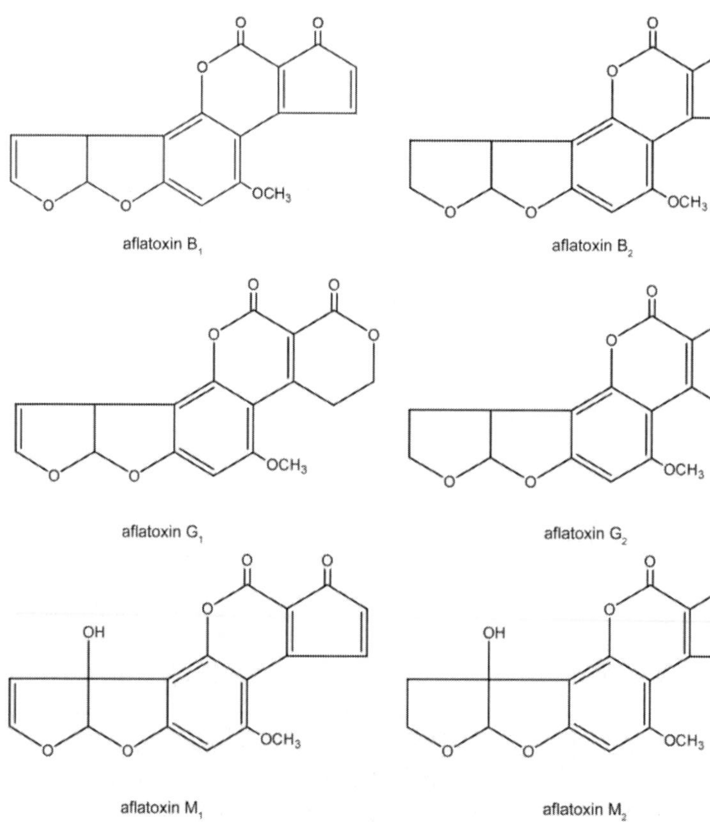

Figure 1. Chemical structure of aflatoxins.

Contamination of agricultural commodities by aflatoxigenic fungi occurs during pre-harvest handling, processing, transport and storage (Saini and Kaur 2012). Aflatoxins have been found as contaminants in agricultural products, especially in peanut, cereals and cereal products (Abbas et al., 2006, Jaksic et al., 2011) as well as animal feeds (Sassahara et al., 2005, Zinedine et al., 2007, Offiah and Adesiyun 2007, Shar et al., 2014).

Studies concerning the presence of aflatoxins in cereals focus mainly on maize culture (Jakic-Dimic et al., 2009, Khatoon et al., 2012, Lattanzio et al., 2007, Lutfullah and Hussain 2012, Rocha et al., 2009, Zinedine et al., 2007), although there are reports on the presence of these toxins in wheat, rice, barley or sorghum (Alexa et al., 2013, Lutfullah and Hussain 2012, Jakic-Dimic et al., 2009, Jaksic et al., 2011, Zinedine et al., 2007, Rajarajan et al., 2013). Frequently in the same tested cereal samples, beside aflatoxins, also such toxins as fumonisins, trichothecenes or zearalenone were observed (Broggi et al., 2007, Kimanya et al., 2008). Particularly strict controls are imposed on cereal products for infants and young children (Aidoo et al., 2011, Alvito et al., 2010, Tam et al., 2006). Aflatoxin derivatives other than those found in cereals may be detected in milk and dairy products (aflatoxins M_1 and M_2) (Figure 1). They are produced in organisms by animals fed contaminated feed (Boudra et al., 2007).

Aflatoxicosis is manifested in different pathomorphological changes, especially in the liver, kidneys, the cardiovascular and nervous systems (Devendran and Balasubramanian 2011). Aflatoxin B_1 is also highly mutagenic, carcinogenic and probably teratogenic to animals (Rocha et al., 2009). The International Agency for Research on Cancer (IARC) classifies this toxin within class 1 of human carcinogens (IARC 1993).

In the European Union, the highest tolerable levels for aflatoxin B_1 and total aflatoxins in cereals for human consumption are 2 and 4 ng/g, respectively, and 20 ng/g for total aflatoxins in poultry feeds (EC 2003, FAO 2004).

3.4.2. Ochratoxin A

This toxin was detected in 1965 as a metabolite of *Aspergillus ochraceus* (van der Merwe et al., 1965; Scott 1965). Shortly afterwards, in the United States ochratoxin A was isolated from maize kernels and recognised as a potentially nephrotoxic compound (Shotwell et al., 1969).

Biosynthesis of ochratoxin A (OA) by certain *Aspergillus* species in warmer and tropical regions worldwide such as the Balkans or Australia has been reported (Reddy and Bhoola 2010). Known species include *Aspergillus ochraceus*, *A. alliaceum*, *A. mellus*, *A. auricomus*, *A. carbonarium*, *A. niger* and *A. glaucus* (Noonim et al., 2008). However, the greatest body of information concerns the toxin-forming capacity of *A. ochraceus* (Reddy and Bhoola 2010). Within *Penicillium*, primarily *Penicillium verrucosum* is responsible for biosynthesis of OA in the temperate and colder climatic zones (0-31°C, optimal 20°C), e.g., in Scandinavia, Central Europe or Canada (Elmholt and Rasmussen 2005). This toxin is commonly related to cereal grain storage and was not found during plant vegetation, although in the case of viticulture it is also produced in the fields (Varga and Kozakiewicz 2006).

Ochratoxin A ($C_{20}H_{18}ClNO_6$) is a dipeptide composed of L-β-phenylalanine and isocoumaric acid (also called ochratoxin α). Biosynthesis of ochratoxin B ($C_{20}H_{19}NO_6$) and ochratoxin C ($C_{22}H_{22}ClNO_6$) by some fungi was reported (Figure 2). While ochratoxin B is hardly toxic (due to the absence of the chlorine substituent in the molecule), ochratoxin C exhibits toxicity comparable to that of ochratoxin A. Usually in natural samples it was found in very small amounts.

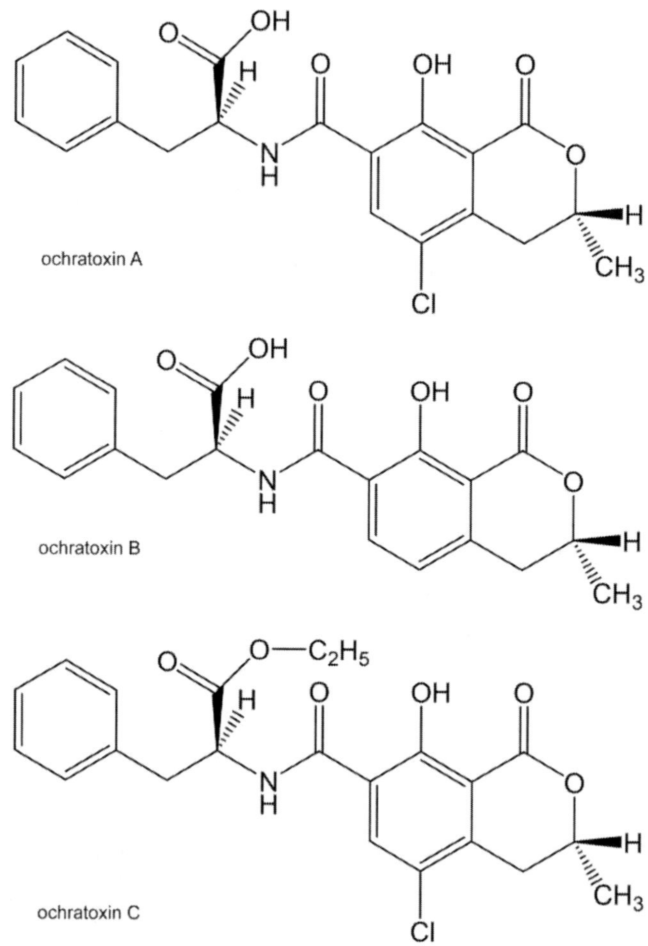

Figure 2. Chemical structure of ochratoxin A, B, C.

Due to the relatively high thermal stability of ochratoxin A, it is not degraded during heating, baking (Paiga et al., 2012) or fermentation processes (Esti et al., 2012), and it is usually detected in final cereal products subjected to technological processing (Mateo et al., 2007).

Inappropriate storage of cereals, under conditions (temperature and humidity) suitable for the development of fungi and the formation of toxic metabolites, as well as the application of inappropriate practices during storage of agricultural products, is responsible for ochratoxin A formation. Natural occurrence of OA is common in starch-rich foodstuffs such as cereals (including wheat, barley, maize, rice, oat, rye) as well as edible legume seeds and related foodstuffs (Llorent-Martínez et al., 2013).

It was observed that OA present in barley used for beer production during malting and/or brewing is not degraded and it is detected in the final product (Anli and Alkis 2010). The level of ochratoxin A contamination depends on the used unmalted barley, maize, rice grits, wheat starch or sorghum groats (sorghum) as fermentable carbohydrate substrates for yeasts (Mateo et al., 2007).

A similar stability was found for ochratoxin A in the milling of cereals for bread and under hydrothermal conditions of its baking, since it was detected at each stage of bread production, as well as in the final product (Paiga et al., 2012). Bread baked on the base of flour contaminated with storage fungi is darker, less porous, slack-baked and slightly sweet in taste. Cereal products for children and infants are not OA free (Araguas et al., 2005).

The amount of this toxin in food and feeds is usually relatively low; however, since it may be accumulated in human and animal organisms, it presents a significant health problem (Zheng et al., 2005). Porcine ochratoxin nephropathy is a disease typical for European countries, with cases of pathological changes in kidneys and the presence of ochratoxin A in blood of these animals reported in many countries (Domijan et al., 2004, Varga et al., 2010). Typical, macroscopic symptoms of porcine ochratoxin nephropathy include the following changes in the kidneys: enlargement, changes in colour (so-called pale kidneys), small vesicles on the surface, as well as cysts and fibrosis of the renal cortex. The disease leads to a reduced filtration capacity and secretion of *p*-aminohippuric acid, limited urine concentration capacity, as well as protein and glucose contents in urine, with progression to renal tubule cell necrosis, while in the advanced condition it is connected with sclerosis and atrophy of renal glomeruli (Abdu et al., 2011).

Similar disease symptoms are observed in humans, particularly in the Balkans (Castegnaro et al., 2006). Exposure of consumers to even very low toxin concentrations may lead to chronic diseases which are very difficult to diagnose and, as a consequence, to kidney and liver cancers (Grollman and Jelakovic 2007).

In 1993 the International Agency for Research on Cancer classified OA as group 2B – possibly carcinogenic to humans (IARC 1993), while several years later, in 1998, the Scientific Committee on Food estimated and introduced a recommendation that in the daily food ration the toxin content should not exceed 5 ng OA kg^{-1} body weight.

3.4.3. Fumonisins

Fumonisins are one of the latest, relatively recently identified and described group of mycotoxins. They were first isolated from the species *Fusarium verticillioides* in 1988 in South Africa (Gelderblom et al., 1988). Beside *Fusarium verticillioides* (formerly *F. moniliforme*), *F. proliferatum*, *F. napiforme*, *F. oxysporum*, *F. dlamini* and *F. nygamai* are also responsible for biosynthesis of fumonisins (Stępień et al., 2011a, b, Waśkiewicz et al., 2013b).

Among 28 identified fumonisin-related analogues, only three of them were found as natural contaminants of food and feed. The above 3 metabolites are: fumonisin B_1 ($C_{34}H_{59}NO_{15}$), the most frequently detected and at the highest concentration levels (70-80% of the total amount of the 3 toxins), followed by fumonisin B_2 ($C_{34}H_{59}NO_{14}$) (15-25%) and by fumonisin B_3 ($C_{34}H_{59}NO_{14}$) (3 to 8%) (Figure 3) (Waśkiewicz et al., 2012b).

Fumonisin B_1 (FB_1) is a white hygroscopic substance, soluble in water, methanol and acetonitrile-water mixture. Mostly food processing has no significant effect on reduction of fumonisin contents. These compounds are quite resistant to temperature below 150°C, more effectively decomposed at higher temperature, but – unfortunately – some fumonisin degradation products are much more harmful than the toxins themselves (Waśkiewicz et al., 2012a).

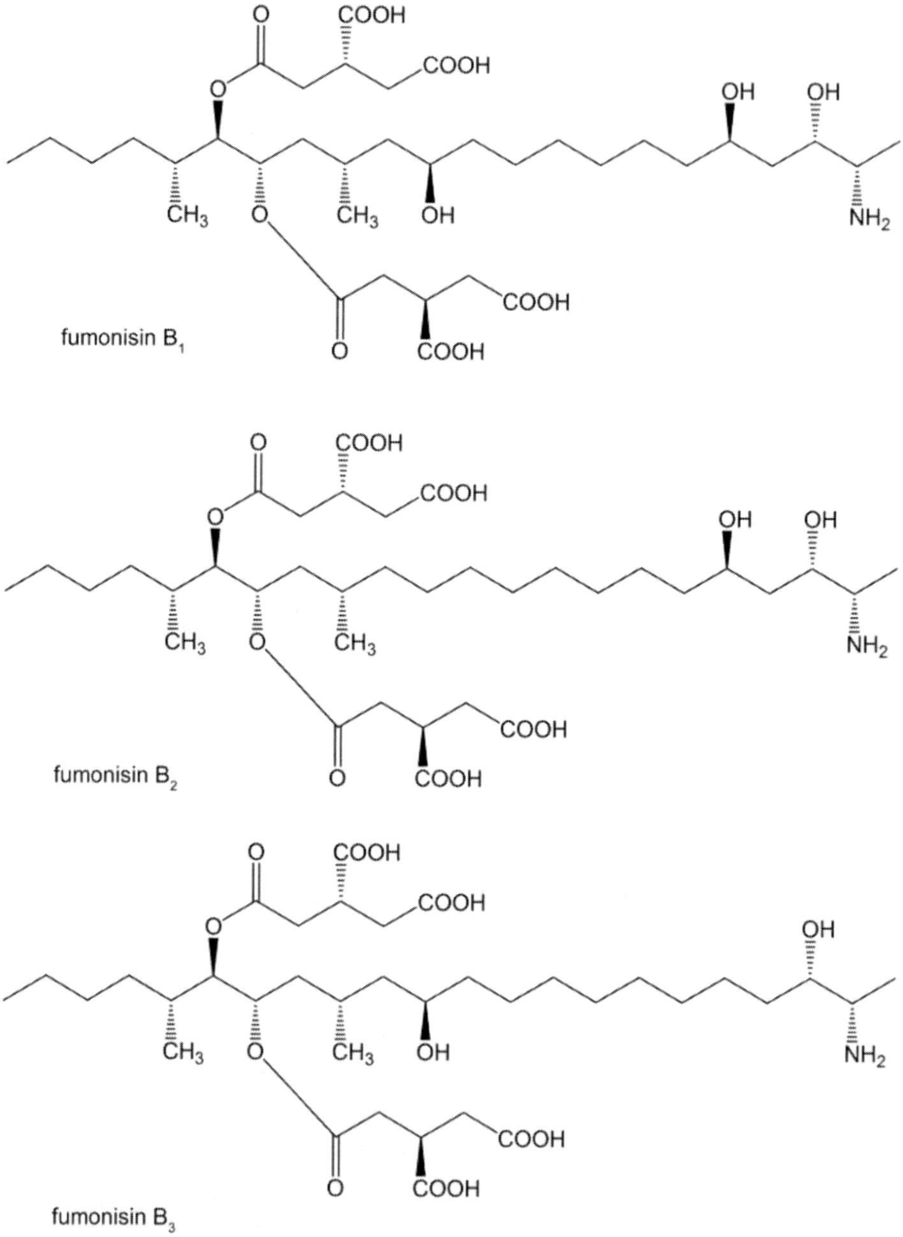

Figure 3. Chemical structure of fumonisin B_1, B_2 and B_3.

Because of easier availability of analytical methods and significant improvement of analytical 'machinery' (for several-fold lowered fumonisin detection limits), recent studies indicate that fumonisins are found almost in all regions where maize is grown, except for colder regions such as North-Eastern Europe and Canada, in which this problem might be neglected. However, individual samples from these regions were found to contain these toxins as well (Waśkiewicz et al., 2012a).

Fungal infection may occur in maize at each stage of development and affect individual parts of the plants – roots, stalks and kernels – while it develops also during harvest and at the early stage of storage, after which the concentration of produced fumonisins does not increase (Waśkiewicz et al., 2012b, Wu et al., 2013).

Fumonisin contamination of maize crops is closely dependent on climatic conditions and geographical location (Domijan et al., 2005, Al-Hazmi 2009, Schjøth et al., 2009, Waśkiewicz et al., 2012, Scussel et al., 2014). Studies were conducted in which contamination of maize and its products with these toxins was analysed for several years, and the analyses indicate an effect of temperature and humidity on pathogen attack (De la Campa et al., 2005). Other conditions are also important, such as origin of plants, drought stress and damage caused by insects (de la Campa et al., 2005).

Studies concerning fumonisin contents in maize and its products have been reported in many countries, e.g., Germany (Zimmer et al., 2008), Poland (Waśkiewicz et al., 2012, Czembor et al., 2014), Italy (Covarelli et al., 2012), Portugal (Lino et al., 2007, Silva et al., 2007), Croatia (Domijan et al., 2005), Turkey (Yazar and Omurtag 2008) as well as India (Sreenivasa et al., 2006), the USA (Yates et al., 2005), China (Wei et al., 2013), Korea (Kim et al., 2014), Argentina (Broggi et al., 2007), Brazil (Ono et al., 2006, Scussel et al., 2014), Nigeria (Adejumo et al., 2007), Zambia (Schjøth et al., 2009), Tanzania (Kimanya et al., 2008), Saudi Arabia (Al-Hazmi 2009) and Benin (Fandohan et al., 2005).

In commercially available, cleaned maize and its products for human consumption (comminuted maize grain, maize meal, maize grits, polenta, semolina, cornflakes and sweet corn) contamination typically does not exceed 1000 $\mu g\ kg^{-1}$, but in some countries it is higher, e.g., in Italy, where the concentrations of fumonisins in polenta samples were higher than 1000 $\mu g\ kg^{-1}$. Considering that in north-eastern Italy polenta is a staple food, it explains why increased risk of oesophagus cancer is observed (Sydenham et al., 1990, Myburg et al., 2009).

Fumonisins typically contaminate maize crops, but their presence in other cereals (Ghiasian et al., 2005) or asparagus plants (Waśkiewicz et al., 2013a) is also observed.

Fumonisin B_1, because of similarity in chemical structure to sphingosine and sphinganine – substrates of ceramide synthetase, which is a key enzyme in biosynthesis of sphingolipids (Riley and Voss 2006, Voss et al., 2007, Waskiewicz et al., 2012a) - is a specific inhibitor of ceramide synthetase, and inhibits formation of sphingolipids, amounts of which decrease in eukaryotic cells (Riley and Voss 2006). An increased amount of sphinganine and a decreased amount of sphingosine because of inhibition of biosynthesis is - on the other hand - an excellent indicator of risk of exposure to fumonisin. Interruption of sphingolipid metabolism explains - to some extent - the important biological effects caused by these toxins, e.g., disturbance in the cell cycle, increased oxidative stress, as well as apoptosis and cell necrosis (Osuchowski and Sharma 2005). Based on numerous data concerning contamination of fumonisins in agricultural products and diseases related to it, the International Research for Research on Cancer in 2002 classified fumonisin B_1 as a substance possibly carcinogenic to humans (class 2B) (IARC 2002).

3.4.4. Zearalenone

Zearalenone, also called F-2 toxin, is the third most commonly detected mycotoxin in plant tissues (Goliński et al., 2009, 2010, Cano-Sancho et al., 2012). This toxin is one of the strongest, non-steroid oestrogens in nature, both in foods and feeds (Danicke et al., 2005).

Biosynthesis of zearalenone by certain *Fusarium* species, mainly *F. graminearum*, *F. culmorum*, *F. semitectum*, *F. equiseti*, and *F. cerealis*, is well known (Ferreira Geraldo et al., 2006, Ezekiel et al., 2008, Wiśniewska et al., 2014), with the highest yields observed at humidity of approximately 16% and temperature below 25°C (Zwierzchowski et al., 2005), being of the most pathogenic, toxin-forming fungi, with the host for them being cereals, predominantly maize.

ZON ($C_{18}H_{22}O_5$) is a macrocyclic lactone (Figure 4), hardly soluble in water and readily soluble in ether, benzene, chloroform, methyl chloride, ethyl acetate, acetonitrile, alcohols, acetone and aqueous solutions of bases. Like most mycotoxins, it is not destroyed or decomposed either in kernel cleaning, milling and storage, or during thermal processing of foods and feeds (Cetin and Bullerman 2005, Boermans and Leung 2007).

Zearalenone is accumulated before harvest, in cereal infested with *Fusarium* growing in the field. A high level of this toxin was also detected in some samples of animal feed of maize origin during inappropriate storage and not as a field problem (Goliński et al., 2010, Wiśniewska et al., 2014). Apart from maize, zearalenone was also found in grain of wheat, barley, oat, sorghum, rice and pea (Seeling and Dänicke, 2005, Waśkiewicz et al., 2008, Goliński et al., 2010, Streit et al., 2012).

Occurrence of zearalenone in cereals, mainly in maize and animal feeds, has been studied in several countries, e.g., Germany (Goertz et al., 2010), Switzerland (Dorn et al., 2011), Italy (Decastelli et al., 2007), Austria (Rodrigues and Naehrer 2012), Belgium (Isebaert et al., 2009, De Boevre et al., 2012), Spain (Cervero et al., 2007, Cano-Sancho et al., 2012, Ibáñez-Vea et al., 2011, 2012a, b), Portugal (Martins et al., 2008, Almeida et al., 2011), the Netherlands (Driehuis et al., 2008), Bulgaria (Manova and Mladenova 2009), Slovakia (Labuda et al., 2005), the Czech Republic (Skládanka et al., 2011), Romania (Banu et al., 2011, Galbenu et al., 2011, Tabuc et al., 2011), Poland (Waśkiewicz et al., 2008, Goliński et al., 2010, Wiśniewska et al., 2014), Croatia (Domijan et al., 2005, Pleadin et al., 2012), Greece (Vlachou et al., 2004) and the USA (Rodrigues and Naehrer 2012), while contamination of agricultural products with the metabolite in Argentina (Broggi et al., 2007), Brazil (Moreno et al., 2009, Almeida-Ferreira et al., 2013), Turkey (Aksoy et al., 2009), Nigeria (Ezekiel et al., 2008), Vietnam (Thieu et al., 2008), Iran (Oveisi et al., 2005, Osboo and Mirabolfathy 2008, Rashedi et al., 2012), Indonesia (Nuryono et al., 2005), Saudi Arabia (Al-Hazmi 2010) and Egypt (El-Desouky and Naguib 2013) has been reported.

The phytoestrogen ZON molecule because of chemical structure similarity – the phenolic ring – exhibits activity and affinity to both oestrogen receptors, ERα and ERβ, present in mammals (Dip et al., 2008). By influencing them it causes several functional changes in the reproductive system by very similar action as other, naturally occurring oestrogens, leading to the development of typical symptoms of hyperoestrogenism (Parveen et al., 2009, Metzler et al., 2010, Gajęcka 2012). Prolonged oestrus, infertility, pseudopregnancies, reduced quality and litter size as well as disturbed lactation are observed (Abid-Essefi et al., 2004). Intensity of observed symptoms and changes in tissues of the reproductive system depend on the age and stages in the reproductive cycle (Wasowicz et al., 2005. Dip et al., 2008, Milicevic et al., 2010).

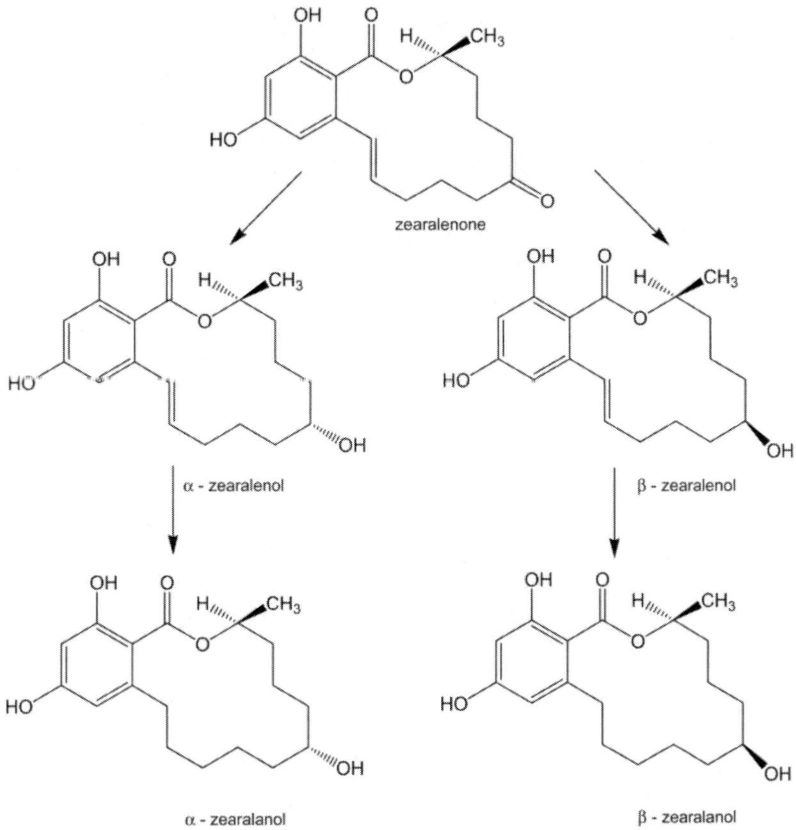

Figure 4. Chemical structure of zearalenone and its toxic derivatives α- and β-zearalenols as well as α- and β-zearalanols.

Zearalenone rarely causes acute poisoning, while it exhibits strong oestrogenic and anabolic properties (Abid-Essefi et al., 2004, Dip et al., 2008, Parveen et al., 2009, Metzler et al., 2010), causing infertility and other related diseases in animals, particularly in pigs and sheep (Gajęcka 2012), in laboratory animals (mice, rats, guinea pigs, hamsters, rabbits) (Pástorová et al., 2004, Ouanes et al., 2005, Zatecka et al., 2014), domestic animals (dogs, cats) (Boermans and Leung 2007, Gajęcka et al., 2013) and also – most importantly – represents a significant hazard for humans.

Rapid absorption of ZON from the alimentary tract is observed, and as early as within 30 minutes after ingestion zearalenone and its metabolites, *α*- and *β*-zearalenol as well as *α*- and *β*-zearalanol (Figure 4), were detected in blood (Zwierzchowski et al., 2005). It needs to be stressed that naturally found zearalenone derivatives also exhibit oestrogenic activity (Seeling and Dänicke 2005).

3.4.5. Trichothecenes

Interest in trichothecenes has developed significantly since detection of the T-2 toxin trichothecene A. Trichothecenes are compounds with the structure of sesquiterpene epoxides, which were divided into four groups: A, B, C and D. Compounds of the first two groups are the most significant, i.e., from group A: T-2 and HT-2 toxins, solaniol and neosolaniol, as well as group B: nivalenol, deoxynivalenol with derivatives 3- and 15-acetyl-deoxynivalenol (Figure 5).

Grain of cereals and other crops is the most commonly colonised by the fungi producing trichothecenes B (deoxynivalenol and its derivatives as well as nivalenol) (Foroud and Eudes 2009). They include *Fusarium culmorum* and *F. graminearum*, the same species that are responsible for biosynthesis of zearalenone (Goliński et al., 2009, 2010, Kushiro 2008). Factors promoting the formation of these mycotoxins include first of all a specific temperature range (21-25°C) and humidity dependent on *Fusarium* species (Welugo 2012, Wiśniewska et al., 2014).

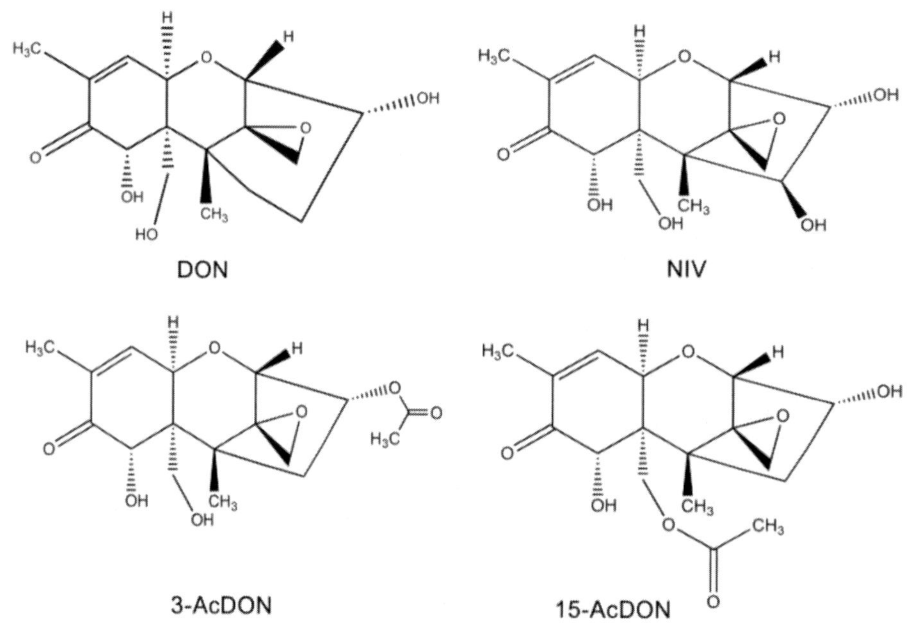

Figure 5. Chemical structure of nivalenol, deoxynivalenol with derivatives 3- and 15-acetyl-deoxynivalenol.

Based on the mycotoxin profile as well as using DNA markers, several species-specific chemotypes were listed in species forming trichothecenes B: chemotypes of nivalenol, 3-acetyl-deoxynivalenol and 15-acetyl-deoxynivalenol (Pasquali et al., 2010, Malihipour et al., 2012). The primary sources of DON are cereals, including wheat, barley, maize and oat (Gonzalez et al., 2008, Yazar and Omurtag 2008, Goliński et al., 2009, 2010, dos Santos et al., 2013, Sampietro et al., 2013, Tajehmiri et al., 2014, Wiśniewska et al., 2014). It was also found in buckwheat, sorghum and in processed plant-origin food such as flour, bread, pasta, beer and malt (Pacin et al., 2010, Yazdanpanah et al., 2014).

One of the most important properties of DON is connected with its resistance to high temperature (170-350°C), which is responsible even for the presence of this toxin in thermally preserved food. High stability of this toxin was shown at a temperature of 170°C during 30-minute exposure (Kushiro 2008, Sobrova et al., 2010).

Toxicity is associated with the presence of three hydroxide groups in the deoxynivalenol chemical structure of the molecule (Nagy et al., 2005). DON and its derivatives may enter the human organism with consumed food, water, through inhalation or through the skin. Animals most sensitive to the action of deoxynivalenol are first of all pigs, with rodents, dogs, cats, poultry and ruminants ranking next (Pestka and Smolinski 2005). The presence of the toxin was detected in blood serum, kidneys and livers of these animals (Doll et al., 2008, Goyarts et al., 2007). Moreover, there is a high risk of DON transmission to hen eggs or animal milk (Keese et al., 2008). The mechanism of deoxynivalenol action at the biochemical level consists in the slowing down of DNA and RNA synthesis as well as protein synthesis at the ribosomal level (Rocha et al., 2005, Bony et al., 2007). This toxin exhibits a haemolytic action, and it also influences nutrient uptake by the animal organism. Consumption of feed containing deoxynivalenol in some animals causes intestinal disorders connected with a reduced uptake of glucose and 5-methyltetrahydrofolic acid. Acute trichothecene poisoning in animals leads to skin irritation, lack of appetite, vomiting, diarrhoea, reduced body weight gain, haemorrhages, leucocytosis, nervous disorders and even death (Pestka and Smolinski 2005, Sobrova et al., 2010, Pestka 2010). Consumption of DON-containing feed may also have a significant effect on animal reproduction due to gestosis occurring in most cases. Deoxynivalenol causes *in vitro* and *in vivo* chromosomal aberration, it is an inhibitor of DNA synthesis, and it also blocks the stage of polypeptide chain elongation during protein synthesis (Zhou et al., 2005, Kolf-Clauw et al., 2009, Pestka 2010). Moreover, its immunosuppressive action was also recorded. In chronic poisoning, apart from the reduced feed intake and body weight gain, certain changes were also observed in blood indices such as a reduced concentration of albumins and globulins in blood serum, reduced levels of calcium and inorganic phosphorus, as well as increased liver weight (Gray and Pestka 2007, Kinser et al., 2005).

The problem of deoxynivalenol presence in crops and foodstuffs is regulated by the European Commission of 28 September 2007 no. 1126/2007, specifying the highest permissible levels of this toxin (Commission Regulation EC 2007). In 1993, the IARC placed DON in Group 3, not classifiable as to its carcinogenicity to humans (IARC 1993). The temporary highest daily dose of deoxynivalenol and its derivatives was established at 1 μg/kg body weight (Pestka and Smolinski 2005), with the same value for DON derivatives (e.g., 3-AcDON or 15-AcDON), since these derivatives are converted *in vivo* to DON (Pinton et al., 2012).

3.4.6. Other Mycotoxins

Beside commonly found mycotoxins (discussed above), less significant toxins are also identified, whose presence in cereal culture is related to the changing profile of fungi of the genus *Fusarium* (Wiśniewska et al., 2013, Czembor et al., 2014). They include enniatins, beauvericin and moniliformin (Figures 6, 7), formed by different fungal species of the genus *Fusarium* (Stępień and Waśkiewicz 2013, Waśkiewicz and Stępień 2012).

Beauvericin (BEA) as well as enniatin analogues (ENNs) A, A_1, B and B_1 belong to cyclic hexadepsipeptide mycotoxins synthesized by *Fusarium* species, mainly *F. avenaceum, F. tricinctum* or *F. poae* (Moretti et al., 2007, Stępień and Waśkiewicz 2013). The occurrence of these mycotoxins is often associated with small-grained cereals, corn and food products (Kokkonen et al., 2010, Meca et al., 2010a, b, Blesa et al., 2012, Zinedine et al., 2011). The concentrations of BEA and ENNs are much lower when compared to zearalenone and deoxynivalenol contents, which results in a non-significant risk for humans and animals.

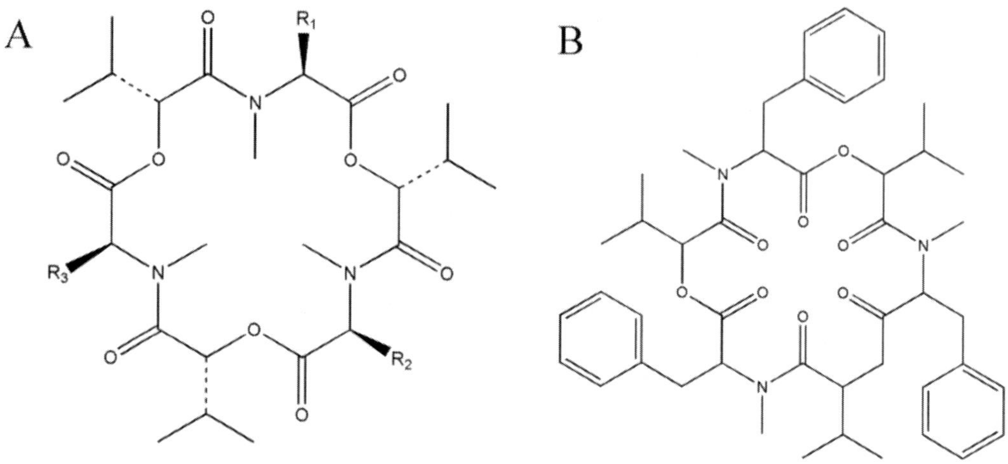

Enniatin	R_1	R_2	R_3
A	$-CH(CH_3)CH_2CH_3$	$-CH(CH_3)CH_2CH_3$	$-CH(CH_3)CH_2CH_3$
A_1	$-CH(CH_3)CH_2CH_3$	$-CH(CH_3)CH_2CH_3$	$-CH(CH_3)_2$
B	$-CH(CH_3)_2$	$-CH(CH_3)_2$	$-CH(CH_3)_2$
B_1	$-CH(CH_3)_2$	$-CH(CH_3)CH_2CH_3$	$-CH(CH_3)CH_2CH_3$

Figure 6. Chemical structure of enniatins and beauvericin.

The primary toxic action of BEA and ENNs is related to their ionophoric properties, thereby influencing and disturbing the physiological ionic balance and pH by dimeric structures transporting monovalent ions across the cell membranes (Kouri et al., 2005, Tonshin et al., 2010).

Beauvericin is toxic to several human cell lines and can induce their apoptosis and DNA fragmentation (Jestoi 2008, Segvic Klaric et al., 2008). Moreover, BEA exerts a negative inotropic effect (decrease in cardiac contraction strength) as well as a negative chronotropic effect (decrease in frequency of cardiac spontaneous beating activity) (Lemmens-Gruber et al., 2000, Peltonen et al., 2010).

Enniatins have been reported to induce a wide range of effects *in vitro*, and have phytotoxic, insecticidal, and antibacterial properties (Jestoi 2008). Other studies have found that enniatins may induce lysosomal as well as mitochondrial modifications and cell cycle disruption, then resulting in apoptotic cell death (Dornetshuber et al., 2007, Meca et al., 2010 a, Wätjen et al., 2009, Gammelsrud et al., 2012).

Similarly, moniliformin (MON) is frequently found in agricultural products in all climatic zones as a metabolite of the fungi *Fusarium avenaceum*, *F. oxysporum* and *F. proliferatum* (Goliński et al., 2010, Irzykowska et al., 2012, Wiśniewska et al., 2014). This toxin was identified e.g., in samples of wheat, oat, maize, rye and triticale, frequently accompanying fumonisins (Waśkiewicz et al., 2010 a, b, 2013). Stability of moniliformin in aqueous solutions is dependent on pH as well as temperature and exposure time. It is assumed that a 1-hour thermal processing at 175°C in a solution at pH 10 completely degrades this toxin (Peltonen et al., 2010, Tittlemier et al., 2014).

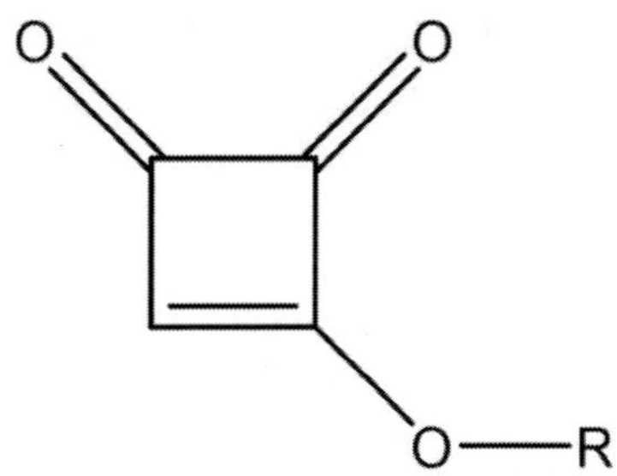

Figure 7. Chemical structure of moniliformin.

A small amount of MON greatly inhibits the oxidative capacity of mitochondrial enzymes, while it may also induce the development of Keshan disease, which attacks mainly the cardiac muscle, leading to cardiac circulatory failure, cardiac rhythm disorders as well as disturbed cardiac contractibility and intravascular clots within cardiac cavities, causing embolism (Ficheux et al., 2012, Sharma et al., 2012).

To date, no strong toxic activity has been shown for this group of mycotoxins, as evidenced by the lack of standards and legal regulations concerning maximum permissible concentrations of these toxins in cereals and their products.

3.5. Prevention of Mycotoxin Formation

Mycotoxin contamination of cereal grain is one of the most important problems in agriculture, investigated more extensively after 1960, when toxic fungal metabolites were identified for the first time.

The best and the most effective method to eliminate mycotoxins from food and feed is to prevent their biosynthesis by appropriate cultivation measures from sowing of cereals up to their harvest, followed by storage under appropriate conditions (Choudhary and Kumari 2010).

Cultivation measures have a considerable effect on a reduction of crop infestation by fungi from the genus *Fusarium*, particularly:

- Crop rotation
- Appropriate fore-crop
- Plant protection preparations preventing weed infestation
- Elimination of sowing material infested by *Fusarium* fungi
- Use of cereal cultivars less susceptible to fusariosis.

Prevention of fungal development during storage of cereal grain may be achieved by:

- Reducing grain moisture content below 13%
- Removal of contamination (dead, broken and damaged kernels, pests)
- Good quality agronomy equipment preventing mechanical damage of kernels
- Regular quality control of stored grain and storage conditions (humidity, temperature, atmosphere).

Processes leading to the reduction, degradation, decomposition, detoxification or elimination of toxins may be divided into physical, chemical and biological (Ni and Streett 2005, Jouany 2007, Halász et al., 2009, Awad et al., 2010). Effectiveness of physical methods such as cleaning, hulling, milling, washing, sorting or heating depends on the level of contamination and toxin distribution in the raw material. Some tested chemical preparations exhibit properties reducing mycotoxin levels. They include e.g., sodium disulfide, ozone, CO_2, hydrogen peroxide, ascorbic acid, sulfur dioxide, formaldehyde and ammonium hydroxide (Abramson et al., 2005, Ni and Streett 2005, Young et al., 2006). It was found that in some cases combinations of physical and chemical methods are more effective; however, due to considerable costs they are not always applied in practice.

It is also admissible to use several materials binding mycotoxins. These include bentonites, synthetic anionites, zeolites, active carbon, released calcium silicates and yeast cell membranes (Bailey et al., 2006, Fairchild et al., 2008, Magnoli et al., 2008). Since sorption on these materials concerns toxins with polar functional groups, particularly aflatoxins and partly zearalenone and vomitoxin, it occurs to some extent in the case of non-polar or weakly polar molecules; thus ochratoxin A and most *Fusarium* toxins may not be eliminated in this manner.

At present the most promising mycotoxin inactivation method is their enzymatic reduction, as a result of which less toxic or non-toxic metabolites are formed. Microbial enzymes disrupt molecules of these metabolites, e.g., in ochratoxin A breaking the peptide bond and in molecules of zearalenone disrupting the lactone ring (Cho et al., 2010, Yu et al., 2011, Tang et al., 2013).

It was also found that lactic or propionic acid bacteria inhibit growth of moulds, causing reduced biosynthesis of mycotoxins (Dalie et al., 2010, Oliveira et al., 2014), which was also found in natural yogurt microflora and yeasts (Masoud and Kaltoft 2006).

It was also shown that the level of mycotoxin contamination in foodstuffs may be reduced, but not eliminated, in food processing, including physical and/or chemical processes such as dry and wet milling, grain cleaning, preservation, roasting, baking, frying or extrusion (Castells et al., 2005, Choudhary and Kumari 2010).

In the case of ochratoxin A and zearalenone, promising detoxification methods are related to the use of ozone (Iacumin et al., 2012), alkali hydrogen peroxide and gamma radiation (Vita et al., 2014). However, both ozone and strongly basic compounds seem to be too drastic for grain treated in this way to be used for food production.

A fumonisin detoxification method using derivatives of phenolic acids naturally found in plants, which as strong antioxidants inhibit the action of fungal enzymes, proved to be effective (Samapundo et al., 2007). Promising results are also reported for the application of essential oils or microbial extracts (Yamamoto-Ribeiro et al., 2013, Xing et al., 2014).

At present, studies are being carried out on substances or groups of compounds (vitamins, proteins, enzymes, fats) which, when added to mycotoxins present in food, reduce their toxicity by decreasing their availability (Karlovsky 2011).

Chemometric methods based on analyses of volatile compounds are an interesting prevention procedure, since in grain samples infested with the fungi *Aspergillus*, *Penicillium* and *Fusarium*, characteristic volatile secondary metabolites are indicators of possible mycotoxin biosynthesis (Wenda-Piesik 2011, Morath et al., 2012). These microorganisms under advantageous conditions may form in stored cereals a large group of compounds, including alcohols, aldehydes, esters, hydrocarbons and heterocyclic compounds (Ting et al., 2010).

Currently studies are also being conducted concerning the potential mycotoxin self-control capacity of plants (Luongo et al., 2005). Such activities include:

- Genetic selection to enhance plant resistance to fungal infection,
- Culture of transgenic plants resistant or less susceptible to colonization by fungi and/or mycotoxin biosynthesis,
- Production of seeds colonised by endophytes – bacteria able to eliminate microscopic fungi,
- Infection of plants with non-toxin-forming fungi in competition with toxigenic fungi responsible for biosynthesis of toxic metabolites.

Identification of optimal conditions for mycotoxin formation over an entire food chain significantly prevents feed and food mycotoxin contamination. Specific programmes and education developed worldwide for food producers, including farmers, are improving the situation. Higher crop quality, its better hygienic status followed by proper storage conditions, introduction of professional monitoring of food products for the presence of mycotoxins and application of adequate food processing technologies facilitate elimination of contaminated batches of products (Kabak 2009, Yang et al., 2014).

Appropriate crop harvesting techniques may minimise the problem of mycotoxins, but will not eliminate it completely. However, it is essential to ensure that mycotoxin permissible levels required for the protection of human and animal health are not exceeded. Thus it is necessary to broaden our knowledge on the safe mycotoxin levels in food for human consumption and in animal feeds, as well as undertake research to improve detection methods (molecular and immunologic), to make them more precise and reliable and at the same time faster and cheaper to execute.

CONCLUSION AND FUTURE PERSPECTIVE

Toxic secondary metabolites of microscopic fungi are as old as fungi themselves, and cause economic, hygiene, health and civilization problems worldwide, but mycotoxins have been the focus of international and multidisciplinary interest for less than 60 years. Development of accurate methods of analysis, improved apparatus with significantly lowered detection limits, followed by official directives and regulations, make the situation better and easier each year. Physical, chemical and biological methods of reducing fungi and their mycotoxins in foods and feeds are promising, but it seems that these simple microorganisms are always "half a step ahead" of us, which suggests that prevention is the most powerful weapon in the extremely difficult war with toxigenic fungi and their mycotoxins. Because of tremendous efforts (including tremendous costs) - there is no doubt about it - the situation now appears much better, with a far brighter perspective than 60 years ago.

References

Abbas, HK; Cartwrihgt, RD; Wie, W; Shier, WT. Aflatoxin and fumonisin contamination of corn (maize, *Zea mays*) hybrids in Arkansas. *Crop Protect.*, 2006, 251-259.

Abdu, S; Ali, A; Ansari, S. Cytotoxic effect of ochratoxin A on the renal corpuscles of rat kidney: could ochratoxin A cause kidney failure? *Histol. Histopathol.*, 2011, 26, 543-549.

Abdulkadar, AHW; Al-Ali, AA; Al-Kildi, AM; Al- Jedah, JH. Mycotoxins in food products available in Qatar. *Food Control*, 2004, 15, 543-548.

Abid-Essefi, S; Ouanes, Z; Hassen, W; Baudrimont, I; Creppy, E; Bacha, H. Cytotoxicity, inhibition of DNA and protein syntheses and oxidative damage in cultured cells exposed to zearalenone. *Toxicol. Vitro*, 2004, 18, 467-474.

Abramson, D; House, JD; Nyachoti, CM. Reduction of deoxynivalenol in barley by treatment with aqueous sodium carbonate and heat. *Mycopathol.*, 2005, 160, 297-301.

Adejumo, TO; Hettwer, U; Karlovsky, P. Survey of maize from south-western Nigeria for zearalenone, alpha- and beta-zearalenols, fumonisin B_1 and enniatins produced by *Fusarium* species. *Food Add. Contam.*, 2007, 24, 993-1000.

Aidoo, KE; Mohamed, SM; Candlish, AA; Tester, RF; Elgerbi, AM. Occurrence of fungi and mycotoxins in some commercial baby foods in North Africa. *Food Nutr. Sci.*, 2011, 2, 751-758.

Aksoy, A; Yavuz, O; Das, YK; Guvenc, D; Muglali, OH. Occurrence of aflatoxin B1, T-2 toxin and zearalenone in compound animal feed. *J. Anim. Vet. Adv.*, 2009, 8, 403-407.

Alexa, E; Dehelean, CA; Poiana, M.-A; Radulov, I; Cimpean, A.-M; Bordean, D.-M; Tulcan, C; Pop, G. The occurrence of mycotoxins in wheat from western Romania and histopathological impact as effect of feed intake. *Chem. Cent. J.*, 2013, 7, 99-110.

Al-Hazmi, NA. Occurrence of fumonisin B_1 in imported and local corn based-snacks collected from Jeddah, Saudi Arabia. *Global J. Biotechnol. Biochem.*, 2009, 4, 193-200.

Almeida, I; Martins, HM; Santos, S; Costa, JM; Bernardo, F. Co-occurrence of mycotoxins in swine feed produced in Portugal. *Mycotoxin Res.*, 2011, 27, 177-181.

Almeida-Ferreira, GC; Barbosa-Tessmann, IP; Sega, R; Machinski Junior, M. Occurrence of zearalenone in wheat- and corn-based products commercialized in the State of Paraná, Brazil. *Braz. J. Microbiol.*, 2013, 44, 371-375.

Alvito, PC; Sizoo, EA; Almeida, CMM; Egmond, HP. Occurrence of aflatoxins and ochrotoxin A in baby food. *Food Anal. Methods*, 2010, 3, 22-30.

Amadi, JE; Adeniyi, DO. Mycotoxin production by fungi isolated from stored grains. *Afr. J. Biotechnol.*, 2009, 8, 1219-1221.

Amezqueta, S; Schorr-Galindo, S; Murillo-Arbizu, M; Gonzalez-Penas, E. OTA - producing fungi in foodstuffs: A review. *Food Control*, 2012, 26, 259-268.

Anli, E; Alkis, M. Ochratoxin A and brewing technology: A review. *J. Inst. Brew.*, 2010, 116, 23-32.

Araguas, C; Gonzalez-Peñas, E; Lopez de Cerain, A. Study on ochratoxin A in cereal-derived products from Spain. *Food Chem.*, 2005, 92, 459-464.

Awad, WA; Ghareeb, K; Bohm, J; Zentek, J. Decontamination and detoxification strategies for the *Fusarium* mycotoxin deoxynivalenol in animal feed and the effectiveness of microbial biodegradation. *Food Add.* Contam., 2010, 27, 510-520.

Aydin, A; Gunsen, U; Demirel, S. Total aflatoxin, Aflatoxin B1 and ochratoxin A levels in Turkish wheat flour. *J. Food Drug Analysis*, 2008, 16, 48-53.

Bailey, CA; Latimer, GW; Barr, AC; Wigle, WL; Haq, AU; Balthrop, JE; Kubena, LF. Efficacy of montmorillonite clay (Novasil plus) for protecting full-term broilers from aflatoxicosis. *J. Appl. Poult. Res.,* 2006, 15, 198-206.

Banu, I; Aprodu, I; Nicolau, A. Occurrence of *Fusarium* mycotoxins (deoxynivalenol and zearalenone) in wheat and high fibre wheat bread in Eastern Romania. *J. Environ. Prot. Ecol.*, 2011, 12, 519-525.

Bennett, JW; Klich, M. Mycotoxins. Clin. Microbiol. Rev., 2003, 16, 497-516.

Blandino, M; Reyneri, A; Vanara, F; Tamietti, G; Pietri, A. Remove from marked records influence of agricultural practices on *Fusarium* infection, fumonisin and deoxynivalenol contamination of maize kernels. *World Mycotoxin J.,* 2009, 2, 409-418.

Blesa, J; Marín, R; Lino, CM; Mañes, J. Evaluation of enniatins A, A1, B, B1 and beauvericin in Portuguese cereal-based foods. *Food Add. Contam.*, 2012, 29, 1727-1735.

Blount, WP. Turkey „X" disease. Turkeys 1961, 9, 55-58.

Bony, S; Olivier-Loiseau, L; Carcelen, M; Devaux, A. Genotoxic potential associated with low levels of the *Fusarium* mycotoxins nivalenol and fusarenon X in a human intestinal cell line. *Toxicol. In Vitro*, 2007, 21, 457-465.

Boermans, HJ; Leung, MCK. Mycotoxins and the pet food industry: Toxicological evidence and risk assessment. *Int. J. Food Microbiol.*, 2007, 119, 95-102.

Boudra, H; Barnouin, J; Dragacci, S; Morgavi, DP. Aflatoxin M_1 and ochratoxin A in raw bulk milk from French dairy herds. *J. Dairy Sci.*, 2007, 90, 3197-3201.

Bouras, N; Kim, YM; Strelkov, SE. Influence of water activity and temperature on growth and mycotoxin production by isolates of Pyrenophora tritici-repentis from wheat. *Int. J. Food Microbiol.*, 2009, 131, 251-255.

Broggi, LE; Pacin, AM; Gasparovic, A; Sacchi, C; Rothermel, A; Gallay, A; Resnik, S. Natural occurrence of aflatoxins, deoxynivalenol, fumonisins and zearalenone in maize from Entre Ríos Province, Argentina. *Mycotoxin Res.*, 2007, 23, 59-64.

Bryden, WL. Mycotoxins in the food chain: human health implications. *Asia Pac. J. Clin. Nutr.*, 2007, 16, 95-101.

Campbell, AW; Thrasher, JD; Gray, MR; Vojdani, A. Mold and mycotoxins: effects on the neurological and immune systems in humans. *Adv. Appl. Microbiol.,* 2004, 55, 375-406.

Cano-Sancho, G; Marin, S; Ramos, AJ; Sanchis, V. Occurrence of zearalenone, an oestrogenic mycotoxin, in Catalonia (Spain) and exposure assessment. *Food Chem. Toxicol.,* 2012, 50, 835-839.

Castegnaro, M; Canadas, D; Vrabcheva, T; Petkova-Bocharova, T; Chernozemsky, IN; Pfohl-Leszkowicz, A. Balkan endemic nephropathy: role of ochratoxins A through biomarkers. *Mol. Nutr. Food Res.*, 2006, 50, 519-529.

Castells, M; Marín, S; Sanchis, V; Ramos, AJ. Fate of mycotoxins in cereals during extrusion cooking: A review. *Food Add. Contam.*, 2005, 22, 150-157.

Cavaliere, C; D'Ascenzo, G; Foglia, P; Pastorini, E; Samperi, R; Laganoe, A. Determination of type B trichothecenes and macrocyclic lactone mycotoxins in field contaminated maize. *Food Chem.*, 2005, 92, 559-568.

Cerveró, MC; Castillo, MA; Montes, R; Hernández, E. Determination of trichothecenes, zearalenone and zearalenols in commercially available corn-based foods in Spain. *Rev. Iberoam. Micol.*, 2008, 24, 52-55.

Cetin, Y; Bullerman, LB. Evaluation of reduced toxicity of zearalenone by extrusion processing as measured by the MTT cell proliferation assay. *J. Agric. Food Chem.,* 2005, 53, 6558-6563.

Cho, KJ; Kang, JS; Cho, WT; Lee, CH; Ha, JK; Song, KB. *In vitro* degradation of zearalenone by *Bacillus subtilis*. *Biotechnol. Lett*., 2010, 32, 1921-1924.

Choudhary, AK; Kumari, P. Management of mycotoxin contamination in preharvest and post harvest crops: present status and future prospects. *J. Phytol.,* 2010, 2, 37-52.

Commission Regulation (EC) 2007. No 1126/2007 of 28 September 2007 amending Regulation (EC) No 1881/2006 setting maximum levels for certain contaminants in foodstuffs as regards *Fusarium* toxins in maize and maize products.

Covarelli, L; Stifano, S; Beccari, G; Raggi, L; Lattanzio, VM; Albertini, E. Characterization of *Fusarium verticillioides* strains isolated from maize in Italy: fumonisin production, pathogenicity and genetic variability. *Food Microbiol.,* 2012, 31, 17-24.

Creppy, EE; Chiarappa, P; Baudrimont, I; Borraci, P; Carratu, MS. Synergistic effects of fumonisin B_1 and ochratoxin A: are *in vitro* cytotoxicity data predictive of *in vivo* acute toxicity? *Toxicol.,* 2004, 201, 115-123.

Czembor, E; Stępień, Ł; Waśkiewicz, A. *Fusarium temperatum* as a new species causing ear rot on maize in Poland. Plant Dis., 2014, 98, 1001.

Danicke, S; Swiech, E; Buraczewska, L; Ueberschar, K-H. Kinetics and metabolism of zearalenone in young female pigs. *J. Anim. Physiol. Anim. Nutr*., 2005, 89, 268-276.

Dalie, DKD; Deschamps, AM; Richard-Forget, F. Lactic acid bacteria – Potential for control of mould growth and mycotoxins: A review. *Food Control,* 2010, 21, 370-380.

De Boevre, M; Di Mavungu, JD; Landschoot, S; Audenaert, K; Eeckhout, M; Maene, P; Haesaert, G; De Saeger, S. *World Mycotoxin J.,* 2012, 5, 207-219.

De la Campa, R; Hooker, DC; Miller, JD; Schaafsma, AW; Hammond, BG. Modeling effects of environment, insect damage, and *Bt* genotypes on fumonisin accumulation in maize in Argentina and the Philippines. *Mycopathol.,* 2005, 159, 539-552.

Decastelli, L; Lai, J; Gramaglia, M; Monaco, A; Nachtmann, C; Oldano, F; Ruffier, M; Sezian, A; Bandirola, C. Aflatoxins occurrence in milk and feed in Northern Italy during 2004-2005. *Food Control*, 2007, 18, 1263-1266.

Devendran, G; Balasubramanian, U. Biochemical and histopathological analysis of aflatoxin induced toxicity in liver and kidney of rat. *Asian J. Plant Sci. Res*., 2011, 1, 61-69.

Dip, R; Lenz, S; Antignac, JP; Le Bizec, B; Gmuender, H; Naegeli, H. Global gene expression profiles induced by phytoestrogens in human breast cancer cells. *Endocr. Relat. Cancer*, 2008, 15, 161-173.

Doll, S; Danicke, S; Valenta, H. Residues of deoxynivalenol (DON) in pig tissue after feeding mash or pellet diets containing low concentrations. *Mol. Nutr. Food Res*., 2008, 52, 727-734.

Domijan, AM; Peraica, M; Ferencić, Z; Cuzić, S; Fuchs, R; Lucić, A; Radić, B. Ochratoxin A-induced apoptosis in rat kidney tissue. *Arh. Hig. Rada Toksikol.,* 2004, 55, 243-248.

Domijan, AM; Peraica, M, Jurjević, Ž; Ivić, D; Cvjetković, B. Fumonisin B_1, fumonisin B_2, zearalenone and ochratoxin A contamination of maize in Croatia. *Food Add. Contam*., 2005, 22, 677-680.

Dorn, B; Forrer, HR; Jenny, E; Wettstein, FE; Bucheli, TD; Vogelgsang, S. *Fusarium* species complex and mycotoxins in grain maize from maize hybrid trials and from grower's fields. *J. Appl. Microbiol*., 2011, 111, 693-706.

Dornetshuber, R; Heffeter, P; Kamyar, M-R; Peterbauer, T; Berger, W; Lemmens-Gruber, R. Enniatin exerts p53-dependent cytostatic and p53-independent cytotoxic activities against human cancer cells. *Chem. Res. Toxicol.,* 2007, 20, 465-473.

Dos Santos, JS; Souzaa, TM; Ono, EYS; Hashimotoc, EH; Bassoi, MC; de Mirandae, MZ; Itanof, EN; Kawamura, O; Hirookaa, EY. Natural occurrence of deoxynivalenol in wheat from Paraná State, Brazil and estimated daily intake by wheat products. *Food Chem.*, 2013, 138, 90-95.

Driehuis, F; Spanjer, MC; Scholten, JM; Giffel, MC. Occurrence of mycotoxins in feedstuffs of Dairy Cows and Estimation of Total Dietary Intakes. *J. Dairy Sci.,* 2008, 91, 4261-4271.

Edwards, SG. Influence of agricultural practices on fusarium infection of cereals and subsequent contamination of grain by trichothecene mycotoxins. *Toxicol. Lett.*, 2004, 153, 29-35.

Ehrlich, KC; Kobbeman, K; Montalbano, BG; Cotty, PJ. Aflatoxin-producing *Aspergillus* species from Thailand. *Int. J. Food Microbiol.*, 2007, 114, 153-159.

El-Desouky, TA; Naguib, K. Occurrence of zearalenone contamination in some cereals in Egypt. *J. Agroalim. Proc. Technol.*, 2013, 19, 445-450.

Elmholt, S; Rasmussen, PH. *Penicillium verrucosum* occurrence and ochratoxin A contents in organically cultivated grain with special reference to ancient wheat types and drying practice. *Mycopathol.,* 2005, 159, 421-432.

Esti, M; Benucci, I; Liburdi, K; Acciaro, G. Monitoring of ochratoxin A fate during alcoholic fermentation of wine-must. *Food Control*, 2012, 27, 53-56.

European Commission (EC). Commission Regulation No. 2174/2003 of 12 December amending directive no 466/2001 concerning maximum levels for aflatoxins in foodstuffs. *Official J. Eur. Comm.*, 2003, L75, 44-45.

Ezekiel, CN; Odebode, AC; Fapohunda, SO. Zearalenone production by naturally occurring *Fusarium* species on maize, wheat and soybeans from Nigeria. *J. Biol. Environ. Sci.*, 2008, 2, 77-82.

Fairchild, AS; Croom, J; Grimes, JL; Hagler, WM. Effect of Astra-Ben 20 on broiler chicks exposed to aflatoxin B1 or T2 toxin. *Int. J. Poult. Sci.,* 2008, 7, 1147-1151.

Fandohan, P; Zoumenou, D; Hounhouigan, DJ; Marasas, WFO; Wingfield, MJ; Hell, K. Fate of aflatoxins and fumonisins during the processing of maize into food products in Benin. *Int. J. Food Microbiol.*, 2005, 98, 249-259.

FAO 2004. Worldwide Regulations for Mycotoxins in Food and Feed in 2003. FAO Food and Nutrition Paper, vol. 81, Food and Agriculture Organization of the United Nations, Roma, Italy.

Ferreira Geraldo, MR; Tessmann, DJ; Kemmelmeier, C. Production of mycotoxins by *Fusarium graminearum* isolated from small cereals (wheat, triticale and barley) affected with scab disease in Southern Brazil. Brazil. *J. Microbiol.*, 2006, 37, 58-63.

Ficheux, AS; Sibiril, Y; Le Garrec, R; Parent-Massin, D. *In vitro* myelotoxicity assessment of the emerging mycotoxins beauvericin, enniatin b and moniliformin on human hematopoietic progenitors. *Toxicon,* 2012, 59, 182-191.

Forgacs, J. Mycotoxicoses-the neglected diseases. *Feedstuffs*, 1962, 34, 124-134.

Forgacs, J; Carll, WT. Preliminary mucotox studies on hemorrhagic disease in poultry. *Vet. Med.*, 1955, 50, 172-177.

Foroud, NA; Eudes, F. Trichothecenes in cereal grains. *Int. J. Mol. Sci.*, 2009, 10, 147-173.

Gajęcka, M. The effect of low-dose experimental zearalenone intoxication on the immunoexpression of estrogen receptors in the ovaries of pre-pubertal bitches. *Polish J. Vet. Sci.*, 2012, 15, 685-691.

Gajęcka, M; Zielonka, Ł; Dąbrowski, M; Mróz, M; Gajęcki, M. The effect of low doses of zearalenone and its metabolites on progesterone and 17β-estradiol concentrations in peripheral blood and body weights of pre-pubertal female Beagle dogs. *Toxicon*, 2013, 76, 260-269.

Galbenu, P; Damiescu, L; Trif, A. Zearalenone occurrence in cereal and cereal-based foodstuffs marketed in Timis country. *Res. J. Agr. Sci.*, 2011, 43, 43-49.

Gammelsrud, A; Solhaug, A; Dendelé, B; Sandberg, WJ; Ivanova, L; Kocbach Bølling, A; Lagadic-Gossmann, D; Refsnes, M; Becher, R; Eriksen, G; Holme, JA. Enniatin B-induced cell death and inflammatory responses in RAW 267.4 murine macrophages. *Toxicol. Appl. Pharmacol.*, 2012, 261, 74-87.

Gelderblom, T; Jaskiewicz, WCA; Marasas, WFO; Thiel, PG; Horak, MJ; Vleggaar, R; Kriek, NJ. Fumonisins-novel mycotoxins with cancer promotion activity produced by *Fusarium moniliforme. Appl. Environ. Microbiol.,* 1988, 54, 1806-1811.

Ghiasian, SA; Rezayat, SM; Kord-Bacheh, P; Maghsood, AH; Yazdanpanah, H; Shephard, GS; van der Westhuizen, L; Vismer, HF; Marasas, WFO. Fumonisin production by *Fusarium* species isolated from freshly harvested corn in Iran. *Mycopathol.,* 2005, 159, 31-40.

Giray, B; Girgin, G; Engin, AB; Aydin, S; Sahin, G. Aflatoxin levels in wheat samples consumed in some regions of Turkey. *Food Control*, 2007, 18, 23-29.

Goertz, A; Zuehlke, S; Spiteller, M; Steiner, U; Dehne, HW; Waalwijk, C; Vries, I; Oerke, EC. *Fusarium* species and mycotoxin profiles on commercial maize hybrids in Germany. *Eur. J. Plant Pathol.*, 2010, 128, 101-111.

Goldblatt, LA. Aflatoxin: Scientific Background, Control, and Implications. Academic Press, Inc., New York, 1969.

Goliński, P; Waśkiewicz, A; Gromadzka, K. Mycotoxins and mycotoxicoses under climatic conditions of Poland. *Polish J. Vet. Sci.,* 2009, 12, 581-588.

Goliński, P; Waśkiewicz, A; Wiśniewska, H; Kiecana, I; Mielniczuk, E; Gromadzka, K; Kostecki, M; Bocianowski, J; Rymaniak, E. Reaction of winter wheat (*Triticum aestivum* L.) cultivars to infection with *Fusarium* spp.: mycotoxin contamination in grain and chaff. Food Add. Contam., 2010, 27, 1015-1024.

Gonzalez, HHL; Molto, GA; Pacin, A; Resnik, SL; Zelaya, MJ; Masana, M; Martinez, EJ; Trichothecenes and mycoflora in wheat harvested in nine locations in Buenos Aires Province, Argentina. Mycopathol., 2008, 165, 105-114.

Goyarts, T; Danicke, S; Valenta, H; Ueberschar, KH. Carry-over of *Fusarium* toxins (deoxynivalenol and zearalenone) from naturally contaminated wheat to pigs. *Food Add. Contam.*, 2007, 24, 369-380.

Gray, JS; Pestka, JJ. Transcriptional regulation of deoxynivalenol-induced IL-8 expression in human monocytes. *Toxicol. Sci.,* 2007, 99, 502-511.

Grollman, AP; Jelakovic, B. Role of environmental toxins in Endemic (Balkan) Nephropathy. *J. Am. Soc. Nephrol.*, 2007, 18, 2817-2823.

Halász, A; Lásztity, R; Abonyi, T; Bata, A. Decontamination of mycotoxin-containing food and feed by biodegradation. *Food Rev. Int.*, 2009, 25, 284-298.

Hedayati, MT; Pasqualotto, AC; Warn, PA; Bowyer, P; Denning, DW. *Aspergillus flavus*: human pathogen, allergen and mycotoxin producer. *Microbiol.,* 2007, 153, 1677-1692.

Hope, JH; Hope, BE. A review of the diagnosis and treatment of ochratoxin A inhalational exposure associated with human illness and kidney disease including focal segmental glomerulosclerosis. *J. Environ. Public Health,* 2012, 1-10.

Iacumin, L; Manzano, M; Comi, G. Prevention of *Aspergillus ochraceus* growth on and Ochratoxin A contamination of sausages using ozonated air. *Food Microbiol.,* 2012, 29, 229-232.

Ibáñez-Vea, M; González-Peñas, E; Lizarraga, E; López de Cerain, A. Co-occurrence of aflatoxins, ochratoxin A and zearalenone in barley from a northern region of Spain. *Food Chem.,* 2012a, 132, 35-42.

Ibáñez-Vea, M; González-Peñas, E; Lizarraga, E; López de Cerain, A. Co-occurrence of mycotoxins in Spanish barley: A statistical overview. *Food Control,* 2012b, 28, 295-298.

Ibáñez-Vea, M; Martinez, R; Gonzalez-Penas, E; Lizarraga, E; Lopez de Cerain, A. Co-occurrence of aflatoxins, ochratoxin A and zearalenone in breakfast cereals. *Food Control,* 2011, 22, 1949-1955.

International Agency for Research on Cancer (IARC). Aflatoxins: Occurring aflatoxins (group 1), aflatoxin M1 (group 2B). In Monographs on the Evaluation of Carcinogenic Risks to Humans, Lyon, France, 1993, 56, 445.

International Agency for Research on Cancer (IARC). Some naturally occurring substances; food items and constituents, heterocyclic aromatic amines and mycotoxins. IARC Monographs on the evaluation of carcinogenic risks to human (Lyon: WHO/IARC), 1993, 56, 397-444, 445-466, 489-521.

International Agency for Research on Cancer (IARC). Toxins derived from *Fusarium moniliforme*: Fumonisin B_1 and B_2 and fusarin C. Lyon; France, 1993.

International Agency for Research on Cancer (IARC). Fumonisin B_1. IARC Monographs on the evaluation of the carcinogenic risks to humans: Some traditional herbal medicines, some mycotoxins, naphthalene and styrene. International Agency for Research on Cancer: Lyon, France, 2002, 82, 301-366.

Irzykowska, L; Bocianowski, J; Waśkiewicz, A; Weber, Z; Karolewski, Z; Goliński, P; Kostecki, M; Irzykowski, W. Genetic variation of *Fusarium oxysporum* isolates forming fumonisin B_1 and moniliformin. *J. Appl. Genet.,* 2012, 53, 237-247.

Iqbal, SA; Khalil, IA; Shah, H. Aflatoxin contents of stored and artificially inoculated cereals and nuts. *Food Chem.*, 2006, 98, 699-703.

Isebaert, S; De Saeger, S; Devreese, R; Verhoeven, R; Maene, P; Heremans, B; Haesaert, G. Mycotoxin-producing *Fusarium* species occurring in winter wheat in Belgium (Flanders) during 2002-2005. *J. Phytopathol.*, 2009, 157, 108-116.

Jakic-Dimic, D; Nesic, K; Petrovic, M. Contamination of cereals with aflatoxins, metabolites of fungi *Aspergillus flavus*. Biotechnol. *Animal Husbandry,* 2009, 25, 1203-1208.

Jaksic, SM; Abramovic, BF; Prunic, BZ; Mihaljev, ZA; Balos, MMZ; Jajic, IM; Despotovic, VN; Bjelica, LJ. Incidence of aflatoxins and fumonisins in cereal food from Serbian market. *J. Agroaliment. Process Technol.*, 2011, 17, 108-112.

Jestoi, M. Emerging fusarium-mycotoxins fusaproliferin, beauvericin, enniatins, and moniliformin: a review. *Crit. Rev. Food Sci. Nutr.*, 2008, 48, 21-49.

Jouany, JP. Methods for preventing, decontaminating and minimizing the toxicity of mycotoxins in feeds. *Anim. Feed Sci. Technol.*, 2007, 137, 342-362.

Kabak, B. The fate of mycotoxins during thermal food processing. *J. Sci. Food Agr.*, 2009, 89, 549-554.

Karlovsky, P. Biological detoxification of the mycotoxin deoxynivalenol and its use in genetically engineered crops and feed additives. *Appl. Microbiol. Biotechnol.*, 2011, 91, 491-504.

Keese, C; Meyer, U; Valenta, H; Schollenberger, M; Starke, A; Weber, IA; Rehage, J, Breves, G; Danicke, S. No carry-over of unmetabolised deoxynivalenol in milk of dairy cows fed high concentrate proportions. *Mol. Nutr. Food Res.*, 2008, 52, 1514-1529.

Kelman, BJ; Robbins, CA; Swenson, LJ; Hardin, BD. Risk from inhaled mycotoxins in indoor office and residential environments. *Int. J. Toxicol.*, 2004, 23, 3-10.

Khatoon, S; Hanif, NQ; Tahira, I; Sultana, N; Sultan,a K; Ayub, N. Natural occurrence of aflatoxins, zearalenone and trichothecenes in maize grown in Pakistan. *Pak. J. Bot.*, 2012, 44, 231-236.

Khoury, A; Atoui, A. Ochratoxin A: general overview and actual molecular status. *Toxins,* 2010, 2, 461-493.

Kim, D-H; Lee, I-H; Do, W-H; Nam, W-S; Li, H; Jang, H-S; Lee, C. Incidence and levels of deoxynivalenol, fumonisins and zearalenone contaminants in animal feeds used in Korea in 2012. *Toxins,* 2014, 6, 20-32.

Kimanya, ME; De Meulenaer, B; Tiisekwa, B; Ndomondo-Sigonda, M; Devlieghere, F; Van Camp, J; Kolsteren, P. Co-occurrence of fumonisins with aflatoxins in home-stored maize for human consumption in rural villages of Tanzania. *Food Add. Contam.,* 2008, 25, 1353-1364.

Kinser, S; Li, MX; Jia, QS; Pestka, JJ. Truncated deoxynivalenol-induced splenic immediate early gene response in mice consuming (n-3) polyunsaturated fatty acids. *J. Nutr. Biochem.*, 2005, 16, 88-95.

Kokkonen, M; Ojala, L; Parikka, P; Jestoi, M. Mycotoxin production of selected *Fusarium* species at different culture conditions. *Int. J. Food Microbiol.,* 2010, 143, 17-25.

Kolf-Clauw, M; Castellote, J; Joly, B; Bourges-Abella, N; Raymond-Letron, I; Pinton, P; Oswald, IP. Development of a pig jejunal explant culture for studying the gastrointestinal toxicity of the mycotoxin deoxynivalenol: Histopathological analysis. *Toxicol. In Vitro,* 2009, 23, 1580-1584.

Kouri, K; Duchen, MR; Lemmens-Gruber, R. Effects of beauvericin on the metabolic state and ionic homeostasis of ventricular myocytes of the guinea pig. *Chem. Res. Toxicol.*, 2005, 18, 1661-1668.

Kushiro, M. Effects of milling and cooking processes on the deoxynivalenol content in wheat. *Int. J. Mol. Sci.*, 2008, 9, 2127-2145.

Labuda, R; Parich, A; Berthiller, F; Tančinová, D. Incidence of trichothecenes and zearalenone in poultry feed mixtures from Slovakia. *Int. J. Food Microbiol.,* 2005, 105, 19-25.

Lattanzio, VMT; Solfrizzo, M; Powers, S; Visconti, A. Simultaneous determination of aflatoxins, ochratoxin A and Fusarium toxins in maize by liquid chromatography/ tandem mass cpecrtometry after multitoxin immunoaffinity cleanup. *Rapid Commun. Mass Spectr.*, 2007, 21, 3253-3261.

Lemmens-Gruber, R; Rachoy, B; Steininger, E; Kouri, K; Saleh, P; Krska, R; Josephs, R; Lemmens, M. The effect of the *Fusarium* metabolite beauvericin on electromechanical and physiological properties in isolated smooth and heart muscle preparations of guinea pigs. *Mycopathol.*, 2000, 149, 5-12.

Lino, CM; Silva, LJG; Pena, A; Fernández, M. Occurrence of fumonisins B_1 and B_2 in broa, typical Portuguese maize bread. *Int. J. Food Microbiol.*, 2007, 118, 79-82.

Mañes, J; Llorent-Martínez, EJ; Ortega-Barrales, P; Fernández-de Córdova, ML; Ruiz-Medina, A. Quantitation of ochratoxin a in cereals and feedstuff using sequential injection analysis with luminescence detection. *Food Control,* 2013, 30, 379-385.

Luongo, L; Galli, M; Corazza, L; Haas, L; Plas, LC; Kohl, J. Potential of fungal antagonists for bio-control of Fusarium spp. in wheat and maize through competition in crop debris. *Biocontrol Sci. Technol.*, 2005, 15, 229-242.

Lutfullah, G; Hussain, A. Studies on contamination level of aflatoxins in some cereals and beans of Pakistan. *Food Control*, 2012, 23, 32-36.

Magnoli, AP; Tallone, L; Rosa, CAR; Dalcero, AM; Chiacchiera, SM; Torres Sanchez, RM. Commercial bentonites as detoxifier of broiler feed contaminated with aflatoxin. *Appl. Clay Sci.*, 2008, 40, 63-71.

Malihipour, A; Gilbert, J; Piercey-Normore, M; Cloutier, S. Molecular phylogenetic analysis, trichothecene chemotype patterns, and variation in aggressiveness of *Fusarium* isolates causing Head Blight in wheat. *Plant Dis.,* 2012, 96, 1016-1025.

Manova, R; Mladenova, R. Incidence of zearalenone and fumonisins in Bulgarian cereal production. *Food Control*, 2009, 20, 362-365.

Marcinkowska, J. Determination of types of fungi import ant in plant pathology [in polish, Oznaczanie rodzajów grzybów ważnych w patologii roślin]. Fundacja Rozwój SGGW, 2003, Warszawa 328.

Marczuk, J; Obremski, K; Lutnicki, K; Gajęcka, M; Gajęcki, M. Zearalenone and deoxynivalenol mycotoxicosis in dairy cattle herds. *Polish J. Vet. Sci.*, 2012, 15, 365-372.

Martins, HM; Marques, M; Almeida, I; Guerra, MM; Bernardo, F. Mycotoxins in feedstuffs in Portugal: An overview. *Mycotoxin Res.*, 2008, 24, 19-23.

Masoud, W; Kaltoft, CH. The effects of yeatsts involved in the fermentation of coffee Arabica in East Africa on growth and ochratoxin A (OTA) production by *Aspergillus ochraceus*. *Int. J. Food Microbiol.*, 2006, 106, 229-234.

Matarese, F; Sarrocco, S; Gruber, S; Seidl-Seiboth, V; Vannacci, G. Biocontrol of *Fusarium* head blight: interactions between *Trichoderma* and mycotoxigenic *Fusarium*. *Microbiol.,* 2012, 158, 98-106.

Mateo, R; Medina, A; Mateo, EM; Mateo, F; Jiménez, M. An overview of ochratoxin A in beer and wine. *Int. J. Food Microbiol.*, 2007, 119, 79-83.

Meca, G; Soriano, JM; Gaspari, A; Ritieni, A; Moretti, A; Mañes, J. Antifungal effects of the bioactive compounds enniatins A, A_1, B, B_1. *Toxicon,* 2010a, 56, 480-485.

Meca, G; Zinedine, A; Blesa, J; Font, G; Mañes, J. Further data on the presence of *Fusarium* emerging mycotoxins enniatins, fusaproliferin and beauvericin in cereals available on the Spanish markets. *Food Chem. Toxicol.,* 2010b, 48, 1412-1416.

Metzler, M; Pfeiffer, E; Hildebrand, AA. Zearalenone and its metabolites as endocrine disrupting chemicals. *World Mycotoxin J.,* 2010, 3, 385-401.

Milicevic, D; Skrinjar, M; Baltic, T. Real and perceived risks for mycotoxin contamination in foods and feeds: challenges for food safety control. *Toxins,* 2010, 2, 572-592.

Moreno, EC; Garcia, GT; Ono, MA; Vizoni, É; Kawamura, O; Hirooka, EY; Ono, EYS. Co-occurrence of mycotoxins in corn samples from the Northern region of Paraná State, Brazil. *Food Chem.,* 2009, 116, 220-226.

Morath, SU; Hung, R; Bennett, JW. Fungal volatile organic compounds: A review with emphasis on their biotechnological potential. *Fungal Biol.* Rev., 2012, 26, 73-83.

Moretti, M; Grunau, A; Minerdi, D; Gehrig, P; Roschitzki, B; Eberl, L; Garibaldi, A; Gullino, ML; Riedel, K. A proteomics approach to study synergistic and antagonistic interactions of the fungal-bacterial consortium *Fusarium oxysporum* wild-type MSA 35. *Proteomics,* 2010, 10, 3292-3320.

Moretti, A; Mulè, G; Ritieni, A; Logrieco, A. Further data on the production of beauvericin, enniatins and fusaproliferin and toxicity to Artemia salina by Fusarium species of Gibberella fujikuroi species complex. *Int. J. Food Microbiol.,* 2007, 118, 158-163.

Mushtaq, M; Sultana, B; Anwar, F; Khan, MZ; Ashrafuzzaman, M. Occurrence of aflatoxins in selected processed foods from Pakistan. *Int. J. Mol. Sci.,* 2012, 13, 8324-8337.

Myburg, RB; Needhi, N; Chuturgoon, AA. The ultrastructural effects and immunolocalisation of fumonisin B_1 on cultured oesophageal cancer cells (SNO). *South Afr. J. Sci.,* 2009, 105, 217-222.

Nagy, CM; Fejer, SN; Berek, L; Molnar, J; Viskolcz, B. Hydrogen bondings in deoxynivalenol (DON) conformations – a density functional study. *J. Mol. Struct.,* 2005, 726, 55-59.

Ni, X; Streett, DA. Modulation of water activity on fungicide effect on Aspergillus niger growth in Sabouraud dextrose agar medium. *Lett. Appl. Microbiol.,* 2005, 41, 428-433.

Noonim, P; Mahakarnchanakul, W; Nielsen, KF; Frisvad, JC; Samson, RA. Isolation, identification and toxigenic potential of ochratoxin A-producing *Aspergillus* species from coffee beans grown in two regions of Thailand. *Int. J. Food Microbiol.*, 2008, 128, 197-202.

Nuryono, N; Noviandi, CT; Böhm, J; Razzari-Fazeli, E. A limited survey of zearalenone in Indonesian maize-based food and feed by ELISA and high performance liquid chromatography. *Food Control*, 2005, 16, 65-71.

Offiah, N; Adesiyun, A. Occurrence of aflatoxins in peanuts, milk, and animal feed in Trinidad. *J. Food Prot.,* 2007, 70, 771-775.

Oliveira, PM; Zannini, E; Arendt, EK. Cereal fungal infection, mycotoxins, and lactic acid bacteria mediated bioprotection: From crop farming to cereal products. *Food Microbiol.*, 2014, 37, 78-95.

Ono, EYS; Biazon, L; da Silva, M; Vizoni, E; Sugiura, Y; Ueno, Y; Hirooka, EY. Fumonisins in corn: correlation with *Fusarium* spp. count, damaged kernels, protein and lipid content. Brazil. *Arch. Biol. Technol.*, 2006, 49, 63-71.

Osboo, RK; Mirabolfathy, M. Natural zearalenone contamination of wheat from Golestan province, Northern Iran. *Iran. J. Plant Pathol.*, 2008, 44, 60-65.

Osuchowski, MF; Sharma, RP. Fumonisin B_1 induces necrotic cell death in BV-2 cells and murine cultured astrocytes and is antiproliferative in BV-2 cells while N2A cells and primary cortical neurons are resistant. *Neurotoxicol.*, 2005, 26, 981-992.

Ouanes, Z; Ayed-Boussema, I; Baati, T; Creppy, EE; Bacha, H. Zearalenone induces chromosome aberrations in mouse bone marrow: preventive effect of 17beta-estradiol, progesterone and Vitamin E. *Mutat. Res.*, 2005, 565, 139-149.

Pacin, A; Ciancio Bovier, E; Cano, G; Taglieri, D; Hernandez Pezzani, C. Effect of the bread making process on wheat flour contaminated by deoxynivalenol and exposure estimate. *Food Control*, 2010, 21, 492-495.

Pardo, E; Malet, M; Marin, S; Sanchis, V; Ramos, AJ. Effects of water activity and temperature on germination and growth profiles of ochratoxigenic *Penicillium verrucosum* isolates on barley meal extract agar. *Int. J. Food Microbiol.*, 2006, 106, 25-31.

Paíga, P; Morais, S; Oliva-Teles, T; Correia, M; Delerue-Matos, C; Duarte, SC; Pena, A; Lino, CM. Extraction of ochratoxin A in bread samples by the QuEChERS methodology. *Food Chem.*, 2012, 135, 2522-2528.

Parveen, M; Zhu, Y, Kiyama, R. Expression profiling of the genes responding to zearalenone and its analogues using estrogen-responsive genes. *FEBS Lett.*, 2009, 583, 2377-2384.

Pasquali, M; Giraud, F; Brochot, C; Cocco, E; Hoffmann, L; Bohn, T. Genetic *Fusarium* chemotyping as a useful tool for predicting nivalenol contamination in winter wheat. *Int. J. Food Microbiol.*, 2010, 137, 246-253.

Pástorová, B; Laciaková, A; Čonková, E; Maraček, I; Danko, J; Pilipčinec, E. Influence of zearalenone on plasma catecholamines in rabbits. *Bull. Vet. Inst. Pulawy*, 2004, 48, 151-153.

Peltonen, K; Jestoi, M; Eriksen, GS. Health effects of moniliformin: a poorly understood *Fusarium* mycotoxin. *World Mycotox. J.*, 2010, 3, 403-414.

Pestka, JJ. Deoxynivalenol: mechanisms of action, human exposure, and toxicological relevance. *Arch. Toxicol.*, 2010, 84, 663-679.

Pestka, JJ; Smolinski, AT. Deoxynivalenol: toxicology and potential effects on humans. *J. Toxicol. Environ. Health B Crit. Rev.*, 2005, 8, 39-69.

Pinton, P; Tsybulskyy, D; Lucioli, J; Laffitte, J; Callu, P; Lyazhri, F; Grosjean, F; Bracarense, AP; Kolf-Clauw, M; Oswald, IP. Toxicity of deoxynivalenol and its acetylated derivatives on the intestine: differential effects on *morphology, barrier function, tight junctions proteins and MAPKinases. Toxicol. Sci.*, 2012, 130, 180-190.

Pleadin, J; Sokolović, M; Perši, N; Zadravec, M; Jaki, V; Vulić, A. Contamination of maize with deoxynivalenol and zearalenone in Croatia. *Food Control*, 2012, 28, 94-98.

Rajarajan, PN; Rajasekaran, KM; Asha Devi, NK. Aflatoxin contamination in agricultural commodities. *Indian J. Pharm. Biol. Res.*, 2013, 1, 148-151.

Ramesh, J; Sarathchandra, G; Sureshkumar, V. Survey of market samples of food grains and grain flour for aflatoxin B_1 contamination. *Int. J. Curr. Microbiol. App. Sci.*, 2013, 2, 184-188.

Rashedi, M; Ashjaazadeh, MA; Sohrabi, HR; Azizi, H; Rahimi, E. Determination of zearalenone contamination in wheat and rice in Chaharmahalva Bakhtyari, *Iran. J. Cell Anim. Biol.*, 2012, 6, 54-56.

Raters, M; Matissek, R. Thermal stability of aflatoxin B_1 and ochratoxin A. *Mycotoxin Res.*, 2008, 4, 130-134.

Reddy, KRN; Salleh, B; Saad, B; Abbas, HK; Abel, CA; Shier, WT. An overview of mycotoxin contamination in foods and its implications for human health. *Toxin Rev.*, 2010, 29, 3-26.

Reddy, KRN; Saritha, P; Reddy, CS; Muralidharan, K. Aflatoxin B1 producing potential of *Aspergillus flavus* strains isolated from stored rice grains. *Afr. J. Biotechnol.*, 2009, 8, 3303-3308.

Reddy, L; Bhoola, K. Ochratoxins-food contaminants: impact on human health. *Toxins*, 2010, 2; 771-779.

Richard, JL. Some major mycotoxins and their mycotoxicoses - An overview. *Int. J. Food Microbiol.*, 2007, 119, 3-10.

Riley, RT; Voss, KA. Differential sensitivity of rat kidney and liver to fumonisin toxicity: organ-specific differences in toxin accumulation and sphingoid base metabolism. *Toxicol. Sci.*, 2006, 92, 335-345.

Rocha, O; Ansari, K; Doohan, FM. Effects of trichothecene mycotoxins on eukaryotic cells: a review. *Food Add. Contam.*, 2005, 22, 369-378.

Rocha, LO; Nakai, VK; Braghini, R; Reis, TA; Kobashigawa, E; Correa, B. Mycoflora and co-occurrence of fumonisins and aflatoxins in freshly harvested corn in different regions of Brasil. *Int. J. Mol. Sci.*, 2009, 10, 5090-5103.

Rodrigues, I; Naehrer, K. 2012. A three-year survey on the worldwide occurrence of mycotoxins in feedstuffs and feed. *Toxins*, 2012, 4, 663-675.

Rojo, FG; Reynoso, MM; Ferez, M; Chulze, SN; Torres, AM. Biological control by Trichoderma species of Fusarium solani causing peanut brown root rot under field conditions. *Crop Protect.*, 2007, 26, 549-555.

Romagnoli, B; Menna, V; Gruppioni, N; Bergamini, C. Aflatoxins in spices, aromatic herbs, herb-teas and medicinal plants marketed in Italy. *Food Control,* 2007, 18, 697-701.

Saini, SS; Kaur, A. Aflatoxin B_1: Toxicity, characteristics and analysis: Mini review. *Global Adv. Res. J. Chem. Material Sci.*, 2012, 1, 63-70.

Sajid, MA; Khan, IA; Rauf, U. *Aspergillus fumigatus* in commercial poultry floks, a serious threat to poultry industry in Pakistan. *J. Anim. Pl. Sci.,* 2006, 16, 79-81.

Samapundo, S; De Meulenear, B; Ostei-Nmoh, D; Lamboni, Y; Debevere, J. Can phenolic compounds be used for the protection of corn from fungal invasion and mycotoxin contamination during storage? *Food Microbiol.*, 2007, 24, 465-473.

Sampietro, DA; Apud, GR; Belizán, MME; Vattuone, MA; Catalán, CAN. Toxigenic potential of *Fusarium graminearum* isolated from maize of northwest Argentina. *Braz. J. Microbiol.,* 2013, 44, 417-422.

Sassahara, M; Pontes Netto, M; Yanaka, EK. Aflatoxin occurrence in foodstuff sopplied to dairy cattle and aflatoxin M1 in raw milk in the North of Parana state. *Food Chem. Toxicol.*, 2005, 43, 981-984.

Schjøth, JE; Visconti, A; Sundheim, L. Fumonisins in maize in relation to climate, planting time and hybrids in two agroecological zones in Zambia. *Mycopathol.,* 2009, 167, 209-219.

Schütze, N; Lehmann, I; Bönisch, U; Simon, JC; Polte, T. Exposure to mycotoxins increases the allergic immune response in a murine asthma model. *Am. J. Respir. Crit. Care Med.*, 2010, 181, 1188-1199.

Scott, DB. Toxigenic fungi isolated from cereal and legume products. *Mycophathol. Mycol. Appl.*, 1965, 25, 213-222.

Scussel, VM; Savi, GD; Freitas Costas, LL; Mendonça Xavier, JJ; Manfio, D; Bittencourt, KO; Aguiar, K; Stein, SM. Fumonisins in corn (*Zea mays* L.) from Southern Brazil. *Food Add. Contam. Part B: Surveillance*, 2014, 7, 151-155.

Seeling, K; Dänicke, S. Relevance of the *Fusarium* toxins deoxynivalenol and zearalenone in ruminant nutrition. A review. *J. Anim. Feed Sci.,* 2005, 14, 3-40.

Segvic Klaric, M; Cvetnic, Z; Pepeljnjak, S; Kosalec, I. Co-occurence of aflatoxins, ochratoxin A, fumonisins and zearalenone in cereals and feed, determined by competitive direct enzyme-linked immunosorbent assay and thin-layer chromatography. *Arh. Hig. Rada Toksikol.*, 2009, 60, 427-434.

Segvic Klaric, M; Pepeljnjak, S; Rozgaj, R. Genotoxicity of fumonisin B1, beauvericin and ochratoxin A in porcine kidney PK15 cells: effects of individual and combined treatment. *Croatica Chem. Acta*, 2008, 81, 139-146.

Shar, ZH; Sumbal, GA; Sherazi, STH; Bhanger, MI; Nizamani, SM. Natural co-occurrence of aflatoxins and deoxynivalenol in poultry feed in Pakistan. Food Add. Contam.: Part B: Surveillance, 2014, DOI: 10.1080/19393210.2013.867904

Sharma, D; Asrani, RK; Ledoux, DR; Rottinghaus, GE; Gupta, VK. Toxic interaction between fumonisin B_1 and moniliformin for cardiac lesions in Japanese Quail. *Avian Dis.,* 2012, 56, 545-554.

Shotwell, LL; Hesseltine, CW; Goulden, ML. Ochratoxin A: occurrence as natural contaminant of a corn sample. *Appl. Microbiol.*, 1969, 17, 765-766.

Silva, LJ; Lino, CM; Pena, A; Moltó, JC. Occurrence of fumonisins B_1 and B_2 in Portuguese maize and maize-based foods intended for human consumption. *Food Add. Contam.*, 2007, 24, 381-390.

Skládanka, J; Nedělník, J; Adam, V; Doležal, P; Moravcová, H; Dohnal, V. Forage as a primary source of mycotoxins in animal diets. *Int. J. Environ. Res. Public Health*, 2011, 8, 37-50.

Sobrova, P; Adam, V; Vasatkova, A; Beklova, M; Zeman, L; Kizek, R. Deoxynivalenol and its toxicity. *Interdisc. Toxicol.*, 2010, 3, 94-99.

Speijers, GJA; Speijers, MHM. Combined toxic effects of mycotoxins. *Toxicol. Lett.,* 2004, 153, 91-98.

Sreenivasa, MY; Sharmila Dass, R; Charith Raj, AP; Janardhana, GR. Molecular detection of fumonisin producing *Fusarium* species of freshly harvested maize kernels using polymerase chain reaction (PCR). Taiwania, 2006, 51, 251-257.

Stępień, Ł; Koczyk, G; Waśkiewicz, A. FUM cluster divergence in fumonisins-producing *Fusarium* species. *Fungal Biol.,* 2011a, 115, 112-123.

Stępień, Ł; Koczyk, G; Waśkiewicz, A. Genetic and phenotypic variation of *Fusarium proliferatum* isolates from different host species. *J. Appl. Gen.*, 2011b, 52, 487-496.

Stępień, Ł; Waśkiewicz, A. Sequence divergence of the enniatin synthase gene in relation to production of beauvericin and enniatins in *Fusarium* species. *Toxins,* 2013, 5, 537-555.

Streit, E; Schatzmayr, G; Tassis, P; Tzika, E; Marin, D; Taranu, I; Tabuc, C; Nicolau, A; Aprodu, I; Puel, O; Oswald, IP. Current situation of mycotoxin contamination and co-occurrence in animal feed-focus on Europe. *Toxins,* 2012, 4, 788-809.

Sydenham, EW; Thiel, PG; Marasas, WFO; Shephard, GS; Van Schalkwyk, DJ; Koch, KR. Natural occurrence of some *Fusarium* mycotoxins in corn from low and high esophageal cancer prevalence areas of the Transkei, Southern Africa. *J. Agric. Food Chem.*, 1990, 38, 1900-1903.

Tabuc, C; Taranu, I; Calin, L. Survey of mould and mycotoxin contamination of cereals in South-Eastern Romania in 2008–2010. *Arch. Zootech.*, 2011, 14, 25-38.

Tajehmiri, A; Aliabadib, MA; Darsanakic, RK. Occurrence of deoxynivalenol in cereals and cereal based products: a short review. *Sci. J. Biol. Sci.*, 2014, 3, 1-5.

Tam, J; Mankotia, M; Mably, M; Pantazopoulos, P; Neil, RJ; Calway, P; Scott, PM. Survey of breakfast and infant cereals for aflatoxins B_1, B_2, G_1 and G_2. *Food Add. Contam.*, 2006, 23, 693-699.

Tang, Y; Xiao, J; Chen, Y; Yu, Y; Xiao, X; Yu, Y; Wu, H. Secretory expression and characterization of a novel peroxiredoxin for zearalenone detoxification in Saccharomyces cerevisiae. *Microbiol. Res.*, 2013, 168, 6-11.

Tangni, EK; Pussemier, L. Ochratoxin A and citrinin loads in stored wheat grains: Impact of grain dust and possible prediction using ergosterol measurement. *Food Add. Contam.*, 2006, 23, 181-189.

Thieu, NQ; Ogle, B; Pettersson, H. Screening of aflatoxins and zearalenone in feedstuffs and complete feeds for pigs in Southern Vietnam. Trop Anim. *Health Prod.*, 2008, 40, 77-83.

Ting, ASY; Mah, SW; Tee, CS. Identification of volatile metabolites from fungal endophytes with biocontrol potential towards *Fusarium oxysporum* F. sp. *cubense* Race 4. *Am. J. Agr. Biol. Sci.*, 2010, 5, 177-182.

Tittlemier, SA; Roscoe, M; Trelka, R; Patrick, SK; Bamforth, JM; Grafenhan, T; Schlichting, L; Fu, BX. Fate of moniliformin during milling of Canadian durum wheat, processing, and cooking of spaghetti. *Can. J. Plant Sci.*, 2014, 94, 555-563.

Tonshin, AA; Teplova, VV; Andersson, MA; Salkinoja-Salonen, MS. The *Fusarium* mycotoxins enniatins and beauvericin cause mitochondrial dysfunction by affecting the mitochondrial volume regulation, oxidative phosphorylation and ion homeostasis. *Toxicol.*, 2010, 276, 49-57.

Toteja, GS; Mukherjee, A; Diwakar, S; Singh, P; Saxena, BN; Sinh, KK; Sinh, AK; Kumar, N; Nagaraja, KV; Bai, G; Prasad, CA; Vanchinathan, S; Roy, R; Parkar, S. Aflatoxin B_1 contamination in wheat grain samples collected from different geographical regions of India: A multicenter study. *J. Food Prot.*, 2006, 69, 1463-1467.

Tritscher, AM; Page, SW. The risk assessment paradigm and its application for trichothecenes. *Toxicol. Lett.*, 2004, 153, 155-163.

Van der Merwe, KJ; Steyne, PS; Fourie, LF; Scott, DB; Theron, JJ. Ochratoxin A, a toxin metabolite produced by *Aspergillus ochraceus* Wilh. *Nature*, 1965, 205, 1112-1113.

Varga, J; Kocsubé, S; Péteri, Z; Vágvölgyi, C; Tóth, B. Chemical, physical and biological approaches to prevent ochratoxin induced toxicoses in humans and animals. *Toxins*, 2010, 2, 1718-1750.

Varga, J; Kozakiewicz, Z. Ochratoxin A in grapes and grape-derived products. Trends Food Sci. Technol., 2006, 17, 72-81.

Vita, DS; Rosa, P; Giuseppe, A. Effect of gamma irradiation on aflatoxins and ochratoxin A reduction in almond samples. *J. Food Res.*, 2014, 3, 113-118.

Vlachou, S; Zoiopoulos, PE; Drosinos, EH. Assessment of some hygienic parameters of animal feeds in Greece. *Anim. Feed Sci. Technol.*, 2004, 117, 331-337.

Voss, KA; Smith, GW; Haschek, WM. Fumonisins: Toxicokinetics, mechanism of action and toxicity. *Anim. Feed Sci. Technol.*, 2007, 137, 299-325.

Wätjen, W; Debbab, A; Hohlfeld, A; Chovolou, Y; Kampkötter, A; Edrada, RA; Ebel, R; Hakiki, A; Mosaddak, M; Totzke, F; Kubbutat, MHG; Proksch, P. Enniatins A1, B and B1 from an endophytic strain of *Fusarium tricinctum* induce apoptotic cell death in H4IIE hepatoma cells accompanied by inhibition of ERK phosphorylation. *Mol. Nutr. Food Res.*, 2009, 53, 431-440.

Waśkiewicz, A; Beszterda, M; Goliński, P. Occurrence of fumonisins in food – an interdisciplinary approach to the problem. *Food Control*, 2012a, 26, 491-499.

Waśkiewicz, A; Beszterda, M; Kostecki, M; Zielonka, Ł; Goliński, P; Gajęcki, M. Deoxynivalenol in the gastrointestinal tract of immature gilts under *per os* toxin application. *Toxins*, 2014, 6, 973-987.

Waśkiewicz, A; Goliński, P; Karolewski, Z; Irzykowska, L; Bocianowski, J; Kostecki, M; Weber, Z. Formation of fumonisins and other secondary metabolites by *Fusarium oxysporum* and *F. proliferatum*: a comparative study. *Food Add. Contam.*, 2010a, 27, 608-615.

Waśkiewicz, A; Irzykowska, L; Bocianowski, J; Karolewski, Z; Kostecki, M; Weber, Z; Goliński, P. Occurrence of *Fusarium* fungi and mycotoxins in marketable asparagus spears. *Polish J. Environ. Stud.*, 2010b, 49, 367-372.

Waśkiewicz, A; Gromadzka, K; Wiśniewska, H; Goliński, P. Accumulation of zearalenone in genotypes of spring wheat after inoculation with *Fusarium culmorum*. *Cereal Res. Commun.*, 2008, 36SB, 401-404.

Waśkiewicz, A; Irzykowska, L; Bocianowski, J; Karolewski, Z; Weber, Z; Goliński, P. Fusariotoxins in asparagus – their biosynthesis and migration. *Food Add. Contam.*, 2013a, 30, 1332-1338.

Waśkiewicz, A; Stępień, Ł. Mycotoxins biosynthesized by plant-derived *Fusarium* isolates. *Arch. Indust. Hyg. Toxicol.*, 2012, 63, 437-446.

Waśkiewicz, A; Stępień, Ł; Wilman, K; Kachlicki, P. Diversity of pea-associated *F. proliferatum* and *F. verticillioides* populations revealed by *FUM1* sequence analysis and fumonisin biosynthesis. *Toxins*, 2013b, 5, 488-503.

Waśkiewicz, A; Wit, M; Goliński, P; Chełkowski, J; Warzecha, R; Ochodzki, P; Wakuliński, W. Kinetics of fumonisin B_1 formation in maize ears inoculated with *Fusarium verticillioides*. *Food Add. Contam.*, 2012b, 29, 1752-1761.

Wasowicz, K; Gajęcka, M; Calka, J; Jakimiuk, E; Gajecki, M. Influence of chronic administration of zearalenone on the processes of apoptosis in the porcine ovary. *Vet. Med.* - Czech, 2005, 50, 531-536.

Wawrzyniak, J; Waśkiewicz, A. Ochratoxin A and citrinin production by Penicillium verrucosum on cereal solid substrates. *Food Add. Contam.*, 2014, 31, 139-148.

Wei, T; Zhu, W; Pang, M; Liu, Y; Dong, J. Natural occurrence of fumonisins B_1 and B_2 in corn in four provinces of China. *Food Add. Contam. Part B: Surveillance,* 2013, 6, 270-274.

Welugo, SN. Factors influencing deoxynivalenol accumulation in small grain cereals. Toxins, 2012, 4, 1157-1180.

Wenda-Piesik, A. Volatile organic compounds emissions by winter wheat plants (*Triticum aestinum* L.) under *Fusarium* spp. infestation and various abiotic conditions. *Polish J. Environ. Stud.*, 2011, 20, 1335-1342.

White, M. Disease facts: Mycotoxicosis. Livestock, 2012, 17, 42.

Wiśniewska, H; Stępień, Ł; Waśkiewicz, A; Beszterda, M; Góral, T; Belter, J. Toxigenic *Fusarium* species infecting wheat heads in Poland. *Cent. Eur. J. Biol.*, 2014, 9, 163-172.

Wu, L; Wang, X; Xu, R; Li, H. Difference between resistant and susceptible maize to systematic colonization as revealed by DsRed-labeled *Fusarium verticillioides*. *The Crop J.*, 2013, 1, 61-69.

Yamamoto-Ribeiro, MMG; Grespan, R; Kohiyama, CY; Ferreira, FD; Mossini, SAG; Silva, EL; de Abreu Filho, BA; Mikcha, JMG; Machinski Jr, M. Effect of Zingiber officinale essential oil on Fusarium verticillioides and fumonisin production. *Food Chem.*, 2013, 141, 3147-3152.

Yang, J; Li, J; Jiang, Y; Duan, X; Qu, H; Yang, B; Chen, F; Sivakumar, D. Natural occurrence, analysis, and prevention of mycotoxins in fruits and their processed products. *Crit. Rev. Food Sci. Nutr.*, 2014, 54, 64-83.

Yates, IE; Widstrom, NW; Bacon, CW; Glenn, A; Hinton, DM; Sparks, D; Jaworski, AJ; Field performance of maize grown from *Fusarium verticillioides* - inoculated seed. *Mycopathol.,* 2005, 159, 65-73.

Yazar, S; Omurtag, G. Fumonisins, trichothecenes and zearalenone in cereals. *Int. J. Mol. Sci.,* 2008, 9, 2062-2090.

Yazdanpanah, H; Shafaati, A; Foroutan, SM; Zarghi, A; Aboul-Fathi, F; Khoddam, A; Shaki, F; Nazari, F. Occurrence of deoxynivalenol in foods for human consumption from Tehran, Iran. *Iran. J. Pharm. Res.*, 2014, 13(Suppl), 87-92.

Yu, J; Cleveland, TE; Nierman, WC; Bennett, JW. *Aspergillus flavus* genomics: gateway to human and animal health, food safety, and crop resistance to diseases. *Rev. Iberoam. Micol.,* 2005, 22, 194-202.

Yu, Y; Qiu, L; Wu, H; Tang, Y; Yu, Y.; Li, X. Degradation of zearalenone by the extracellular extracts of Acinetobacter so. SM04 liquid culture. *Biodegr.,* 2011, 22, 613-622.

Zain, ME. Impact of mycotoxins on humans and animals. *J. Saudi Chem. Soc.,* 2011, 15, 129-144.

Zatecka, E; Ded, L; Elzeinova, F; Kubatova, A; Dorosh, A; Margaryan, H; Dostalova, P; Korenkova, V; Hoskova, K; Peknicova, J. 2014. Effect of zearalenone on reproductive parameters and expression of selected testicular genes in mice. *Reproduct. Toxicol.*, 2014, 45, 20-30.

Zhang, JX; Jin, Y; Rudd, JC; Bockelman, HE. New Fusarium Head Blight resistant spring wheat germplasm identified in the USDA national small grains collection. *Crop Sci.,* 2008, 48, 224-235.

Zheng, Z; Hanneken, J; Houchins, D; King, RS; Lee, P; Richard, JL. Validation of an ELISA test kit for the detection of ochratoxin A in several food commodities by comparison with HPLC. *Mycopathol.,* 2005, 159, 265-272.

Zhou, HR; Islam, Z; Pestka, JJ. Induction of competing apoptotic and survival signaling pathways in the macrophage by the ribotoxic trichothecene deoxynivalenol. *Toxicol. Sci.*, 2005, 87, 113-122.

Zimmer, I; Usleber, E; Klaffke, H; Weber, R; Majerus, P; Otteneder, H; Gareis, M; Dietrich, R; Märtlbauer, E. Fumonisin intake of the German consumer. *Mycotoxin Res.*, 2008, 24, 40-52.

Zinedine, A; Juan, C; Soriano, JM; Idrissi, L; Manes, J. Limited survey for the occurrence of aflatoxins in cereals and poultry feeds from Rabat, Morocco. *Int. J. Food Microbiol.*, 2007, 115, 124-127.

Zinedine, A; Meca, G; Mañes, J; Font, G. Further data on the occurrence of Fusarium emerging mycotoxins enniatins (A, A1, B, B1), fusaproliferin and beauvericin in raw cereals commercialized in Morocco. *Food Control*, 2011, 22, 1-5.

Zrari, TJO. Detection of aflatoxin from some *Aspergillus* sp. isolated from wheat seeds. *J. Life Sci.*, 2013, 7, 1041-1047.

Zwierzchowski, W; Przybyłowicz, M; Obremski, K; Zielonka, Ł; Skorska-Wyszyńska, E; Gajęcka, M; Polak, M; Jakimiuk, E; Jana, B; Rybarczyk, L; Gajęcki, M. Level of zearalenone in blood serum and lesions in ovarian follicles of sexually immature gilts in the course of zearalenone mycotoxicosis. *Polish J. Vet. Sci.,* 2005, 8, 209-218.

Xing, F; Hua, H; Selvaraj, JN; Yuan, Y; Zhao, Y; Zhou, L; Liu, Y. Degradation of fumonisin B_1 by cinnamon essential oil. *Food Control*, 2014, 38, 37-40.

Yang, J; Li, J; Jiang, Y; Duan, X; Qu, H; Yang, B; Chen, F; Sivakumar, D. Natural occurrence, analysis, and prevention of mycotoxins in fruits and their processed products. *Crit. Rev. Food Sci. Nutr.*, 2014, 54, 64-83.

Young, JC; Zhu, H; Zhou, T. Degradation of trichothecene mycotoxins by aqueous ozone. *Food Chem. Toxicol.*, 2006, 44, 417-424.

In: Mycotoxins
Editor: Cesar Rios

ISBN: 978-1-63483-544-2

Chapter 6

GUT-DERIVED BIOMARKERS FOR 8-KETOTRICHOTHECENE MYCOTOXINS

Yuseok Moon[1,2,*]
[1]Laboratory of Mucosal Exposome and Biomodulation,
Department of Biomedical Sciences,
Pusan National University School of Medicine, Yangsan, Korea,
[2]Immunoregulatory Therapeutics Group in Brain Busan 21 Project, Busan, Korea

ABSTRACT

Deoxynivalenol (DON) and related 8-ketotrichothecene mycotoxins are extensively distributed in the cereal-based food and feed stuffs worldwide. Monitoring the oral exposures is crucial in assessing acute and chronic toxicities of the mixed type of 8-ketotrichothecenes mainly important in grain-based staple. For the efficient mucosal biomonitoring, development of effective gut-derived biomarkers is a very important procedure of the toxicological risk assessment of 8-ketotrichothecenes. In the luminal parts of mucosa, the 8-ketotrichothecenes are mixed with dietary components and gut microbiota, all of which interact together, leading to complex events in the body. For instance, de-epoxidation of DON by gut microbes can lead to more fecal secretion of less toxic DOM-1 (metabolic inactivation). By contrast, most of plant-derived DON-glucosides are also easily catabolized to free DON by gut microbes, which produces more burden to the body (metabolic activation). We can use these 8-ketotrichothecene metabolites as efficient mucosal exposure biomarkers from luminal parts of the exposed host. Moreover, the diversity and presence of the 8-ketotrichothecene-metabolizing microbes in the gut are another important determinant of the susceptibility to the toxicity, provided as useful susceptibility biomarkers. In terms of the mucosal biological responses to 8-ketotrichothecenes in food, mucosal effector biomarkers are important in estimating

[*] Corresponding author: Yuseok Moon, Department of Biomedical Sciences, Pusan National University School of medicine, Yangsan, 626-870 Korea; Telephone: 82-51-510-8094; Fax: 82-55-382-8090; Email: moon@pnu.edu.

the gut-related diseases including gastroenteritis and igA nephritis in human and animal population. Enterocyte chemokines as the early effector mucosal biomarker and mucosal igA as the chronic biomarker have been investigated. In particular, mucosal igA is very sensitive effector biomarker which can be prominently correlated with exposure at dose as low as the regulatory limit in most countries. Moreover, anorexia monitoring using murine models are recently developed to monitor 8-ketotrichothcene-induced feed refusal at the gut-brain axis. These behavioral responses are mechanistically linked to alteration of eating disorder-linked hormones from the intestinal neuroendocrine cells, which can be useful biomarkers in some species. Using these gut-derived biomarkers, reporter readout systems and more realistic gut mimetic models have been developed and would be useful in monitoring the potent risk of 8-ketotrichothecene mycotoxins with minimal use of experimental animals.

Keywords: 8-ketotrichothecenes, biomarkers, gut microbiota, metabolism

1. INTRODUCTION

Trichothecne mycotoxins are one of the extensively contaminated mycotoxin family in grains, their processed food as well as the non-food environment including soil, water, and indoor air. In the agricultural setting, type B 8-ketotrichothecenes including deoxynivalenol (DON), nivalenol (NIV), and their metabolites from *Fusarium* species are the main type of 8-ketotrichothecene subgroup based on the chemical structure. In particular, high occurrence of the type B trichothecenes has been extensively associated with phytopathogen infections in the grain-based crops including *Fusarium* head blight in wheat and *Gibberella* ear rot in maize. Moreover, more than 50% of 8-ketotrichothecene occurrence are found to be associated with multi-toxin production in according to global survey. Three chemical ecotypes of 8-ketotrichothecenes in the agronomic field distribution have been classified by David Miller (Miller 2002, Miller 2008). DON shows co-occurrence with 10-20% of 3-acetyl- or 15-Acetyl-deoxynivalenol. NIV also can exist with 10-20% 4-acetyl NIV. These occasional co-occurrence of multiple 8-ketotrichothecenes pose risk of mixture exposure in human and animals, indicating the need to investigate the mammalian biomarkers to prevalent 8-ketotrichothecene mycotoxins in food and other human and animal environment. Toxicologically, DON is one of most prevalent mycotoxins in the world crop and its absorption is very fast, and it can be detected in serum within 5-15 minutes after ingestion in mice. Therefore, half-lives of distribution and clearance are very short as 20.4 and 11.8 hours, respectively (Pestka et al. 2008). However, since most 8-ketotrichothecene exposure is mediated by daily oral intake of the grain-based staple, the gastrointestinal tract can be the most susceptible part of body under early and extended chronic exposure to the mucosal 8-ketotrichothecenes at relatively high local concentrated doses. The present review thus focused on the impact of trichothecnes on the gastrointestinal physiology and related biomarkers for better biomonitoring and toxicological risk assessment of 8-ketotrichothecenes.

2. Mucosal Metabolic Susceptibility Biomarkers and Toxin Metabolites as Exposure Biomarkers

2.1. Gut Microbial Metabolic Inactivation-Related Biomarkers

Gut microflora hold metabolic enzymes system in response to luminal xenobiotics, which can be used as mucosal exposome for food-borne gut diseases (Fiocchi 2012). In particular, DON is metabolized by gut microbe-secreting metabolic enzymes before intestinal absorption, converted into less toxic de-epoxide metabolite, DOM-1 (de-epoxyde-=oxynivalenol). Susceptibility to microbial metabolism of DON in GI tract can be different depending on species or each microbial composition in host. Generally, swine is the most susceptible species than other domestic animals such as poultry and ruminants in terms of general toxicity of DON. In case of ruminants, high resistance to DON-induced toxicity contributes to excretion of de-epoxide metabolites by microbes in ruminants (Prelusky et al. 1987, Swanson et al. 1986). Mechanistically, the ruminant is more resistant to DON toxicity than other animals because of its metabolic inactivation (de-epoxidation) by the ruminant gut microflora. An example is *Eubacterum* strain BBSH797 (a strain producing de-epoxidase) from ruminant gut which is now commercially available as an additive in feed (Awad et al. 2010, Schatzmayr et al. 2006). *Eubacterium sp*. DSM 11798 from poultry GI tract also has detoxification mechanism for DON and is used as the feed additive poultry (Awad et al. 2006, Awad et al. 2004). These fecal microbial biomarkers are evaluated by measuring the conversion rates form DON to DON-1 using *in vitro* culture of individual stool (mixture of luminal bacteria). Since most of DOM-1 in the lumen is excreted and generally detected in feces (Nagl et al. 2012), DOM-1 is the main DON metabolite as a crucial exposure biomarker in the fecal samples. Although additionally DOM-1-glucoside is also detected (Lattanzio et al. 2011), it is excluded from the biomarker due to no appropriate standard available.

2.2. Gut Microbial Metabolic Activation-Related Biomarkers

Crops also have metabolic inactivation mechanisms in response to DON since the toxin can attack the plant ribosomes as well. Therefore, they convert DON into less toxic DON-3-β-glucoside via glycosylation reaction in the plant cells. For instance, barley and maize have 7-29% and 5-46% DON-3-β-glucoside, respectively (Berthiller et al. 2009). Although this conjugated mycotoxin in itself is not easy to be absorbed into mammalian circulation, human and some animals have gut microflora which hydrolyze the masked mycotoxin, leading to increase of free form of DON, a form of high absorption (Berthiller et al. 2009). Therefore, the susceptibility would be increased by microbial degradation of DON-glucosidase in GI tract. It means if degradation level is high, real body burden with the masked DON will be increased. In according to Berthiller's study (Berthiller et al. 2011), *E. Casseliflavus*, *E. Faecalis* and *E. Gallinarum* showed 1-8% of hydrolysis rate of DON-3-glucoside, but *E. Cloacae, E. Durans, E. Faecium, E. Mundtii, L. Plantarum,* and *B. Adolesentis* showed 62%

of conversion. Mainly, *Lactobacullus, Enterococcus, Enterobacter, Bifidobacterim* have high activity of this enzyme, and its susceptibility can be evaluated through difference in the enzyme activity between species or metagenomics analysis. A metagenomics analysis focused on Japanese showed that *Bacteriodes, Eubacterium and Ruminococcus* have high diversity and dominant position in adults whereas *Bidobacterium* and *Lactobacillus* are dominant in infant (Hattori and Taylor 2009). Therefore, body burden of DON will be bigger at infant than adult because strain which have high activity of DON-3-glucosidase in infant.

3. Gastrointestinal Effector Biomarkers

3.1. Gastrointestinal Immune-Hematological Effector Biomarkers

In terms of mechanism of action, DON causes acute sepsis-like syndrome, which as closely related to global translational inhibition by interfering with the ribosomal action or ribosomal cleavage. In terms of clinical diagnosis, there are anemia, leukopenia and thrombocytopenia in response to DON exposure, but related hematological factors are hard to identify as specific to DON toxicity since these index are also elevated in other sepsis-like disorders. Recently, heptoglobin was known to be specifically increased by DON exposure and has been verified as an effector biomarker (Kim et al. 2008). Heptoglobin is an important protein in acute hemolysis, playing important roles in adsorbing and recycling of the free hemoglobin (Kim et al. 2008, Mikami et al. 2011). Although heptoglobin and serum amyloid A were increased for 1-8 days in mini-pig model (Mikami et al. 2011), extended-time exposure is required for their usefulness as biomarkers of chronic exposure. Moreover, enterocyte cytokines can be considered as diagnostic biomarkers of mucosal immune toxicity. Epithelial interleukin-8 (IL-8), a representative chemokine, is detected after exposure for 3 hour in murine exposure (Mikami et al. 2011). It is known as an early effector biomarkers of immune activation via mrna stabilization in NF-κB-independent way in response to DON exposure (Moon 2014, Moon et al. 2007, Park et al. 2013). Although immunoglobulin M, A and G is decreased during the early time of exposure (Kim et al. 2008), igA production is significantly increased in mucosa and spleen after dietary exposure for 8 weeks and excessive increase of igA in blood forms immune complex, is deposited in glomerulus, and leads to igA nephritis symptoms, which is similar to human Berger's disease as the autoimmune diseases (Dewa et al. 2011, Goyarts et al. 2006, Pestka et al. 1989). Only 1-10 ppm of dietary DON exposure in animal experiments can increase igA as a sub-chronic effector biomarker of hypersensitivity. Since these exposure levels are almost close to the regulation limits (1 ppm) of EU and USA, it is necessary to re-establish safe levels of DON exposure. Considering the doses (1-10 ppm) triggering igA produced and renal deposition in experimental animal models, the safe levels after extrapolation to human exposure would be much lower than the present regulation levels. Mucosal or serum igA and igA-immune complex thus are expected very sensitive effector biomarker of chronic exposure in association with mucosal stress-related systemic diseases including igA nephritis.

3.2. Gastrointestinal Neuroendocrine Effector Biomarkers

As speculated from the nomenclature of synonym vomitoxin for DON, it was originally found to trigger vomiting and anorexia in swine, rapidly leading to weight loss. This toxico-pathologic pattern can be commonly observed in murine model for 8-ketotrichothecenes (Flannery et al. 2011, Wu et al. 2012). Generally, B6C3F1 mouse model can recover feed refusal in 16 hour later, but in case of fusarenon-X and nivalenol-exposed mice, feed refusal lasts 48-96 hours (Wu et al. 2012). Mechanistically, mucosal or systemic inflammatory factors (IL-1β, IL-6., TNF-α, cyclooxygenase-2 (COX-2), microsomal prostaglandin synthase1 (mpges-1)) triggering feed refusal are also increased by 8-ketotrichothecenes (Girardet et al. 2011). However, the anorexia is not generally associated with systemic inflammatory factors. Instead, serotonin (monoamine 5-hydroxy-trytamine, 5-HT) is well-known neuro-endocrine biomarker of DON exposure and related anorexia since antagonists of serotonin receptor administered to pig suppressed feed refusal (Prelusky 1993, Prelusky 1994, Prelusky et al. 1997). Although serotonin and other related metabolites can be effector biomarkers, they are temporarily elevated in response to short-staying DON and thus they are not enough to represent extended or chronic exposure to 8-ketotrichothecenes. Serotonin receptor independently mediates feed refusal by DON, and its upstream modulators including peptide $PYY_{3\text{-}36}$ (Wu et al. 2013) identified as a hormone secreted at digestive endocrine cells regulates the receptor. Taken together, serotonin and $PYY_{3\text{-}36}$ can be used as effector biomarkers of feed refusal to response to type B trichothecenes.

4. Applications and Perspectives

Gut microbe-mediated metabolically active and inactive products of 8-ketotrichothecenes are crucial exposure biomarkers to indicate the luminal crosstalk between toxins and host. Moreover, these luminal microbes as susceptibility markers may determine personalized host responses and give crucial information for reducing the exposure risk for susceptible groups. Indeed, the biological effector biomarkers can be applicable to develop accurate biomonitoring protocols and reporter-based assay along with the chemical-based analysis methods. Recently, human enterocyte-based biomarker reporter system was developed to assess the early mucosal actions of mucosal toxicants as the animal alternate monitoring (Yang et al. 2010), which may evade from the ethical issues. Moreover, development of biomimetic models close to physiological conditions would substantially contribute to reduction of animal use in the toxicological assays for mycotoxicity in the future.

Acknowledgment

This work was carried out with the support of "Cooperative Research Program for Agriculture Science & Technology Development (Project No. PJ01093206)" Rural Development Administration, Republic of Korea and Brain Busan 21 Project, Busan, Korea

REFERENCES

Awad, W. A., Bohm, J., Razzazi-Fazeli, E., Ghareeb, K. & Zentek, J. (2006). Effect of addition of a probiotic microorganism to broiler diets contaminated with deoxynivalenol on performance and histological alterations of intestinal villi of broiler chickens. *Poult Sci*, Jun., *85*, 974-979.

Awad, W. A., Bohm, J., Razzazi-Fazeli, E., Hulan, H. W. & Zentek, J. (2004). Effects of deoxynivalenol on general performance and electrophysiological properties of intestinal mucosa of broiler chickens. *Poult Sci*, Dec., *83*, 1964-1972.

Awad, W. A., Ghareeb, K., Bohm, J. & Zentek, J. (2010). Decontamination and detoxification strategies for the fusarium mycotoxin deoxynivalenol in animal feed and the effectiveness of microbial biodegradation. *Food Addit Contam Part A Chem Anal Control Expo Risk Assess*, Apr., *27*, 510-520.

Berthiller, F., Dall'asta, C., Corradini, R., Marchelli, R., Sulyok, M., Krska, R., Adam, G. & Schuhmacher, R. (2009). Occurrence of deoxynivalenol and its 3-beta-d-glucoside in wheat and maize. *Food Addit Contam Part A Chem Anal Control Expo Risk Assess*, Apr., *26*, 507-511.

Berthiller, F., Krska, R., Domig, K. J., Kneifel, W., Juge, N., Schuhmacher, R. & Adam, G. (2011). Hydrolytic fate of deoxynivalenol-3-glucoside during digestion. *Toxicol Lett*, Oct 30., *206*, 264-267.

Dewa, Y., Kemmochi, S., Kawai, M., Saegusa, Y., Harada, T., Shimamoto, K., Mitsumori, K., Kumagai, S., Sugita-Konishi, Y. & Shibutani, M. (2011). Rapid deposition of glomerular igA in BALB/c mice by nivalenol and its modifying effect on high igA strain (HIGA) mice. *Exp Toxicol Pathol*, Jan., *63*, 17-24.

Fiocchi, C. (2012). Towards a 'cure' for IBD. *Dig Dis*, *30*, 428-433.

Flannery, B. M., Wu, W. & Pestka, J. J. (2011). Characterization of deoxynivalenol-induced anorexia using mouse bioassay. *Food Chem Toxicol*, Aug., *49*, 1863-1869.

Girardet, C., Bonnet, M. S., Jdir, R., Sadoud, M., Thirion, S., Tardivel, C., Roux, J., Lebrun, B., Mounien, L., Trouslard, J., Jean, A., Dallaporta, M. & Troadec, J. D. (2011). Central inflammation and sickness-like behavior induced by the food contaminant deoxynivalenol: a PGE2-independent mechanism. *Toxicol Sci*, Nov., *124*, 179-191.

Goyarts, T., Danicke, S., Tiemann, U. & Rothkotter, H. J. (2006). Effect of the fusarium toxin deoxynivalenol (DON) on igA, igM and igG concentrations and proliferation of porcine blood lymphocytes. *Toxicol In Vitro*, Sep., *20*, 858-867.

Hattori, M. & Taylor, T. D. (2009). The human intestinal microbiome: a new frontier of human biology. *Dna Res*, Feb., *16*, 1-12.

Kim, E. J., Jeong, S. H., Cho, J. H., Ku, H. O., Pyo, H. M., Kang, H. G. & Choi, K. H. (2008). Plasma haptoglobin and immunoglobulins as diagnostic indicators of deoxynivalenol intoxication. *J Vet Sci*, Sep., *9*, 257-266.

Lattanzio, V. M., Solfrizzo, M., De Girolamo, A., Chulze, S. N., Torres, A. M. & Visconti, A. (2011). LC-Ms/Ms characterization of the urinary excretion profile of the mycotoxin deoxynivalenol in human and rat. *J Chromatogr B Analyt Technol Biomed Life Sci*, Apr 1., *879*, 707-715.

Mikami, O., Kubo, M., Murata, H., Muneta, Y., Nakajima, Y., Miyazaki, S., Tanimura, N. & Katsuda, K. (2011). The effects of acute exposure to deoxynivalenol on some inflammatory parameters in miniature pigs. *J Vet Med Sci*, May., *73*, 665-671.

Miller, J. D. (2002). Aspects of the ecology of fusarium toxins in cereals. *Adv Exp Med Biol*, *504*, 19-27.

Miller, J. D. (2008). Mycotoxins in small grains and maize: old problems, new challenges. *Food Addit Contam Part A Chem Anal Control Expo Risk Assess*, Feb., *25*, 219-230.

Moon, Y. (2014). Ribosomal alteration-derived signals for cytokine induction in mucosal and systemic inflammation: noncanonical pathways by ribosomal inactivation. *Mediators Inflamm*, 201*4*, 708193.

Moon, Y., Yang, H. & Lee, S. H. (2007). Modulation of early growth response gene 1 and interleukin-8 expression by ribotoxin deoxynivalenol (vomitoxin) via ERK1/2 in human epithelial intestine 407 cells. *Biochem Biophys Res Commun*, Oct 19., *362*, 256-262.

Nagl, V., Schwartz, H., Krska, R., Moll, W. D., Knasmuller, S., Ritzmann, M., Adam, G. & Berthiller, F. (2012). Metabolism of the masked mycotoxin deoxynivalenol-3-glucoside in rats. *Toxicol Lett*, Sep 18., *213*, 367-373.

Park, S. H., Do, K. H., Choi, H. J., Kim, J., Kim, K. H., Park, J., Oh, C. G. & Moon, Y. (2013). Novel regulatory action of ribosomal inactivation on epithelial nod2-linked proinflammatory signals in two convergent atf3-associated pathways. *J Immunol*, Nov 15., *191*, 5170-5181.

Pestka, J. J., Islam, Z. & Amuzie, C. J. (2008). Immunochemical assessment of deoxynivalenol tissue distribution following oral exposure in the mouse. *Toxicol Lett*, May 5., *178*, 83-87.

Pestka, J. J., Moorman, M. A. & Warner, R. L. (1989). Dysregulation of igA production and igA nephropathy induced by the trichothecene vomitoxin. *Food Chem Toxicol*, Jun., *27*, 361-368.

Prelusky, D. B. (1993). The effect of low-level deoxynivalenol on neurotransmitter levels measured in pig cerebral spinal fluid. *J Environ Sci Health B*, Dec., *28*, 731-761.

Prelusky, D. B. (1994). The effect of deoxynivalenol on serotoninergic neurotransmitter levels in pig blood. *J Environ Sci Health B*, Nov., *29*, 1203-1218.

Prelusky, D. B., Rotter, B. A., Thompson, B. K. & Trenholm, H. L. (1997). Effect of the appetite stimulant cyproheptadine on deoxynivalenol-induced reductions in feed consumption and weight gain in the mouse. *J Environ Sci Health B*, May., *32*, 429-448.

Prelusky, D. B., Veira, D. M., Trenholm, H. L. & Foster, B. C. (1987). Metabolic fate and elimination in milk, urine and bile of deoxynivalenol following administration to lactating sheep. *J Environ Sci Health B*, Apr., *22*, 125-148.

Schatzmayr, G., Zehner, F., Taubel, M., Schatzmayr, D., Klimitsch, A., Loibner, A. P. & Binder, E. M. (2006). Microbiologicals for deactivating mycotoxins. *Mol Nutr Food Res*, May., *50*, 543-551.

Swanson, S. P., Dahlem, A. M., Rood, H. D., Jr., Cote, L. M., Buck, W. B. & Yoshizawa, T. (1986). Gas chromatographic analysis of milk for deoxynivalenol and its metabolite DOM-1. *J Assoc Off Anal Chem*, Jan-Feb., *69*, 41-43.

Wu, W., Bates, M. A., Bursian, S. J., Flannery, B., Zhou, H. R., Link, J. E., Zhang, H. & Pestka, J. J. (2013). Peptide YY3-36 and 5-hydroxytryptamine mediate emesis induction by trichothecene deoxynivalenol (vomitoxin). *Toxicol Sci*, May., *133*, 186-195.

Wu, W., Flannery, B. M., Sugita-Konishi, Y., Watanabe, M., Zhang, H. & Pestka, J. J. (2012). Comparison of murine anorectic responses to the 8-ketotrichothecenes 3-acetyldeoxynivalenol, 15-acetyldeoxynivalenol, fusarenon X and nivalenol. *Food Chem Toxicol*, Jun., *50*, 2056-2061.

Yang, H., Park, S. H., Choi, H. J., Do, K. H., Kim, J., An, T. J., Lee, S. H. & Moon, Y. (2010). Mechanism-based alternative monitoring of endoplasmic reticulum stress by 8-keto-trichothecene mycotoxins using human intestinal epithelial cell line. *Toxicol Lett*, Oct 20., *198*, 317-323.

In: Mycotoxins
Editor: Cesar Rios

ISBN: 978-1-63483-544-2

Chapter 7

APPLICATION OF METAGENOMICS TO THE IDENTIFICATION OF POTENTIAL TOXIGENIC FUNGI

Salvador Mirete*

Department of Molecular Evolution. Centro de Astrobiología (CSIC-INTA), Madrid, Spain

ABSTRACT

Mycotoxins are harmful secondary metabolites mainly produced by members of the genera *Fusarium*, *Aspergillus* and *Penicillium*. The identification of these fungi has been traditionally carried out through procedures that require their growth under standard laboratory conditions. Nevertheless, in the recent years a significant progress has been made in the development of metagenomics, a culture independent approach that aims at studying the overall microbial genomes present in a particular habitat. Metagenomic analysis of uncultured fungal communities can be used for the identification of potential toxigenic fungi and also to assess their distribution, which may have important implications for their management. Thus, molecular methods based on the polymerase chain reaction (PCR) have been used to access the uncultured fraction of fungal communities in different environments as diverse as soils and air masses. Here, different studies using a sequence-based metagenomics approach and identifying potential toxigenic fungi from diverse habitats are reviewed with special emphasis on the methods as well as on the genetic markers employed.

INTRODUCTION

Mycotoxins are secondary metabolites produced by diverse fungal species that are not essential for survival but can lead to toxic responses in animals and humans [1]. Fungal species capable of producing these harmful secondary metabolites can be found in diverse

* Corresponding author: E-mail: miretecs@cab.inta-csic.es.

environments such as soils where they can thrive and produce a number of mycotoxins. The most relevant mycotoxin producing fungi involved in the food chain are members of the genera *Fusarium*, *Aspergillus* and *Penicillium* which can be found in soils as well as in plant- and animal-based foods and commodities [2, 3, 4]. Thus, among the most important fungi found in agricultural soils are mycotoxigenic species of *Fusarium* spp., which can produce important and chemically diverse mycotoxins such as trichothecenes, zearalenone and fumonisins. Other relevant mycotoxigenic fungi with agricultural significance that may be found in diverse environments are members of *Aspergillus* spp. producing aflatoxins, ochratoxins, patulin and citrinin, and also fungi belonging to *Penicillium* spp. producing ochratoxin, patulin and citrinin. The identification and distribution of these toxigenic fungi in soils and other environments have been traditionally derived from their cultivation using standard media, and may not represent the overall fungal diversity present within a particular environment. This is so because an important fraction of fungi can not be cultivated on standard laboratory media or preserved in culture collections [5, 6]. In order to circumvent this limitation, the potential genetic diversity of the different toxigenic fungi present in the environment can be studied through metagenomics since it does not require the cultivation of microorganisms but instead, it is a culture-independent approach based on the analysis of genomic DNA isolated from all the microbial community [7]. Thus, in the recent years, several laboratories worldwide have focused on the study of fungal communities through the sequence-based metagenomics approach by using a collection of sequences retrieved directly from environmental DNA using the polymerase chain reaction (PCR) amplification and diverse sequencing methods. However, it is worth to note that the number of different sequences from potential toxigenic fungi retrieved by using the sequence-based metagenomics approach and the number of different isolates obtained through cultivation can overlap with each other (Figure 1).

In this review I will address the different sequence-based methods using metagenomic DNA that have lead to the detection of potential mycotoxin producing species, mainly members of the genera *Fusarium*, *Aspergillus* and *Penicillium*. In the different studies reported here, members of these fungi have been detected but their toxin production remains unknown as they were not cultured. Nevertheless, their early detection within a particular environment may help to delineate subsequent diagnostic strategies.

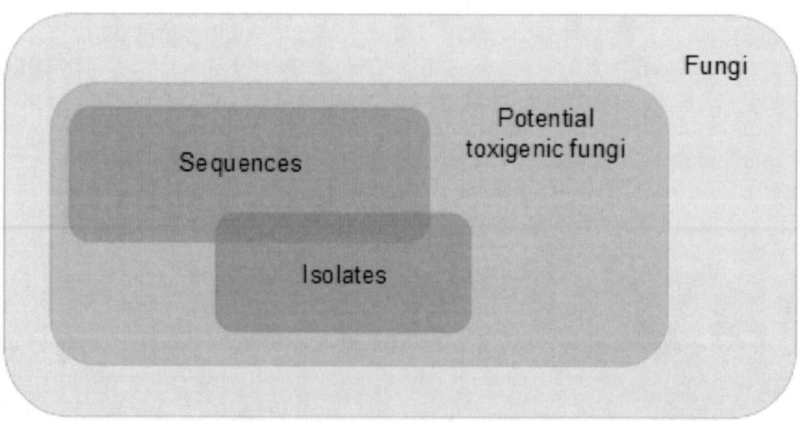

Figure 1. Venn diagram showing the distribution of the different fungal groups that can be found within an environmental sample.

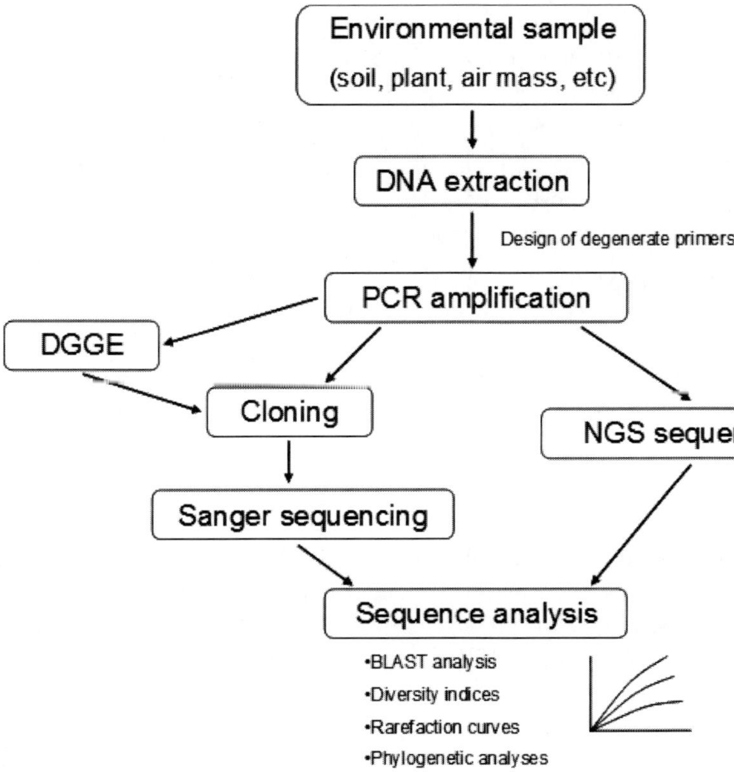

Figure 2. Schematic overview of the main molecular strategies that have allowed the detection of potential toxigenic fungi from the environment. PCR: polymerase chain reaction; DGGE: denaturing gradient gel electrophoresis; NGS: next-generation sequencing.

DNA Extraction and PCR Amplification from Environmental Samples

To analyze the metagenome, it is necessary the isolation of high quality DNA from the environmental samples. The process of DNA extraction can be considered a critical step because nucleic acids of environmental origin are usually coextracted with humic substances and other inhibitors of subsequent molecular procedures employed in metagenomic analysis such as PCR amplification and cloning [8]. Different DNA extraction protocols have been developed so far and diverse commercial kits are available, which can be chosen depending on the use for downstream applications and on the properties of the microorganisms present in the samples [9, 10, 11, 12]. Once the environmental DNA is conveniently isolated, diverse culture-independent methods have been used, most of them relying on the PCR amplification of target sequences as shown in Figure 2.

A successful PCR amplification of environmental samples depends greatly on the choice of the nucleotide sequence to be amplified as well as on the proper design of the PCR primers, and as a result, enough phylogenetic resolution and specificity should be obtained. In fungal species, different primers have been used based on the nucleotide sequence of diverse genes such as the 18S and 28S rRNA genes, the elongation factor 1-α gene (EF-1α) and the β-tubulin gene sequences. A problem that stems from the use of primers based on gene sequences is that they may be highly conserved and thus may provide low phylogenetic signal. In contrast, noncoding spacer regions are more variable and include the internal transcribed sequence (ITS), located between the 18S and 28S genes, and the nuclear ribosomal DNA intergenic spacer (IGS), which separates rDNA repeat units. From these two sequences, the IGS represents the most rapidly evolving spacer region present within the rDNA array, thus providing a better phylogenetic resolution [13]. For example, within *Fusarium* species, genetic markers based on the IGS and the EF-1α gene have been used to detect variability among closely related species and at the intraspecific level since they seem to have the most phylogenetic signal [14, 15, 16]. Here, it is interesting to note that the same mycotoxin can be produced by different species within a fungal genus, and that there may be both producing and non producing strains within the same fungal species. For example, different *Fusarium* species such as members of *F. verticillioides* and *F. proliferatum* are capable of producing fumonisins, but non producing strains have also been found within *F. verticillioides* [16]. The use of gene sequences involved in the biosynthesis of mycotoxins can also be considered for the environmental detection of potential toxigenic fungi. Thus, previous studies carried out using PCR amplification, southern blot and RT-PCR revealed that fumonisin biosynthetic genes (*FUM*) are absent in those members within the *Gibberella fujikuroi* species complex incapable of producing fumonisins [17, 18]. A similar result was obtained for the *TRI5* gene involved in the trichothecene biosynthesis [19].

PCR-DGGE

The application of molecular techniques such as the PCR amplification of environmental DNA in combination with denaturing gradient gel electrophoresis (DGGE) has been widely used in the field of microbial ecology. In this approach, the amplicons obtained during PCR amplification can be distinguished by their nucleotide sequences on polyacrylamide gels by using denaturing gradients during electrophoresis [20]. Specific bands of interest obtained during PCR-DGGE can be excised, purified and cloned into appropriate vectors to sequence the insert. For the application of this methodology to study fungal communities, PCR primers based on the 18S rRNA gene sequence have been widely used [21]. For example, these primers have been of great value for studying fungal groups difficult to cultivate from wood [22] and revealed the presence of *Aspergillus oryzae* at the initial stage of a vinegar fermentation process [27]. This approach was applied to DNA extracted from asparagus and orchid samples and, as a result, showed the presence of different *Fusarium* species in planta when using primers based on the EF1-α gene [23] and the ITS region [24], respectively. In addition, the diversity of fungi and bacteria that may play a role as microbial antagonists of the colonization of toxigenic *Fusarium* spp. in maize stalks was assessed by PCR-DGGE using the bacterial 16S rRNA gene and the ITS sequence, respectively [25]. This technique was also used to study the dynamics of potential ochratoxin A (OTA) producing fungi in coffee samples by using the 28S rDNA as genetic marker [26].

CLONE LIBRARIES

PCR amplification of target sequences carried out directly on fungal DNA extracted from environmental samples and with specific primers can be subsequently cloned into bacterial strains using appropriate vectors such as pUC19, TOPO TA and pGEM. The nucleotide sequences of the PCR amplification products are individually cloned and are flanked by universal primers that can be used for Sanger sequencing. The resulting DNA sequences are then analysed in silico for the identification of closest matches in public databases such as the National Center for Biotechnology Information (NCBI) using BLAST [28], the FUSARIUM-ID sequence database [29] or the ribosomal RNA sequence databases, Silva [30] and the Ribosomal Data Poject [31]. A rarefaction curve and richness estimators such as Chao and the abundance-based coverage estimator (ACE) [32] can be accomplished in order to assess the number of operational taxonomic units (OTUs) as well as the sampling effort necessary to cover all the fungal diversity within a particular community. Diverse parameters such as the Simpson's index (D), Shannon index (H'), Shannon evenness (E) values can be calculated to assess the fungal diversity found in a particular sample and can serve to compare them with data obtained from cultivation or from other environmental samples.

The use of clone libraries served to illustrate the high abundance and diversity of novel fungal sequences retrieved from forest soils where up to 412 unique OTUs were recovered from a total of 863 fungal ITS sequences sampled [33]. The construction of clone libraries based on the PCR amplification of the EF1-α gene of *Fusarium* species present in different soil samples revealed a total of 663 different clones belonging to eight species including *F. solani*, *F. oxysporum* and *F. incarnatum-equiseti* species complexes, a higher diversity than that obtained by cultivation, with a total of 44 different isolates [34]. This approach was also applied to assess the diversity of fungi in air particulate matter through the amplification of the ITS sequence, and as a result members of *Aspergillus* spp. as well as *Penicillium* spp. were found [35]. Other study carried out on arable soils and using the ITS- and partial 28S rDNA-region as genetic markers revealed the presence of diverse *Fusarium* species including *F. merismoides*, *F. incarnatum* and *F. solani* [36].

NEXT-GENERATION SEQUENCING (NGS) METHODS

Next-generation sequencing (NGS) also referred to as massively parallel sequencing methods have been employed for the identification of fungi in diverse environmental samples. One of the main advantages of using these sequencing methods relies on that it does not require the construction of clone libraries after PCR amplification. Also, the number of DNA sequences obtained using NGS methods is higher than that obtained through Sanger sequencing which can lead to a better identification of the overall fungal diversity present, but can also require an exhaustive bioinformatic analysis. The same DNA sequence analyses conducted for sequences retrieved from clone libraries can also be carried out with those sequences obtained using massively parallel sequencing. In addition, there is dedicated software that was conceived to deal with the wealth of data retrieved using high-throughput sequencing such as the Qiime [37] and the Mothur [38] software pipelines.

There are diverse commercially available technologies of NGS including Roche/454, Illumina/Solexa, Life/APG Solid 3 and the recent technology developed by Pacific Biosciences (PacBio). A limitation of using these sequencing methods is the short length of the reads which can reach from 50 bp to 330 bp in the case of Solid 3 and 454 plattforms, respectively, although PacBio-based methods can provide longer reads [39]. From these methods, 454 pyrosequencing have been widely used for the identification of fungal members found in diverse environments such as in six different forest soils where approximately 600-1000 OTUs were found in each soil sample [40]. The genetic marker of choice in most environmental studies was the ITS sequence because it can provide enough phylogenetic signal, and thus it has been proposed to be used as the primary fungal barcode marker [41]. Nevertheless, some caution is needed when using this sequence for phylogenetic analyses since intragenomic variation has been found when using ITS sequences as genetic markers [42]. Other fungal sequences used, in a lesser extent, in 454 pyrosequencing of fungal communities include the 18S rRNA gene sequence, which has been used to assess the influence of pH on fungal communities composition in an arable soil [43].

The detection of potential toxigenic fungi through 454 pyrosequencing has been carried out in diverse soils and environments. For example, the application of this methodology to the study of fungal communities in a native tallgrass prairie revealed the presence of members belonging to the three main potential toxigenic genera *Fusarium*, *Aspergillus* and *Penicillium* [44]. Also this method allowed the detection of diverse fungal species present in pea field soils where a total of 264 sequences affiliated with members of *F. merismoides* [45]. Members of *Fusarium* spp. were also detected by pyrosequencing in a potato monoculture soil and its presence was linked to yield decline and disease incidence [46]. Different works have reported the use of pyrosequencing to detect potential human pathogenic species such as *A. fumigatus* and *F. oxysporum* in compost [47], and *A. fumigatus* and *P. brevicompactum* in the atmosphere [48], which, in addition, can also represent potential toxigenic species.

CONCLUSION

The use of sequence-based metagenomics allows the detection of fungal isolates present in natural ecosystems and other areas of interest impacted by human activity. The identification of uncultured fungi using suitable genetic markers and the environment where they are retrieved may be considered essential factors to control the presence of mycotoxins in diverse environments, including arable soils destined to the production of food for human and animal consumption. Thus, the use of specific PCR primers designed from genes related to toxin biosynthesis will undoubtedly provide further diagnostic resolution. Also, it is remarkable that the combination of the sequence-based methods reviewed here and culture-based methods carried out on standard laboratory conditions should provide a clearer picture of the potential toxigenic fungi present in the environment.

Acknowledgements

I would like to thank María Teresa González Jaén and José Eduardo González Pastor for critical reading of this manuscript. This work was funded by the Spanish Ministry of Science and Innovation (CGL2012-39627-C03-02).

References

[1] Bennett, JW; Bentley, R. What's in a name? Microbial secondary metabolism. *Adv Appl Microbiol*, 1989, 34, 1-28.
[2] Elmholt, S. Mycotoxins in the soil environment. In: Karlovsky, P. (ed). *Secondary Metabolites in Soil Ecology., Springer*, 2008,167-203.
[3] Sweeney, MJ; Dobson, ADW. Mycotoxin production by *Aspergillus*, *Fusarium* and *Penicillium* species. *Int J Food Microbiol*, 1998, 43(3), 141-158.
[4] Logrieco, A; Bottalico, A; Mulé, G; Moretti, A; Perrone, G. Epidemiology of toxigenic fungi and their associated mycotoxins for some Mediterranean crops. *Eur J Plant Pathol*, 2003, 109(7), 645-667.
[5] Bridge, P; Spooner, B. *Soil fungi: diversity and detection. Plant Soil*, 2001, 232(1-2), 147-154.
[6] Hawksworth, DL. The fungal dimension of biodiversity: magnitude, significance, and conservation. *Mycol Res*, 1991, 95(6), 641-655.
[7] Handelsman, J. Metagenomics: application of genomics to uncultured microorganisms. *Microbiol Mol Biol Rev*, 2004, 68(4), 669-685.
[8] Daniel, R. The metagenomics of soil. *Nat Rev Microbiol*, 2005, 3(6), 470-478.
[9] Simon, C; Daniel, R. Metagenomic analyses: past and future trends. *Appl Environ Microb*, 2011, 77(4), 1153-1161.
[10] Ekkers, DM; Cretoiu, MS; Kielak, AM; Elsas, JD. The great screen anomaly--a new frontier in product discovery through functional metagenomics. *Appl Microbiol Biotechnol*, 2012, 93(3), 1005-1020.
[11] González-Pastor, JE; Mirete, S. Novel metal resistance genes from microorganisms: a functional metagenomic approach. In: Streit, WR; Daniel, R. (eds). Metagenomics: Methods and protocols. 1st Edition ed. *Springer*, 2010.
[12] López-Pérez, M; Mirete, S. Discovery of novel antibiotic resistance genes through metagenomics. *Recent Adv DNA Gene Seq*, 2014, 8(1), 15-19.
[13] Hillis, DM; Dixon, MT. Ribosomal DNA: molecular evolution and phylogenetic inference. *Q Rev Biol*, 1991, 66(4), 411-453.
[14] O'Donnell, K; Gueidan, C; Sink, S; Johnston, PR; Crous, PW; Glenn, A; et al. A two-locus DNA sequence database for typing plant and human pathogens within the *Fusarium oxysporum* species complex. *Fungal Genet Biol*, 2009, 46(12), 936-948.
[15] Mirete, S; Patiño, B; Jurado, M; Vázquez, C; González-Jaén, MT. Structural variation and dynamics of the nuclear ribosomal intergenic spacer region in key members of the *Gibberella fujikuroi* species complex. *Genome*, 2013, 56(4), 205-213.

[16] Mirete, S; Vázquez, C; Mulé, G; Jurado, M; González-Jaén, MT. Differentiation of *Fusarium verticillioides* from banana fruits by IGS and EF-1α sequence analyses. *Eur J Plant Pathol*, 2004, 110(5-6), 515-523.

[17] González-Jaén, MT; Mirete, S; Patiño, B; López-Errasquin, E; Vázquez, C. Genetic markers for the analysis of variability and for production of specific diagnostic sequences in fumonisin-producing strains of *Fusarium verticillioides*. *Eur J Plant Pathol*, 2004, 110(5-6), 525-532.

[18] Proctor, RH; Plattner, RD; Brown, DW; Seo, J-A; Lee, Y-W. Discontinuous distribution of fumonisin biosynthetic genes in the *Gibberella fujikuroi* species complex. *Mycol Res*, 2004, 108(07), 815-822.

[19] Desjardins, AE; Hohn, TM; McCormick, SP. Trichothecene biosynthesis in *Fusarium* species: chemistry, genetics, and significance. *Microbiol Rev*, 1993, 57(3), 595-604.

[20] Muyzer, G; Smalla, K. Application of denaturing gradient gel electrophoresis (DGGE) and temperature gradient gel electrophoresis (TGGE) in microbial ecology. *Anton Leeuw Int J G*, 1998, 73(1), 127-141.

[21] Hoshino, YT; Morimoto, S. Comparison of 18S rDNA primers for estimating fungal diversity in agricultural soils using polymerase chain reaction-denaturing gradient gel electrophoresis. *Soil Sci Plant Nutr*, 2008, 54(5), 701-710.

[22] Vainio, EJ; Hantula, J. Direct analysis of wood-inhabiting fungi using denaturing gradient gel electrophoresis of amplified ribosomal DNA. *Mycol Res*, 2000, 104(08), 927-936.

[23] Yergeau, E; Filion, M; Vujanovic, V; St-Arnaud, M. A PCR-denaturing gradient gel electrophoresis approach to assess *Fusarium* diversity in asparagus. *J Microbiol Meth*, 2005, 60(2), 143-154.

[24] Tao, G; Liu, ZY; Hyde, KD; Lui, XZ; Yu, ZN. Whole rDNA analysis reveals novel and endophytic fungi in *Bletilla ochracea* (*Orchidaceae*). *Fungal Divers*, 2008, 33(1), 101-122.

[25] Köhl, J; Lombaers, C; Moretti, A; Bandyopadhyay, R; Somma, S; Kastelein, P. Analysis of microbial taxonomical groups present in maize stalks suppressive to colonization by toxigenic *Fusarium* spp.: A strategy for the identification of potential antagonists. *Biol Control*, 2015, 83(0), 20-28.

[26] Durand, N; El Sheikha, AF; Suarez-Quiros, M-L; Oscar, G-R; Nganou, ND; Fontana-Tachon, A; et al. Application of PCR-DGGE to the study of dynamics and biodiversity of yeasts and potentially OTA producing fungi during coffee processing. *Food Control*, 2013, 34(2), 466-471.

[27] Haruta, S; Ueno, S; Egawa, I; Hashiguchi, K; Fujii, A; Nagano, M; et al. Succession of bacterial and fungal communities during a traditional pot fermentation of rice vinegar assessed by PCR-mediated denaturing gradient gel electrophoresis. *Int J Food Microbiol*, 2006, 109(1), 79-87.

[28] Altschul, SF; Gish, W; Miller, W; Myers, EW; Lipman, DJ. Basic local alignment search tool. *J Mol Biol*, 1990, 215, 403-410.

[29] Geiser, DM; del Mar Jiménez-Gasco, M; Kang, S; Makalowska, I; Veeraraghavan, N; Ward, TJ; et al. *In Molecular Diversity and PCR-detection of Toxigenic Fusarium Species and Ochratoxigenic Fungi; Springer*, 2004, 473-479.

[30] Quast, C; Pruesse, E; Yilmaz, P; Gerken, J; Schweer, T; Yarza, P; et al. The SILVA ribosomal RNA gene database project: improved data processing and web-based tools. *Nucleic Acids Res*, 2013, 41(D1), D590-D596.

[31] Cole, JR; Wang, Q; Fish, JA; Chai, B; McGarrell, DM; Sun, Y; et al. Ribosomal Database Project: data and tools for high throughput rRNA analysis. *Nucleic Acids Res*, 2014, 42(Database issue), D633-642.

[32] Chao, A; Chazdon, RL; Colwell, RK; Shen, TJ. A new statistical approach for assessing similarity of species composition with incidence and abundance data. *Ecol Lett*, 2005, 8(2), 148-159.

[33] O'Brien, HE; Parrent, JL; Jackson, JA; Moncalvo, J-M; Vilgalys, R. Fungal community analysis by large-scale sequencing of environmental samples. *Appl Environ Microb*, 2005, 71(9), 5544-5550.

[34] Edel-Hermann, V; Gautheron, N; Mounier, A; Steinberg, C. *Fusarium* diversity in soil using a specific molecular approach and a cultural approach. *J Microbiol Meth*, 2015, 111, 64-71.

[35] Fröhlich-Nowoisky, J; Pickersgill, DA; Després, VR; Pöschl, U. High diversity of fungi in air particulate matter. *P Natl Acad Sci USA*, 2009, 106(31), 12814-12819.

[36] Klaubauf, S; Inselsbacher, E; Zechmeister-Boltenstern, S; Wanek, W; Gottsberger, R; Strauss, J; et al. Molecular diversity of fungal communities in agricultural soils from Lower Austria. *Fungal Divers*, 2010, 44(1), 65-75.

[37] Caporaso, JG; Kuczynski, J; Stombaugh, J; Bittinger, K; Bushman, FD; Costello, EK; et al. QIIME allows analysis of high-throughput community sequencing data. *Nat Methods*, 2010, 7(5), 335-336.

[38] Schloss, PD; Westcott, SL; Ryabin, T; Hall, JR; Hartmann, M; Hollister, EB; et al. Introducing mothur: open-source, platform-independent, community-supported software for describing and comparing microbial communities. *Appl Environ Microb*, 2009, 75(23), 7537-7541.

[39] Metzker, ML. Sequencing technologies - the next generation. *Nat Rev Genet*, 2010, 11(1), 31-46.

[40] Buee, M; Reich, M; Murat, C; Morin, E; Nilsson, RH; Uroz, S; et al. 454 Pyrosequencing analyses of forest soils reveal an unexpectedly high fungal diversity. *New Phytol*, 2009, 184(2), 449-456.

[41] Schoch, CL; Seifert, KA; Huhndorf, S; Robert, V; Spouge, JL; Levesque, CA; et al. Nuclear ribosomal internal transcribed spacer (ITS) region as a universal DNA barcode marker for *Fungi*. *P Natl Acad Sci USA*, 2012, 109(16), 6241-6246.

[42] O'Donnell, K; Cigelnik, E. Two divergent intragenomic rDNA ITS2 types within a monophyletic lineage of the fungus *Fusarium* are nonorthologous. *Mol Phylogenet Evol*, 1997, 7(1), 103-116.

[43] Rousk, J; Bååth, E; Brookes, PC; Lauber, CL; Lozupone, C; Caporaso, JG; et al. Soil bacterial and fungal communities across a pH gradient in an arable soil. *ISME J*, 2010, 4(10), 1340-1351.

[44] Jumpponen, A; Jones, KL; Blair, J. Vertical distribution of fungal communities in tallgrass prairie soil. *Mycologia*, 2010.

[45] Xu, L; Ravnskov, S; Larsen, J; Nilsson, RH; Nicolaisen, M. Soil fungal community structure along a soil health gradient in pea fields examined using deep amplicon sequencing. *Soil Biol Biochem*, 2012, 46, 26-32.

[46] Liu, X; Zhang, J; Gu, T; Zhang, W; Shen, Q; Yin, S; et al. Microbial community diversities and taxa abundances in soils along a seven-year gradient of potato monoculture using high throughput pyrosequencing approach. *PLoS One*, 2014, 9(1), e86610.

[47] De Gannes, V; Eudoxie, G; Hickey, WJ. Insights into fungal communities in composts revealed by 454-pyrosequencing: implications for human health and safety. *Front Microbiol*, 2013, 4.

[48] Yamamoto, N; Bibby, K; Qian, J; Hospodsky, D; Rismani-Yazdi, H; Nazaroff, WW; et al. Particle-size distributions and seasonal diversity of allergenic and pathogenic fungi in outdoor air. *ISME J*, 2012, 6(10), 1801-1811.

In: Mycotoxins
Editor: Cesar Rios

ISBN: 978-1-63483-544-2

Chapter 8

OCCUPATIONAL EXPOSURE TO AFLATOXIN B1: A NEW OR A FORGOTTEN RISK FACTOR?

***Susana Viegas*[1,2], *Raquel Sabino*[1,3] *and Carla Viegas*[1,4]**

[1]Environment & Health RG, Lisbon School of Health Technology, Polytechnic Institute of Lisbon, Portugal

[2]Centro de Investigação e Estudos em Saúde Pública, Escola Nacional de Saúde Pública, ENSP, Universidade Nova de Lisboa, Portugal

[3]Mycology Laboratory, Instituto Nacional de Saúde Dr. Ricardo Jorge, Portugal

[4]Environmental Health Institute, Faculty of Medicine from Lisbon University, Portugal

ABSTRACT

Aflatoxin B1 (AFB1) is normally predominant in food cultures and products and considered by different International Agencies as a genotoxic and potent hepatocarcinogen agent. However, despite the fact that the fungi producing this compound being detected in some work environments, AFB1 is rarely studied in workplaces and considered a possible risk factor. The aim of this chapter is to claim attention for the need of considering AFB1 presence in some specific occupational settings. Additionally, this chapter has the intention to explain the importance of performing exposure assessment, allowing the identification of the most suitable preventive and protective measures to avoid exposure. The data analyzed and reported in this review suggest that AFB1 is still not recognized as a risk factor, which explains the reduced number of published papers on this topic. Moreover, this chapter describes some of the most important aspects to consider in the risk assessment process, such as: new climatic scenarios that promote fungi growth and mycotoxins' production, the probable simultaneous presence of other mycotoxins, and also the challenge of choosing the most suitable method to perform exposure assessment.

Keywords: AflatoxinB1, occupational exposure, exposure assessment, risk assessment

Introduction

Aflatoxins are secondary metabolites produced under certain environmental conditions (temperature, water activity, substrate composition and pH or modified atmospheres) by the fungal species *Aspergillus flavus* and *Aspergillus parasiticus* (Bhatnagar et al., 2006). *Aspergillus flavus* and *A. parasiticus* are closely related fungi which can contaminate primary agricultural products in the fields, during the harvest process, in the storage period, and during their processing (Diener et al., 1987). There are some characteristics that helps distinguish between these two species: *A. parasiticus* appears to be adapted to a soil environment, being prominent in peanuts, whereas *A. flavus* seems to be adapted to the aerial and foliar environment, being dominant in maize, cottonseeds, and tree nuts (Diener et al., 1987). However, researchers have frequently failed to distinguish between these two species in their studies (Kurtzman et al., 1987) and have addressed this problem through DNA comparisons.

In both species of *Aspergillus* section *Flavi* (*flavus* and *parasiticus*), there are strains that are nonaflatoxigenic (Smith and Moss, 1985) and the detection of mycotoxins is a relevant issue to consider regarding risk assessment.

Eighteen different types of aflatoxins have been identified so far, of which the major members are aflatoxins B1, B2, G1 and G2. It is now generally accepted that *A. flavus* usually produces only aflatoxin B1 (AFB1) and aflatoxin B2 (AFB2), but it is also capable of synthesizing cyclopiazonic acid (CPA), a mycotoxin present in the batch of contaminated groundnuts that killed approximately 100,000 turkeys in 1960 (Turkey 'X' disease) (Smith, 1997). On the other hand, *A. parasiticus* often produces the four primary aflatoxins: AFB1, B2, G1 and G2 (D'Mello and Macdonald, 1997; Diener et al., 1987). However, recent studies demonstrated that certain strains of *A. flavus* can also be able to produce AFG1 and AFG2 (Cotty and Cardwell, 1999).

There are many theories about why these fungi produce aflatoxins, but none has been clearly demonstrated. Thus, several hypotheses have been proposed: aflatoxins may be a defense response produced by fungi to stress, a way to protect fungi from UV damage, products of fungi primary metabolism necessary to enhance the fungal fitness or to provide protection from predators for reproductive structures such as conidia and sclerotia (Cary and Ehrlich, 2006; Magan and Aldred, 2007). However, it is well known that aflatoxins are synthesized by a polyketide metabolic pathway and that the genes from both *A. flavus* and *A. parasiticus* linked to this aflatoxin biosynthetic pathway are clustered (Bhatnagar et al., 2004; Chang et al., 2004).

Aflatoxin B1 (AFB1) is normally predominant in food cultures and products. It is bioactivated by cytochrome P450, a group of enzymes present in the liver in high amounts, related with the bioactivation and metabolism of several xenobiotics and also of endogenous compounds (Nebert and Dalton, 2006).

The CYP450 enzymes activate AFB1 into a unstable metabolite (aflatoxin-8,9-epoxide) that is able to react with cellular macromolecules such as DNA (causing genotoxicity) and proteins (causing cytotoxicity), leading to the formation of covalent adducts (Autrup et al., 1991; Doi et al., 2002; Brera et al., 2002; Dash et al., 2007; Diaz et al., 2010).

Aflatoxin B1 is the most potent hepatocarcinogen known in mammals and it isclassified by the International Agency of Research on Cancer as Group 1 carcinogen. It also has toxic, mutagenic and teratogenic properties (IARC, 1993).

In a study of 71 Dutch oil-press workers exposed to aflatoxins, Hayes et al. (1984) showed that the mortality from liver cancer was higher in those workers than the expected norm. Similarly, in a registry-based analysis of occupational risks for primary liver cancer in Sweden, a significant increase incidence of this disease was observed in both male and female workers in grain mills (McLaughlin et al., 1987).

Several epidemiological and laboratory studies have shown that the human respiratory system is also a target for AFB1, as a carcinogenicity agent (Hayes et al., 1984; Donnelly et al., 1996; Kelly et al., 1997; Massey et al., 2000).

More recently, Yang and colleagues (2012, 2013) demonstrated that AFB1 is also activated at the lung by CYP2A13, to subsequently form AFB1–DNA adducts, inducing cytotoxicity and apoptosis via the mitochondrial signaling pathway.

Since aflatoxins are considered as genotoxic and carcinogenic, their presence in food should be reduced to as low as reasonably achievable (ALARA), with no tolerable daily intake (TDI) indicated.

2. Occupational Exposure to Aflatoxin B1

Although dietary exposure to AFB1 has been extensively documented, there are still few studies dedicated to the problem of occupational exposure to AFB1

The requisites for mycotoxin exposure are the presence of mould and, as already mentioned previously, the presence of favorable conditions for mycotoxin production. Mycotoxin production commonly requires higher available water content in the environment/substrate than the one needed for mold growth (Corey, 1987).

At workplaces, the natural background concentration of mould in the air is often exceeded, especially when organic products are handled. As good examples of this situation are the waste handling or building restoration after water damage, where the presence and exposure to mould is evident (Mayer et al., 2008; Malta-Vacas et al., 2012; Tawniczek-Wałczyk et al., 2012; Johanning et al., 2014; Viegas et al., 2014).

Occupational exposure to aflatoxins can also occur during the processing and handling of contaminated grains, particularly from animal feed. Airborne concentrations at the workplace are typically in the ng/m3-range, but higher concentrations (up to $\mu g/m^3$) have been reported (IARC, 2002).

Some previous publications and reports suggest the potentially high risks of occupational exposure to AFB1 through inhalation, being the most probable exposure route present in the occupational settings (Dvorackova et al., 1976; Baxter et al., 1981; Hayes et al., 1984; Popendorf et al., 1985; Dvorackova et al., 1986).

However, under the majority of conditions, the level of exposure to mycotoxins by inhalation is lower than the exposition by oral intake. Nevertheless, workers can be occasionally exposed to high airborne levels during certain operations and this can cause different and more sever health effects (Halstensen et al., 2004; Mayer et al., 2008; Degen, 2008).

Workers exposed to aflatoxins by inhalation, particularly in the form of airborne dust, are also more prone to ingest (by absorbance through the oral mucosa) and inhale the AFB1 released during tasks involving storing, loading, handling or milling contaminated materials such as grain, waste, feed and others (Sorenson et al., 1981; Jargot and Meline, 2013). Additionally, the possibility of exposure by dermal absorption is also particularly relevant in occupational settings where the use of short clothes is allowed and where the particulate matter settles on skin frequently (Degen, 2008; Mayer et al., 2008).

Therefore, and considering all these information, occupational exposure to aflatoxins can be a real problem during a high number of tasks, such as: bulk handling of agricultural foodstuffs (nuts, grain, maize, coffee and others already reported such as onions and spices), animal and feed production, brewing/malting, waste management, composting, food production and horticulture (IARC, 2002).

Moreover, in all these settings there are specific tasks more associated to dust aerossolization, which can really contribute to promote the exposure to AFB1 (Brera et al., 2002).

3. Occupational Settings with Reported Exposure to Aflatoxin B1

To answer to possible questions regarding the possibility of exposure to AFB1 in occupational settings, we have made an exhaustive search for papers available in scientific databases, written in English and related with occupational exposure to AFB1.

The articles considered were all the ones that, besides following the previous conditions, also showed quantitative data regarding exposure assessment (Table 1). Sixteen articles were considered and analyzed, fourteen of which were published based on agricultural activities (production or handling of products) and animal production (swine and poultry) and two other based on different occupational settings, such as the waste management setting and the textile industry (Table 1).

4. Assessment of Occupational Exposure to Aflatoxin B1

Due to the ease and low ability required for detecting fungi, they are often used as an indirect indicator of mycotoxin presence in both agricultural and occupational settings. However, it is essential to identify the collected isolates at the strain level because mycotoxins are only produced by specific strains of each species (Thrane et al., 2004). Conventional methods for fungal identification, such as culture and microscopy, do not discriminate among closely related species and are limited in regard to obtain cultures of all species present in a given sample. Molecular techniques such as polymerase chain reaction (PCR) and DNA hybridization have provided significant advances in rapid identification and quantification of specific fungal DNA, irrespective of fungal viability (Halstensen, 2008).

Table 1. Overview of studies that report occupational exposure to AFB1

Authors	Occupational setting/Country	Title	Year of publication	Samples	Results	Observations
A. Hussein; M. M. Taha; S. Beshir; E. M. Shahy; W. Shaheen; M. Elhamshary	Wheat handlers, Egypt	Carcinogenic effects of aflatoxin B1 among wheat handlers	2015	Serum	AFB1/Alb results were significantly higher among workers employed as bakers compared to mill workers and controls	
S. Viegas; L. Veiga; P. Figueiredo; A. Almeida; E. Carolino; C. Viegas	Waste management, Portugal	Assessment of Workers' Exposure to Aflatoxin B1 in a Portuguese Waste Industry	2015	Serum	All the workers showed detectable levels of AFB1 with values ranging from 2.5 to 25.9 ng ml−1 and with a median value of 9.9 ± 5.4 ng ml−1. In the control group, the AFB1 values were all below LOD.	A control group was used
A. Malik; S. Ali; M. Shahid; R. Bhargava	Food-grain industry, India	Occupational exposure to Aspergillus and aflatoxins among food-grain workers in India	2014	Serum	Aflatoxins were detected in 32·6% of the food-grain workers and 9·1% of non food grain workers ($P < 0{\cdot}01$).	A control group was used
H. Lai; X. Mo;Y. Yang; K. He; J. Xiao; Chao Liu & Jiansi Chen & Yuan Lin	Sugar production and papermaking using sugarcane, China	Association between aflatoxin B1 occupational airway exposure and risk of hepatocellular carcinoma: a case-control study	2014	Dust samples from sugarcane bagasse warehouse and the presser and paper production workshops. Serum (AFB1 albumin adducts)	Dust Samples (7.2 ± 1.30, 8.0 ± 1.23 and 8.6 ± 1.82 µg kg−1, respectively). Serum samples AFB1 albumin adducts were detected in 102 (56.35%) workshop employees, with values ranging from 8 to 212 pg/mg albumin (mean value 38.51 ± 44.80 pg/mg albumin). In contrast, only 12 (5.9%) controls had detectable levels of AFB1 albumin adducts.	A control group was used

Table 1. (Continued)

Authors	Occupational setting/Country	Title	Year of publication	Samples	Results	Observations
A. Saad-Hussein;AA. El-Ghor; MM. Noshy; MM. Taha; EA. Abdel-Shafy	Flour mill, Egypt	Hepatotoxic Effect of Aflatoxin B1 and Polymorphisms in Glutathione S-Transferase in Millers Occupationally Exposed to Aspergillus	2013	Serum (AFB1 albumin adducts)	AFB1/Alb (ng/g) = Workers Mean = 0.06 SD = 0.04/Controls Mean 0.04 SD 0.01, $P < 0.0001$ (with statistical difference).	A control group was used
A. Saad-Hussein; S. Beshir; G. Moubarz; S. Elserougy; and M.M. Ibrahim	Textile industry, India	Effect of Occupational Exposure to Aflatoxins on Some Liver Tumor Markers in Textile Workers	2013	Urine (AFM1)	Urinary AFM1 mean levels were significantly higher in the pre-spinning, spinning and weaving groups compared to controls (554.2 ± 346.2 pg/ml, 459.1 ± 781.6 pg/ml, 296.5 ± 336.5 pg/ml and 68.5 ± 136.8 pg/ml, respectively) and in pre-spinning compared to the other two textile groups.	A control group was used
S. Viegas; L. Veiga; P. Figueiredo; A. Almeida;E. Carolino; R. Sabino; C. Veríssimo; C. Viegas	Swine production, Portugal	Occupational Exposure to Aflatoxin B 1 in Swine Production and Possible Contamination Sources	2013	Serum	Twenty-one workers (75%) showed detectable levels of AFB1 with values ranging from <1 ng/ml to 8.94 ng/ml and with a mean value of 1.91 ± 1.68 ng/ml (Table 2). In the control group, the AFB1 values were all below LOD.	A control group was used
F. Oluwafemi, T. Odebiyi, A. Kolapo	Feed mills, Nigeria	Occupational aflatoxin exposure among feed mill workers in Nigeria	2012	Serum	73.4-189.2 ng/ml and Aflatoxin B1 was not detected in the blood samples of the control group	A control group was used

Authors	Occupational setting/Country	Title	Year of publication	Samples	Results	Observations
S. Viegas; L. Veiga; J. Malta-Vacas; R.Sabino; P. Figueredo; A. Almeida; C. Viegas; E. Carolino	Poultry, Portugal	Occupational Exposure to Aflatoxin (AFB1) in Poultry Production	2012	Serum	Eighteen poultry workers (59%) had detectable levels of AFB1 with values ranging from <1 ng/ml to4.23 ng/ml and with a mean value of 2 ± 0.98ng/ml. AFB1 was not detected in the serum sampled from any of the controls.	A control group was used
M. I. Selim	Farms, USA	Significance of Aflatoxins in Rural and Global Health: Concern for Agricultural Workers	2010	Airborne dust	0.04 to 4,349 ng/m3	
Z. Wanga; T. Chai; C. Quan; H. Duan; Yao M et al.	Poultry, China	Simultaneous detection of airborne Aflatoxin, Ochratoxin and Zearalenone in a poultry house by immunoaffinity clean-up and high-performance liquid chromatography	2008	Airborne dust	0.08 ng AFB1/m³	
C. Brera; R. Caputi; M. Miraglia;I. Iavicoli;A. Salerno; G. Carelli	Warehouse of gree coffee, nutmeg, black pepper and cocoa bean, Italy	Exposure assessment to mycotoxins in workplaces: aflatoxins and	2002	Airborne dust	Total dust - <LOD-0.080 ng (aflatoxins); airborne particles - 0.002–0.045 ng/m3 (AFB1).	
S.K. Ghosh; M.R. Desai; G.L. Pandya; K. Venkaiah	Rice and maize processing plants, India	Airborne aflatoxin in the grain processing industries in India.	1997	Airborne dust	Concentrations of total airborne aflatoxin in the respirable dusts were 26 pG/m3 and 19 pg/m3, respectively, for the workplace and the storage area of the rice mill, while the concentrations of total	

Table 1. (Continued)

Authors	Occupational setting/Country	Title	Year of publication	Samples	Results	Observations
					dust were 12 pg/m3 for the workplace and 11 pg/m3 for the storage area.	
J.L. Autrup; J. Schmidt; T. Seremet; H. Autrup	Feed industry, Denmark	Determination of exposure to aflatoxins among Danish workers in animal-feed production through the analysis of aflatoxin B1 adducts to serum albumin.	1991	Settled dust and serum (AFB1 albumin adducts)	After four weeks of work 7 of 45 workers had detectable levels of aflatoxin B1), a finding suggesting occupational exposure to aflatoxin B1.	Workers were sampled before starting working after vacations and after four weeks of work
W.G. Sorenson; W. Jones et al.	Peanut shelling plant/ testing	Aflatoxin in respirable airborne peanut dust.	1984	Airborne dust	0.4-7.6 ng AFB1/m^3	
W.A. Burg; O.L. Shotwell; B.E. Saltzman	Corn handling/USA	Measurements of airborne aflatoxins during the handling of contaminated corn.	1981	Airborne dust	DUST AFB1 - 12.5 to 204.3 ppb, with an average of 138 ppb. AFB2 content ranged from 1.1 to 41.6 ppb, with an average of 24.6 ppb. Air - 7 mg/m3 to 417 mg/m3.	

However, it was already reported in some occupational studies the simultaneous exposure to the mycotoxin AFB1 and to no aflatoxigenic strains from *A. flavus* complex, as well (Viegas et al., 2015). This can occur because mycotoxins can be present in the environment long time after fungal structures disappearance (Halstensen, 2008). Due to this important fact, and to confirm exposures levels to AFB1, mycotoxins and/or mycotoxin metabolites should be detected in environmental and/or biological samples using biomarkers (Hooper et al., 2008).

A detailed exposure assessment is needed when evaluating health effects related with workplace exposures. This is one of the four steps included in the risk assessment process. The process starts with one preliminary step known as hazard identification. This is the first step and where normally the workplace evaluation fails, since AFB1 is not recognized as an occupational hazard and, because of that, it is not routinely studied. The next step is the dose-response relation and in this stage there are also some difficulties, since the most common available data is related only with exposure by ingestion to very low concentrations and during a long time frame. The third step corresponds to the exposure assessment and, at the end, and considering all the information collected and obtained, the risk characterization is performed.

Although several scientific reports have been published in order to propose the best methodologies for the exposure assessment framework, harmonization is far from being achieved. Nevertheless, an approach that uses the environment and/or biological monitoring as tools for risks assessment analysis is the most common scenario.

Some of the papers considered presented data regarding environment monitoring, using settled dust or airborne dust to estimate exposure to AFB1. The most used method is the airborne dust and usually involves the use of personal dust sampling equipment that typically consists of a portable pump that aspirates air from the breathing zone through a sampling cassette which collects airborne dust on a filter. The sampling equipment is carried by the worker during work in order to sample dust that is representative for the workers exposure (Halstensen, 2008).

In the specific case of aflatoxins, the biological monitoring and more precisely, the usage of biomarkers of internal dose (exposure) is common. When compared to the environmental monitoring, the biological monitoring using metabolites or the original molecule provides additional information that can be effective in improving the occupational risk assessment at the individual and/or group level. Besides the total uptake of the chemical by different routes of absorption, this information includes the assessment of the integrated, an estimate of current, recent or past exposure, the degree of metabolism, the early biological effects and the ability of an organism to tolerate and respond to chemical insult (Manno et al., 2009).

Biological monitoring can be done by measuring the levels of the produced metabolites or the free aflatoxins in biological samples, including serum plasma, urine, milk, and feces. Such measurements may provide information on not only in regard to aflatoxin intake, but also to the degree of individual absorption and metabolism rate (Groopman, 1994). It takes into consideration all the possible routes of absorption related with exposure (inhalation, skin, ingestion) that can be occurring in occupational environment or during leisure activities. It accounts for individual differences in absorption rate due to variations in workload or to co-exposure to additional components of complex mixtures, and in metabolic capabilities, or even due to either genetically determined or acquired changes in gene expression and enzyme activity (Manini et al., 2007).

Regarding the analyzed articles, the most often resource used to perform exposure assessment was the biologic monitoring (67%), showing that the authors chosen to use a biomarker of internal dose that quantify the AFB1 levels or the albumin adducts in the serum. Additionally, there is one paper that uses urine instead of serum to measure a different AFB1 metabolite (AFM1) (Saad-Hussein et al., 2013).

The use of a control group is frequently an option when using biomarkers of internal dose (Table 1), enabling to presume exclusion of exposure by diet of the considered group of workers. This option has been suggested and applied by several authors (Mayer et al., 2003; Degen, 2011; Viegas et al., 2012, 2013, 2015).

Recent advances in analytical chemistry have made possible measuring trace levels of specific chemicals or their metabolites in almost any biological fluid or tissues and have, therefore, contributed significantly to the increased usage of biomonitoring methods in exposure assessment process (Manno et al., 2010). This may explain the fact that only since 2010 were published articles presenting data on chemicals/metabolites collected from biologic fluids (serum in the majority of the cases). This can be related with the fact that only more recently specific analytical resources, such as ELISA, have been available and with acceptable limits of detection. This and other now available resources turn the mycotoxins biomonitoring a more easy and affordable method. However, specific points must be considered when working on this area: biomarkers data need interpretation based on deep toxicological knowledge and ethical issues should be addressed, as generally required in human studies (Manini et al., 2007).

5. Exposure Prevention

The most adequate measure to avoid exposure is the control of handled materials' contamination by mycotoxins. In settings such as food and feed, where several laws limit the mycotoxins concentration, this can be easier to achieve but in others, such as in the waste management and in textile production, it can be a real challenge to restrain this type of contamination (Viegas et al., 2015). In those cases no exposure limits are available to be used in order to avoid or control exposure.

All the analyzed papers considered in this review were developed in occupational settings where the exposure to organic dust is also a reality and the presence of particles is probably contributing significantly for the exposure, acting as a carrier of AFB1 to the workers breathing zone and mouth. This point has already been well documented in several studies (Autrup et al., 1999; Brera et al., 2002; Viegas et al., 2013). Following the previous, it is possible to consider that preventing exposure to particles will also contribute to reduce exposure to AFB1 (Viegas et al., 2015). Exhaust ventilation and the use of masks are some of the most common measures applied. Additionally, previous studies have shown, as mentioned already, that skin is permeable to AFB1 and other mycotoxins (Kemppainen et al., 1988; Boonen et al., 2012) and this exposure route can have a significant role in some occupational settings, particularly if the workers are sweating due to tasks involving high metabolic rates or if they use short clothes. In this case is important to protect skin and to follow strict hygienic procedures after work.

6. Additional Aspects to Consider in the Risk Assessment

The risk assessment approach applied for non-genotoxic chemicals is similar among different regulatory agencies: insertion of an uncertainty (safety) factor permits the derivation of permissible exposure levels at which no relevant human cancer risks are anticipated. However, for genotoxic carcinogens, particularly DNA reactive agents such as AFB1, there are no threshold values and all regulations should be based on the ALARA principle, technical feasibility, and other socio-political considerations (Bolt et al., 2004). So, this aspect supports the need of the appliance of preventive measures aiming to avoid exposure.

Moreover, the most common scenario in occupational settings is the co-exposure to several risk factors. This aspect has to be considered in the risk assessment process because the toxicity and the health effects can change when dealing with a co-exposure to two or more chemical agents and also with the co-exposure to chemicals and biologic agents, such as fungi. In the case of mycotoxins' exposure, there are some papers reporting on the ingestion of food contaminated with several mycotoxins, however, only a very small number of studies explore the co-exposure in occupational settings (Halstensen, 2008). Considering the co-exposure scenario, the additive and synergistic effects reported in experimental studies regarding the interaction between mycotoxins and the significant risk that represents to human health should be considered (Speijers and Speijers, 2004). Therefore, research studies are needed to develop work on co-exposure to mycotoxins in different occupational settings, based also on the fact that exposure can occur by different exposure routes (Inhalation and ingestion, for instance). Recently, a Portuguese study developed in swine production showed that occupational exposure to AFB1 by inhalation was occurring and the exposure to ocrathoxin A was also a reality, but in this specific case, related with the consumption of contaminated food (Viegas et al., 2015).

Climate changes and their effects on fungal distribution and activity are another aspect of concern regarding the Human exposure to AFB1. These effects are hard to predict because they are mediated in many different ways, including: fungal physiology, reproduction and survival, host physiology, spatial and temporal distribution of hosts, resource availability and competition (Boddy et al., 2014). In the case of mycotoxins there are many factors involved in their production by fungi but the climate is the most important. Production of these compounds, for example, on different crops, is highly susceptible to environmental factors, such as temperature and available moisture in the pre- and/or post-harvest period (Milani, 2013). Thus, when changes in the weather occur, mycotoxins production will be affected. Mycotoxins are therefore climate-dependent and related to plant and storage-associated problems. The mycotoxins production can also be affected by non-infectious factors, such as the bioavailability of micronutrients and insect damage, that are also determined by climate conditions (Milani, 2013). Moreover, the impact of climate changes has been identified as an emerging concern for food and feed safety (Miraglia et al., 2009) and European Food Safety Agency Emerging Risks Unit has identified changing patterns in mycotoxin production in cereals, wheat, maize and rice related to climate changes as a potential area of concern. The study had concluded that aflatoxin in maize is an emerging problem in Europe (available on EFSA website on: http://www.efsa.europa.eu/en/supporting/pub/223e.htm).

Additionally, some authors have stated that European countries with temperate climates, such as Portugal, Spain and Italy have the biggest risk regarding exposure to fungi and mycotoxins (Paterson and Lima, 2010). As an example, with climate changes, the climate of these countries will probably become warmer reaching temperatures of 33ºC, which is a temperature very close to the optimal for aflatoxin' productions. This situation may already occur with current crops. Recognized aflatoxin-susceptible plants (e.g., peanuts, maize) are grown in an increasingly high rate to exploit the new climate conditions (Paterson and Lima, 2010). Therefore, exposure in agricultural settings can increase significantly in the near future.

In conclusion,and by the number of articles found in our research, there is still a long way until AFB1 be recognized as a risk factor present in some specific occupational settings. This failing implicates that exposure will not be studied with the frequency and detail needed to perform a correct risk assessment. Therefore, occupational health professionals should be informed and trained to identify, assess exposure and to recognize the most suitable technical measures to avoid contamination and exposure.

Funding

The authors are grateful to Environment and Health Research Group from Lisbon School of Health Technology. Raquel Sabino was financially supported by a fellowship from Fundação para a Ciência e Tecnologia Portugal (contract SFRH/BPD/72775/2010).

References

Autrup, J; Schmidt, J; Seremet, T; Autrup, H. Determination of exposure to aflatoxins among Danish workers in animal feed production through the analysis of aflatoxin 81 adducts to serum albumin. *Scand J Work Environ Health.* 1991; 17: 436–40.

Baxter, CS; Wey, HE; Burg, WR. A prospective analysis of the potential risk associated with inhalation of aflatoxin contaminated grain dusts. *Food Cosmet Toxicol.* 1981; 19: 765–9.

Bhatnagar, D; Payne, GA; Cleveland, TE; Robens, JF. Mycotoxins: current issues in U.S.A. In: Meeting the mycotoxin menace. H Barug, HP van Egmond, R Lopez-Garcia, WA van Osenbruggen, A Visconti. Wageningen Academic Publisher, Wageningen, the Netherlands, 2004; 17-47.

Bhatnagar, D; Cary, JW; Ehrlich, K; Yu, JJ; Cleveland, TE. Understanding the genetics of regulation of aflatoxin production and *Aspergillus flavus* development. *Mycopathologia,* 2006; 162, 155-166.

Boddy, L; Beuntgen, U; Egli, S; Gange, A; Eheegaard, E; Kirk, PM; Mohammad, A; Kauserud, H. Climate variation effects on fungal fruiting. *Fungal Ecology.* 2014; 10: 20-33.

Bolt, HM; Foth, H; Hengstler, JG; Degen, GH. Carcinogenicity categorization of chemicals—new aspects to be considered in a European perspective. *Toxicology Letters.* 2004; 151: 29–41.

Boonen, J; Malysheva, SV; Taevernier, L; Di Mavungu, JD; De Saeger, S; De Spiegeleer, B. Human skin penetration of selected model mycotoxins. *Toxicology.* 2012; 301: 21–32.

Brera C, Caputi R, Miraglia, M; Iavicoli, I; Salernoet, A. Exposure assessment to mycotoxins in workplaces: aflatoxins and ochratoxin A occurrence in airborne dusts and human sera. *Microchem J,* 2002; 73: 167–73.

Cary, JW; Ehrlich, KC. Aflatoxigenicity in Aspergillus: molecular genetics, phylogenetic relationships and evolutionary implications. *Mycopathologia*. 2006; 162, 167-177.

Chang, PK; Yu, J; Yu, JH. aflT, a MFS transporter-encoding gene located in the aflatoxin gene cluster, does not have a significant role in aflatoxin secretion. *Fungal Genetics and Biology*. 2004; 41, 911-920.

Corey, JEI. Relationships of water activity to fungal growth. In Beuchat LR (ed.) Food and beverage mycology. *Van Nostrand Reinhold International Company Limited,* 1987 New York US.

Cotty, PJ; Cardwell, KF. Divergence of West African and north American communities of Aspergillus section Flavi. *Applied and Environmental Microbiology*. 1999; 65, 2264-2266.

Dash, B; Afriyie-Gyawu, E; Huebneret, HJ. et al. Determinants of the variability of aflatoxin-albumin adduct in levels in Ghanaians. *Toxicol Environ Health A*. 2007; 70: 58–66.

Degen, GH. The challenge to assess workplace related risks from mycotoxin exposure (Guest editorial). *Mycotoxin Res*. 2008; 24: i–ii.

Degen, GH. Tools for investigating workplace-related risks from mycotoxin exposure. *World Mycotoxin J*. 2011; 4: 315–27.

Diaz, GJ; Murcia, HW; Cepeda, SM. Cytochrome P450 enzymes involved in the metabolism of aflatoxin B1 in chickens and quail. *Poult Sci*. 2010; 89: 2461–9.

Diener, UL; Cole, RJ; Sanders, TH; Payne, GA; Lee, LS; Klich, MA. Epidemiology of aflatoxin formation by Aspergillus flavus. *Annual Review of Phytopathology*. 1987; 25, 249-270.

D'Mello, JPF; Macdonald, AMC. Mycotoxins. *Animal Feed Science and Technology*. 1997; 69, 155-166.

Doi, AM; Patterson, PE; Gallagher, EP. Variability in aflatoxin B(1)-macromolecular binding and relationship to biotransformation enzyme expression in human prenatal and adult liver. *Toxicol Appl Pharmacol*. 2002; 181: 48–59.

Donnelly, PJ; Devereux, TR; Foley, JF; Maronpot, RR; Anderson, MW; Massey, TE. Activation of K-ras in aflatoxin B1-induced lung tumors from AC3F1 (A/J x C3H/HeJ) mice. Carcinogenesis. 1996; 17: 1735–40.

Dvorackova, I. Aflatoxin inhalation and alveolar cell carcinoma. *Br Med J*. 1976; 1: 691–3.

Dvorackova, I; Pichova, V. Pulmonary interstitial fibrosis with evidence of aflatoxin B1 in lung tissue. *J Toxicol Environ Health*. 1986; 18: 153–7.

Groopman, JD. Molecular dosimetry methods for assessing human aflatoxin exposures. In Eaton DL, Groopman JD, editors. The toxicology of aflatoxins: human health, veterinary, and agricultural significance. *San Diego, CA: Academic Press*. 1994: 259–80.

Halstensen, AS. Species-specific fungal DNA in airborne dust as surrogate for occupational mycotoxin exposure? *Int J Mol Sci*. 2008; 9: 2543–58.

Hayes, RB; Van Nieuwenhuize, JP; Raatgever, JW; Ten Kate, FJ. Aflatoxin exposures in the industrial setting: an epidemiological study of mortality. *Food Chem Toxicol*. 1984; 22: 39–43.

Hooper, 0; Bolton, V; Gray, ME. Fungal mycotoxins can be detected in tissue and body fluids of patients with a history of exposure to toxin producing molds 2006. Available from:http://www.reaitimelab.com/documents /MycotoxinPosterMayl 620 7.pdf.

IARC. Some naturally occurring substances: food items and constituents, heterocyclic aromatic amines and mycotoxins. *IARC Monogr Eval Carcinogen Risks Hum.* 1993; 56: 1–599.

IARC. Some traditional herbal medicines, some mycotoxins, naphthalene and styrene. *IARC Monogr Eval Carcinogen Risks Hum.* 2002; 82: 1–556.

Jargot, D; Melin, S. Characterization and validation of sampling and analytical methods for mycotoxins in workplace air. *Environ Sci Process Impacts.* 2013; 15: 633–44. doi:10.1039/c2em30566f

Johanning, E; Auger Philip, P; Morey, R; Yang, CS; Olmsted, E. Review of health hazards and prevention measures for response and recovery workers and volunteers after natural disasters, flooding, and water damage: mold and dampness. *Environ Health Prev Med* 2014; 19:93–99.

Kelly, JD; Eaton, DL; Guengerich, FP; et al. Aflatoxin B1 activation in human lung. *Toxicol Appl Pharmacol.* 1997; 144: 88–95.

Kemppainen, BW; Riley, RT; Pace, JG. Skin absorption as a route of exposure for aflatoxin and trichothecenes. *Toxin Rev.* 1988; 7: 95–120.

Kurtzman, CP; Horn, BW; Hesseltine, CW. Aspergillus nomius, a new aflatoxin-producing species related to Aspergillus flavus and Aspergillus tamarii. *Antonie van Leeuwenhoek,* 1987; 53, 147-158.

Magan, N; Aldred, D. Post-harvest control strategies: minimizing mycotoxins in the food chain. *International Journal of Food Microbiology.* 2007; 119, 131-13.

Malta-Vacas, J; Viegas, S; Sabino, R; Viegas, C. Fungal and microbial volatile organic compounds exposure assessment in a waste sorting plant. *Journal of Toxicology and Environmental Health* 2012; 75: 1410-1417.

Massey, TE; Smith, GB; Tam, AS. Mechanisms of aflatoxinB1 lung tumorigenesis. *Exp Lung Res.* 2000; 26: 673–83.

Manno, M; Viau, C; Cocker, J; Colosio, C; Lowry, L; Mutti, A; Nordberg, M; Wangh, S. Biomonitoring for occupational health risk assessment (BOHRA) *Toxicology Letters.* 2010; 192 (1): 3–16.

Manini, P; Palma, G; Mutti, A. Exposure assessment at the workplace: Implications of biological variability. *Toxicology Letters.* 2007; 168: 210–218.

Mayer, S; Engelhart, S; Kolk, A; Blome, H. The significance of mycotoxins in the framework of assessing workplace related risks. *Mycotoxin Res.* 2008; 24: 151-164.

McLaughlin, JK; Malker, HSR; Malker, BK. Registry based analysis of occupational risk for primary liver cancer in Sweden. *Cancer Res.* 1987; 47: 287–93.

Milani, JM. Ecological conditions affecting mycotoxin production in cereals: a review. *Veterinarni Medicina.* 2013; 58 (8): 405–411.

Miraglia, M; Marvin, HJP; Kleter, GA; Battilani, P; Brera, C; Coni, E; Cubadda, F;Croci, L; De Santis, B; Dekkers, S; Filippi, L;. Hutjes, RWA; Noordam, NY; Pisante, M; Piva, FG; Prandini, A; Toti, L; Van den Born, JG; Vespermann, A. Climate change and food safety: An emerging issue with special focus on Europe. *Food and Chemical Toxicology.* 2009; 47: 1009–1021.

Murphy, PA; Hendrich, S; Landgren, C; Bryant, CM. Food mycotoxins: an update. *Journal of Food Science.* 2006; 71, R51-R65.

Nebert, DW; Dalton, TP. The role of cytochrome P450 enzymes in endogenous signaling pathways and environmental carcinogenesis. *Nature Reviews Cancer* 6, 947-960 (December 2006) | doi:10.1038/nrc2015.

Russell, R; Paterson, M, Lima, N. How will climate change affect mycotoxins in food? *Food Research International.* 2010; 43: 1902–1914.

Popendorf, W; Donham, KJ; Easton, DN et al. A synopsis of agricultural respiratory hazards. *Am Ind Hyg Assoc* J. 1985; 46: 154–61.

Saad-Hussein, A; El-Ghor, AA; Noshy, MM; Taha, MM; Abdel-Shafy, EA. Hepatotoxic Effect of Aflatoxin B1 and Polymorphisms in Glutathione S-Transferase in Millers Occupationally Exposed to Aspergillus. *Journal of Applied Sciences Research.* 2013; 9(3): 1724-1730.

Viegas, C; Gomes, AQ; Abegão, J; Sabino, R; Graça, T; Viegas, S. Assessment of Fungal Contamination in Waste Sorting and Incineration — Case Study in Portugal. *Journal of Toxicology and Environmental Health, Part A: Current Issues,* 2014; 77:1-3: 57-68.

Viegas, S; Veiga, L; Malta-Vacas, J; Sabino, R; Figueredo, P; Almeida, A; Viegas, C; Carolino, E. Occupational exposure to aflatoxin (AFB1) in poultry production. *Journal of Toxicology and Environmental Health, Part A,* 2012; 75:1330–1340.

Viegas, S; Veiga, L; Verissimo, C; Sabino, R; Figueredo, P; Almeida, A; Carolino, E; Viegas, C. Occupational Exposure to Aflatoxin B1 in Swine Production and Possible Contamination Sources, *Journal of Toxicology and Environmental Health, Part A: Current Issues.* 2013; 76:15, 944-951.

Viegas, S; Veiga, L; Figueiredo, P; Almeida, A; Carolino, E; Viegas, C. Assessment of Workers' Exposure to Aflatoxin B1 in a Portuguese Waste Industry. *Annals of Occupational Hygiene* 2015; 59(2): 173–181.

Viegas, S; Veiga, L; Almeida, A; Figueiredo, P; Carolino, E. Occupational exposure to aflatoxin B1 and ochratoxin A: Co-exposure in swine production. - International Symposium on Occupational Safety and Hygiene SHO2014, Arezes, P. et al., Portuguese Society of Occupational Safety and Hygiene. 2014: 424-426.

Tawniczek-Wałczyk, A; Gołofit-Szymczak, M; Cyprowski, M; Górny, RL. Exposure to harmful microbiological agents during the handling of biomass for power production purposes. *Medycyna Pracy.* 2012; 63:395–407.

Thrane, U; Adler, A; Clasen, P; Galvano, F; Langseth, W; Lew, H; Logrieco, A; Nielsen, KF; Ritieni, A. Diversity in metabolite production by Fusarium langsethiae, Fusarium poae, and Fusarium sporotrichioides. *Int. J. Food Microbiol.* 2004; 95: 257–66.

Smith, JE. Aflatoxins. In: Handbook of Plant and Fungal Toxicants. JPF D'Mello. *Boca Raton, FL,* 1997: 269-285.

Smith, JE; Moss MO. Mycotoxins - Formation, analysis and significance. Editor. *John Wiley & Sons, Chicester*, Great Britain, 1985: 148.

Sorenson, WG; Jones, W; Simpson, J; Davidson, JI. Aflatoxin in respirable airborne peanut dust. *J Toxicol Environm Health* 1984; 14: 525-533.

Yang, XJ; Lu, HY; Li, ZY; Bian, Q; Qiu, LL; Li, Z; Liu, Q; Li, J; Wang, XJ; Wang, S. Cytochrome P450 2A13 mediates aflatoxin B1-induced cytotoxicity and apoptosis in human bronchial epithelial cells. *Toxicology.* 2012; 300: 138–48.

Yang, X; Zhang, Z; Wang, X; et al. Cytochrome P450 2A13 enhances the sensitivity of human bronchial epithelial cells to aflatoxin B1-induced DNA damage. *Toxicol Appl Pharmacol.* 2013; 270: 114–21.

In: Mycotoxins
Editor: Cesar Rios
ISBN: 978-1-63483-544-2

Chapter 9

A DECADE OF MYCOTOXINS RESEARCH IN AFRICA: A REVIEW

T. I. Ekwomadu and M. Mwanza
Faculty of Agriculture Science and Technology, North-West University,
Mmabatho, South Africa

ABSTRACT

Mycotoxins are prevalently toxic compounds produced as secondary metabolites by various fungi and excreted on agricultural crops at pre and post-harvest stages of food and feed production under a wide range of climatic conditions. Majority of mycotoxins are produced by fungi that belong to the genera *Aspergillus*, *Fusarium* and *Penicillium*. Most of the known hosts include maize, peanut, wheat, rice, sorghum, barley and various tree nuts which are staple foods in most parts of Africa. Exposure to mycotoxins is mostly by direct ingestion of contaminated agricultural crops or by a carry-over of mycotoxins into animal tissues, milk and eggs and can also be inhaled or absorbed through the skin. Mycotoxins exert various toxic effects from relatively mild G.I disturbances to severe organ damage on exposed humans and animals. Risks of mycotoxin contamination of food and feed in Africa is increasing due to climatic conditions, agronomic practices and the socio-economic and food security status of the public (inhabitants).

Climatic conditions such as hot and humid environment favour the proliferation of fungal and mycotoxin contamination of food and feeds. Farming practices in Africa sustain fungal and mycotoxin contamination of food as well as feeds. The socio-economic status of greater parts of the public predisposes them to the consumption of contaminated food either directly or at various points in the food chain. The resulting health risk of mycotoxin contamination depends on the toxicological property of the particular compound (carcinogenicity, mutagenicity, teratogenicity, genotoxicity, immunosuppression, nephrotoxicity and hepatotoxicity) as well as the extent of exposure. Mycotoxins have assumed great economic importance worldwide because of their influence on human health and livestock and many countries have set regulatory limits in order to reduce their intake. In Africa, fifteen countries which accounts for about 59% of the population of the continent, were considered to have mycotoxin regulations in 2003.

Mycotoxin control strategies should also include the adoption of appropriate agronomic practices at the pre and post-harvest prevention stages and also the physical, biological and chemical methods of decontamination at the post-harvest stage. Strategies for the diversion of contaminated products to lower risk uses and laboratory safety measures to protect researchers from the toxic effects of mycotoxin should also be included. There should also be producer /consumer education and accurate and affordable detection methods of mycotoxin in developing African countries. The aim of this review is to highlight the occurrence of dietary mycotoxins in Africa as well as the advances in analytical methods of mycotoxin extraction and detection over the last decade (2003-2013).

Keywords: Mycotoxin, Africa, decade

1. INTRODUCTION

Mycotoxins (from "*myco*" fungus and toxin) are natural, chemically diverse, fungal products that have harmful effects on exposed humans and animals in a variety of ways [1]. They are non-volatile, relatively low molecular weight compounds,[1] synthesized mainly by the secondary metabolism of some filamentous fungi [2]. Agricultural crops are susceptible to contamination by these mycotoxins either while growing or when stored. The existence of mycotoxins came into the lime light in 1960 when more than 100,000 turkeys died in the United Kingdom as a result of the Turkey *X* disease after consuming aflatoxin-contaminated peanut meal [3]. Currently, some 300 to 400 mycotoxins are known, of which, just about a dozen regularly receive attention as threats to human and animal health [2].

Exposure to mycotoxins is mostly by ingestion, but can also occur through inhalation and dermal contact [2, 4, 5]. The health effects of mycotoxins are referred to as mycotoxicoses. They are referred to as primary mycotoxicoses when provoked by consuming contaminated food or feeds or secondary mycotoxicoses when produced by eating meat, eggs or milk from animals that ingested feed contaminated with mycotoxins [1]. Exposure to mycotoxins can produce acute and chronic toxicities or both on the central nervous, pulmonary, digestive and cardiovascular systems. Mycotoxins may also exhibit some detrimental properties which include carcinogenesis, mutagenesis, teratogenesis, estrogenesis and immunosuppression in both animals and man [6]. This will eventually lead to a variety of biological effects such as liver and kidney damage, CNS effects and reproductive effects. Mycotoxins (aflatoxins) have also been suggested to have a synergistic effect with the hepatitis B virus in the etiology of liver cancer and could interact with some other important diseases in the African continent such as malaria, kwashiorkor and HIV/AIDS [7].

The majority of agro-economically important mycotoxins are produced by *Aspergillus*, *Fusarium* and *Penicillium* fungal species, although other genera such as *Claviceps* also produce potent toxins [4, 8-10]. These genera are ubiquitous in the environment and can exploit any nutritive-rich matrix given the appropriate conditions [4, 10]. The Food and Agriculture Organisation of the United Nations, has estimated that over one quarter of the world's crops, including many basic foods are significantly contaminated by these mycotoxin-producing fungi [4]. They produce mycotoxins which include aflatoxin (AF), ochratoxin (OT), fumonisins, zearalenone (ZEA) and trichothecenes (TCT) such as deoxynivalenol (DON), T-2 toxin and nivalenol [11]. Tropical conditions such as high temperatures, humidity and unseasonal rains lead to the proliferation of fungal and the production of mycotoxins on various agricultural crops especially while growing in the field [12]. If optimal conditions of temperature, humidity and a suitable substrate prevail, the production of mycotoxins can also occur during storage and/or processing of agricultural commodities. These climatic conditions as well as poor harvesting practices, improper storage and processing are characteristic in most parts of Africa [7]. Food-borne mycotoxins likely to be of greatest importance in African countries and other developing countries in the tropics are the fumonisins and the aflatoxins [13].

Mycotoxins have attracted worldwide concerns due to their impact on human and animal health. This has given rise to the establishment of specific regulatory limits in order to reduce their intake. Fifteen countries in Africa had specific mycotoxin regulations in 2003. These countries make up approximately 59 percent of inhabitants of the continent but for the majority of countries in Africa specific mycotoxin regulations (probably) do not exist [14]. The establishment of mycotoxin regulations will have little or no effect in terms of the protection of health in these countries where many farmers grow agricultural produce for their own consumption (subsistence farming), which is the situation in many African countries.

Furthermore, analytical methods for rapid, sensitive and precise determination of these mycotoxins in agricultural produce are greatly needed. Apart from this need, it is also necessary to regularly identify fungal species responsible for their production in order to properly asses the toxicological risk of exposure to both human and animals and also to ensure that the regulatory limits are met [15, 16].

2. Factors Influencing the Occurrence of Mycotoxins in African Food Commodities

Mycotoxin contamination is inevitable and unpredictable, which makes it a serious challenge to food safety [17]. It is thus unreasonable to expect that the supply of food can ever be made entirely mycotoxin free. Production of mycotoxins occurs during secondary metabolism as they are not necessary for fungal growth and play no biochemical function [1]. They are produced from one or more primary metabolites [18]. The incidence of mycotoxins from fungi is governed entirely by the existence of conditions that favour the growth of fungus concerned [18]. Formation of mycotoxins on a given food crop involves basically the mycotoxigenic fungi and suitable nutrient factors (intrinsic) of the substrate and environmental factors (extrinsic) surrounding the food crop [19]. However, the ultimate level of mycotoxin that accumulates on food and feed material is due to overall interaction of all these factors.

2.1. Intrinsic Factors

Intrinsic factors include moisture and relative humidity, crop variety and its level of growth and maturity, type of nutrient and their concentrations as well as the pH of the substrate.

a) Moisture and relative humidity: Important factors in the growth and production of mycotoxins are moisture content and relative humidity equilibrium of the substrate. High moisture content of 14% and higher in maize, and high relative humidity are essential for fungal spore germination and proliferation. This can lead to mycotoxin contamination [20].
b) Crop variety: Naturally some varieties of field crops are relatively resistant, whereas others are susceptible to fungal development and subsequent mycotoxin production. For example, yellow maize is more susceptible to *Fusarium* infection than white maize while white maize is a better substrate for fumonisin production [21]. The development of crop varieties that resist the growth of toxigenic fungi or that would completely inhibit the production of toxin even under bad storage conditions would be an ideal solution.
c) Level of growth and maturity: This is very important since it has been confirmed that there is higher incidence of mycotoxin invasion in over-mature grains, that is, grains that stayed longer in the field before their harvest, than on immature and mature ones from the same plant at harvest [19].
d) pH: The optimum production of mycotoxin, for example, fumonisins occurs under relatively high oxygen tension and low pH (approx.2) due to organic acids produced by starch metabolism of well-rotted maize [18].

2.2. Extrinsic Factors

Extrinsic factors occur when the fungus invades the food grain or feed material. Extrinsic factors such as temperature, light, heat, UV or gamma irradiation, microbial interaction, time and aeration (oxygen) play important roles in influencing the growth and formation of mycotoxins on these materials.

a) Temperature: It plays important roles in fungal growth and the development of mycotoxins. Fungi usually grow at a wide range of temperatures, but mycotoxigenic fungi are mostly mesophilic. Elevated temperature in maize increases the ability of *A. flavus* to proliferate and produce aflatoxin [21].
b) Irradiation: It is one of the methods of increasing the storage life of agricultural crops. However, it also increases the susceptibility of the product to certain storage fungi such as *A. flavus* that lead to mycotoxin contamination. Previous studies have revealed that levels of aflatoxin were significantly higher in some irradiated cereals, including maize, than in non-irradiated samples of same food [19].
c) Microbial Interaction: Under natural conditions, production of Aflatoxins by *Aspergillus* species is affected by the presence of other fungi on agricultural commodities. *A. flavus* is frequently associated with many other microorganisms particularly in stored grains. The microbial interaction between fungi and other microorganism, for substrate under favourable environmental conditions will restrict or reduce the quantity of aflatoxin formation by what is referred to as competitive exclusion process. Depending on the population density, the atoxigenic strain competes for the same site of invasion of the crop plant and nutrient resource as that

of toxigenic strain, thereby reducing the toxigenic strain population and mycotoxin accumulation [22].

3. Analysis of Health Risks of Dietary Mycotoxins

Most outbreaks of food-borne mycotoxin poisoning occur in sub-Saharan Africa where maize and groundnuts are dietary staples [7]. Also, the globalisation of food trade with stringent mycotoxin standards on exported foods means that developing countries are likely to export their best-quality foods while keeping contaminated food domestically. This will inadvertently result in higher risk of exposure in those countries [23]. The resulting health risks of mycotoxin poisoning depends not only on the toxicological property of the particular compound but also on its concentration and the extent of exposure, as it is said that the dose makes the poison [24]. Animals (and humans) vary widely in their response to a given dose of a mycotoxin based not only on the age, sex, nutritional or gestational status but also on their ability to detoxify [24, 25].

Thus the ability of a mycotoxin to cause detrimental effects will not only depend on the nature and concentration of the compound but also on the metabolic status of the organism involved, the route of exposure and the potential for metabolic detoxification [24]. Generally, the greater the amount of toxin, the more severe the effect, although mycotoxins have different threshold levels at which they are toxic [24]. This threshold level (lowest concentration) at which they are toxic may vary from one animal species to another and even from one individual to another within a population. Often the young are more sensitive than adults and such characteristics as diet, pregnancy, general state of health and drug interactions can affect the sensitivity of mycotoxin [2]. It was believed that ruminants are relatively tolerant to the adverse effects of mycotoxins, presumably because of the ability of rumen microflora to detoxify mycotoxins. Further studies have shown that some of the rumen metabolites are even more toxic than parent mycotoxins, for example, conversion of zearalenone to α-zearanol [26]. In addition, mycotoxins can impair the functions of rumen by exerting antimicrobial effects on rumen microflora and also, the increased rate of passage of feed through the rumen may possibly overwhelm the ability of the rumen to completely denature the toxins [27, 28].

Considering the opportunistic nature of filamentous fungi, there is almost no food or feedstuff that has not at one time or another been found to be contaminated with some mycotoxins. Mycotoxins are considered one of the most important contaminants of food and feed. As mentioned earlier, the FAO has estimated that more than 25% of the world's agricultural produce is contaminated with mycotoxins [15, 29]. While mycotoxins occur in many varieties of fungi, not all fungi produce mycotoxin. One strain of given specie may be capable of producing mycotoxin whereas another strain of that specie is not toxigenic [30]. Some fungi have the ability to produce more than one mycotoxin and some mycotoxins are produced by more than one fungal species. Usually more than one mycotoxin can be detected on a contaminated substrate [8, 30]. They have been reported on cereals, legumes, nuts, dried seafood products, spices and various other African food commodities [31-36], (Table 1).

The most important mycotoxins both from the public health and agronomic perspectives include aflatoxins (AF), ochratoxins (OT), fumonisins (FBs) zearalenone (ZEA) and trichothecenes (TCT) such as deoxynivalenol (DON), T-2 toxin and nivalenol [11, 37]. However, food borne mycotoxins likely to be of greatest importance in tropical developing countries such as Africa are the fumonisins and aflatoxins [7, 38, 39] with others such as ochratoxins and zearalenone and trichothecenes occurring at varying levels and different regions. The presence of mycotoxins in food is often overlooked due to several reasons which include; public ignorance about their existence and lack of regulatory mechanisms [7]. However, human exposure to mycotoxins can also increase during chronic food shortage due to drought, war, famine, economic instability and other hardships [7, 10]. Nevertheless, mycotoxins have been found to be carcinogenic (aflatoxins, ochratoxin A and fumonisins), mutagenic (aflatoxins), teratogenic (ochratoxin A), estrogenic (zearalenone), hemorrhagic (trichothescenes), immunotoxic (aflatoxins and ochratoxin A), nephrotoxic (ochratoxin A), dermotoxic (trichothescenes) and neurotoxic (patulin) [40]. Each mycotoxin group exerts its toxic effects by a different mechanism. For example, some intercalate with DNA, inhibit sphingolipid biosynthesis and some interfere with protein synthesis [41]. Many researchers have carried out studies on the occurrence of mycotoxins on African food commodities. The data is reported in Table 1.

Table 1. Occurrence of Mycotoxins on African food/feed commodities

Country	Food/Commodity	Mycotoxin	Freq.	Conc.	Method	Ref.
Botswana	Raw peanut	Aflatoxin	78%	12-329μg/kg		[42]
	Opaque beer	FB_1	6.5%	47-1316 μg/l		[43]
	Malt	ZEA	56%	102-2213 μg/l		[43]
Nigeria	Pre-harvest maize	Aflatoxin	65%	3-138 μg/kg		[44]
	Dried yam chips	AFB_1		27.1ppb		[45]
	Mellon seeds	AFB_1	32.2%	Above5 μg/kg		[46]
	Feed	Aflatoxin	94%	115ng/g	HPLC	[47]
		Fumonisin	51%	46ng/g	HPLC	[47]
	Garri	Aflatoxins	17.5%	0.12-5.71 μg/kg	TLC	[48]
	Maize	ZEA	37%	<50-196 μg/kg	ELISA	[49]
		T-2	36%	<7.5-29 μg/kg	ELISA	[49]
	Feed	Ochratoxin	100%	12ng/g	HPLC	[47]
	Rice	Ochratoxin	66.7%	134-341 μg/kg	TLC/ HPLC	[50]
	Rice	DON	23.8%	107.9 μg/kg	TLC/HPLC	[50]
		Aflatoxins	100	28-372 μg/kg	TLC/ HPLC	[50]
	Yam	AFB_1	85.7%	nd -0.27 μg/kg	HPLC	[51]
	Garri	AFB_1	72.2%	nd-0.69μg/kg	HPLC	[51]
	Semovita	AFB_1	33.3%	nd-0.17μg/kg	HPLC	[51]
	White yam flour	AFB_1	57%	0.02-3.2	HPLC	[52]
	Breast milk	AFM_1	82%	nd-92.14μg/kg	HPLC	[51]

Country	Food/Commodity	Mycotoxin	Freq.	Conc.	Method	Ref.
	Rice	AFB_1	90.4%	nd-0.30μg/kg	HPLC	[51]
	Bush mango seeds	AFB_1		0.2-4.0 μg/kg	TLC	[53]
		AFG1		0.3-4.2 μg/kg	TLC	[53]
	Diary milk	AFM_1	5.9%	0.2-0.4 μg/l	TLC	[54]
	Stored sweet potato flour	AFB_1	100%	0.0023 μg/kg	ELISA	[55]
		AFB_2	100%	0.0014 μg/kg	ELISA	[55]
	Peanut cake	Aflatoxin	90%	>20μg/kg	LC/ESI-MS/MS	[35]
	Coconut candy snack	α- Zearalenol	12.5%	Up to 54μg/kg	QuEChERS-HPLC-MS/MS	[56]
	Cocoa seeds	Aflatoxins	88.2%	0.03-4.97 μg/kg	HPLC	[57]
		OTA	88.2%	0.7-525.4 μg/kg	HPLC	[57]
	W. Food (Rice + Milk)	AFM_1		4.6 ng/ml	HPLC	[58]
		AFB_2		8.290 ng/ml	HPLC	[58]
	Cocoa seeds	FB_1	100%	18.7-206 μg/kg	HPLC	[57]
		ZEA	23.5%	24.2-83.6 μg/kg	HPLC	[57]
	Akpalata	AFB_1	100%	8.5-95 μg/kg	TLC	[59]
	Offor	AFB_1	100%	8.0-90 μg/kg	TLC	[59]
	Achi	AFB_1	100%	4.5-56 μg/kg	TLC	[59]
	Ukpo	AFB_1	100%	4.0-50 μg/kg	TLC	[59]
	Fresh meat	AFB_1 + AFB_2	100%	0.0217-0.085 μg/kg	TLC	[60]
	Sun-dried meat	AFB_1 + AFB_2	100%	0.0029-0.0758 μg/kg	TLC	[60]
	Maize	DON	22%	9.6-745.1 μg/kg	HPLC/MS	[61]
		DAS	9%	1.0-51.0 μg/kg	HPLC	[61]
		AcDON	17%	0.7-72.4 μg/kg	HPLC	[61]
	G.nut-Maize based snack	AFB_1	100%	9.3-12 μg/kg	LC-MS/MS	[62]
	G.nut-Maize based snack	AFM_1	100%	124-130 μg/kg	LC-MS/MS	[62]

Table 1. (Continued)

Country	Food/Commodity	Mycotoxin	Freq.	Conc.	Method	Ref.
	W.Food(Maize + Milk)	AFB_1		181.6 ng/ml	HPLC	[58]
	W.Food (Maize + Soya)	AFB_1		4,806 ng/ml	HPLC	[58]
	G.nut based snacks	AFB_1	30%	52-362 μg/kg	LC-MS/MS	[62]
	G.nut based snacks	AFM_1	20%	256-410 μg/kg	LC-MS/MS	[62]
	Maize based snacks	AFB_1	62.5%	1-69.5 μg/kg	LC-MS/MS	[62]
	Maize based snacks	AFM_1	37.5%	2-2.2 μg/kg	LC-MS/MS	[62]
	Maize based snacks	FB_1	100%	44-339 μg/kg	LC-MS/MS	[62]
	Yam chips	AFB_1		14 μg/kg	TLC	[63]
	G.nut	AFB_1		24 μg/kg	TLC	[63]
	Pawpaw fruits	Aflatoxins	80%		TLC	[64]
	Breadfruit	AFB_1	100%	40.06-48.59 μg/kg	TLC	[65]
	Grains	AFB_1	100%	17.10-82.12 μg/kg	TLC	[65]
	Rotten tomato fruits	Aflatoxins	100%		TLC	[66]
	Breast milk	AFM_1	82%	3.49-35mg/l	HPLC	[51]
	Lactating mother's food	AFB_1	79%	0.16-0.33 μg /g	HPLC	[51]
Senegal	Peanut oil	AFB_1	85%	About 40ppb		[67]
South Africa	Trad. Brewed beer	Aflatoxin	0.3%	300 μg/l	HPLC	[68]
		ZEA	45%	2.6-426 μg/l	HPLC	[68]
	Dietry/Med.plant	Fumonisin	16%	8.6-1553 μg/kg	HPLC/LC-MS/MS	[69]
	Diary milk	AFM_1	94.5%	0.02-1.5μg/kg	HPLC	[70]
	Maize	Fumonisin		222-1,142ppb	RAPP	[71]
	Feed	Fumonisin	57%	454ng/g	HPLC	[47]
		ZEA	29%	86ng/g	HPLC	[47]
		Aflatoxin	6%	0.2ng/g	HPLC	[47]
		B-trichoth	87%	1469ng/g	HPLC	[47]
	Cowpea seeds	FB_1	100%	0.12-0.61 μg /g	HPLC	[72]
	Maize	Fumonisins	60%	0-2.16ppm	HPLC/TLC	[73]
	Bambara nuts	Fumonisins	100%	0.9-6.01ppm	HPLC/TLC	[73]
	Maize	Fumonisin Bs	100%	64-1035 ppb	HPLC/TLC	[74]
		AFB_1	70%	0-741ppb	HPLC/TLC	[75]
		ZEA	90%	0-135ppb	HPLC/TLC	[74]
	Beer	OTA	27%	>0.1ppb	HPLC	[76]
	Barley & barley pdts				HPLC	[76]

Country	Food/Commodity	Mycotoxin	Freq.	Conc.	Method	Ref.
	Whole wheat	DON		8 ppm	TLC-VICAM	[77]
	Breakfast cereals	OTA		4 ppb	TLC-VICAM	[77]
		ZEA		4 ppb	TLC-VICAM	[77]
	Cotton seed feed	AFB_1		26-164 µg/kg	VICAM	[78]
	Feed	Aflatoxins	30%	0.2-71.8 µg/kg	LC-MS/MS	[79]
		Fumonisins	100%	104-2999 µg/kg	LC-MS/MS	[79]
		DON	99%	124-2352 µg/kg	LC-MS/MS	[79]
	P.nut	Aflatoxins	90%	2.1-73.5µg/kg	HPLC	[80]
Kenya	Market maize	Aflatoxin		1- 46,400 ppb	VICAM-IAC	[81]
	Maize	Aflatoxin		Up to 160 µg/kg	ELIZA	[39]
	Feed	Aflatoxin	78%	52ng/g	HPLC	[47]
		Fumonisin	56%	67ng/g	HPLC	[47]
		Ochratoxin	50%	2ng/g	HPLC	[47]
	Lager beer	DON	100%	1.6-6.4 µg/l	ELISA	[82]
	Maize & maize pdts	Aflatoxins	83%	0.11-4593.93ppb	TLC	[83]
	Feed	Aflatoxins	95%	5.13-1123ppb	TLC	[83]
	Homegrown maize	Aflatoxins		354.53ppb	VICAM –IAC	[84]
	Maize	Aflatoxins	46.4%	0-58,000	IAC	[85]
	Maize	Aflatoxins		nd-48,000ppb	VICAM-IAC	[86]
	Maize	Aflatoxins	20.1%	>100 ppb	VICAM-1AC	[87]
	Wheat	DON	68%	up to 302 µg/kg	ELISA	[88]
	Feed	AFB_1	81%	up to 280 ppb	ELISA	[89]
	Dairy milk	AFM_1	99%	up to 600 ppt	ELISA	[89]
Ethiopia	Wheat	DON		40-490ppb	HPLC	[90]
	Cereals	OTA		5.1-2106ppb	HPLC	[90]
Tanzania	Cassava pdt	Aflatoxins	100%	>1.2ppb	ELISA	[91]
	Maize	Aflatoxins	18%	158 µg/kg	HPLC	[92]
		Fumonisins	52%	11,048 µg/kg	HPLC	[92]
Cameroon	Peanut	ZEA	57%	70 µg/kg	HPLC	[93]
	Breast milk	AFM_1	4.8%	0.005-0.625 µg/l	HPLC	[94]
	Diary milk	AFM_1	15.9%	0.006-0.527µg/l	HPLC	[94]
	Cocoa bean	OTA		0-48.02ppb	HPLC	[95]
	Cassava chips	Aflatoxins	25%	5.2-14.5 µg/kg	ELISA	[96]

Table 1. (Continued)

Country	Food/Commodity	Mycotoxin	Freq.	Conc.	Method	Ref.
	Maize	FB_1		6717 μg/kg	ELISA	[32]
		DON		380 μg/kg	ELISA	[32]
		ZEN		56 μg/kg	ELISA	[32]
	G.nut	AFB_1	97%	47 μg /kg	LC-ESI-MS/MS)	[97]
	Maize	AFB_1	30%	4 μg /kg	LC-ESI-MS/MS)	[97]
		FB_1	100%	2-2313 μg /kg	LC-ESI-MS/MS)	[97]
	P. nut meal(feed)	AFB_1		1.2-234.2 μg /kg	VICAM	[98]
	Broiler feed	AFB_1		1-383.2 μg /kg	VICAM	[98]
Coted'Ivoire	Peanut	Fumonisin	86%	0.6-64 μg /kg	HPLC	[99]
		Ochratoxin	86%		HPLC	[99]
		Aflatoxin	86%		HPLC	[99]
	Maize	OTA	100%	3-1738ng/g	HPLC	[100]
Zimbabwe	Sorghum	Fumonisins	61%	8-187ng/g	HPLC	[101]
	P.nut	Aflatoxins	17%	6.6-622.1 ng/g	HPLC	[101]
	Breast milk	AFM_1	9%	≤ 50 ng/l	ELISA	[102]
Morocco	Rice	OTA	90%	0.02-32.4ng/g	IAC, LC-FLD	[103]
	Dried raisins	OTA	30%	0.05-4.95ng/g	IAC, LC-FLD	[103]
	Paprika	AFB_1	95%	1.1-15.4ppb	TLC	[104]
	Diary milk	AFM_1	88.8%	0.001-0.117 μg /l	IAC,LC-FLD	[105]
	Cumin	AFB_1	25%	0.8-6.7 ppb	TLC	[104]
	Black pepper	AFB_1	15%	0.7-1.3ppb	TLC	[104]
	White paper	AFB_1	10%	2.8-3.7ppb	TLC	[104]
	Rice	OTA	26%	0.08-47ng/g	LC-FD	[106]
	Red paprika	AFB_1		2.88μg /kg	HPLC	[107]
	Corn	ZEA		14 μg /kg	HPLC	[107]
	Diary milk	AFM_1	27%	10-100 ng/l	IAC, LC-FD	[108]
	Rice	FB_2		176 μg /kg	LC-MS/MS	[109]
		NIV		97 μg /kg	LC-MS/MS	[109]
Egypt	Feed	ZEA	19%	9 ng/g	HPLC	[47]
		Fumonisin	81%	266 ng/g	HPLC	[47]
		Aflatoxin	19%	1 ng/g	HPLC	[47]
	Infant formula	AFM_1	43.2%	0.3-21.8 ng/l	ELISA	[110]
	Breast milk	AFM_1	69.6%	7.3-328.6 ng/l	ELISA	[110]
	Diary milk	AFM_1	49%	0.01-0.25μg/ kg	ELISA	[111]
	Breast milk	AFM_1	36%	0.5-5μg /kg	HPLC	[112]

Country	Food/Commodity	Mycotoxin	Freq.	Conc.	Method	Ref.
	Breast milk	AFM_1	100%	0.0014µg /kg	HPLC	[113]
	Breast milk	OTA		1.890 ng/l	HPLC	[114]
	Diary milk	AFM_1	38%	23-73 ng/l	ELISA	[115]
	Cheese	AFM_1	40%	52-87.6 ng/g	ELISA	[115]
Sudan	Diary milk	AFM_1	95.45%	0.22-6.9µg /kg	HPLC	[116]
	Breast milk	AFM_1	54.3%	0.401-2.561ng/g	HPLC	[117]
	Feed	Fumonisin	11	23ng/g	HPLC	[47]
		Aflatoxin	54	90ng/g	HPLC	[47]
		Ochratoxin	67	15ng/g	HPLC	[47]
	P. nut butter	AFB_1	100%	29-128ppb	VICAM	[118]
	Sesame oil	AFB_1	43.75%	0.2-0.8 µg/kg	HPLC	[119]
	G. nut oil	AFB_1	3.57%	0.6 µg/kg	HPLC	[119]
	P.nut butter	Aflatoxins	100%	26.7- 853 µg/kg	HPLC	[120]
	Feed	Aflatoxins	64.29%	130.63 µg/kg	HPLC	[121]
Sierra Leone	Breast milk	AFM1	23%	0.003-336 µg/kg		[97]
Ghana	Feed	Aflatoxin	72%	26ng/g	HPLC	[47]
		ZEA	11%	178ng/g	HPLC	[47]
		Fumonisin	89%	500ng/g	HPLC	[47]
	Breast milk	AFM_1	80%	0.194ng/g		[102]
Algeria	Feed	Fumonisin	64%	977ng/g	HPLC	[47]
		B-Trichothecen	0%	0ng/g	HPLC	[47]
	Wheat & derived pdts	AFB_1	56.6%	0.13-37.42 µg/kg	RF-HPLC	[122]
B/Faso	Maize	Fumonisin		10-3,100ng/g	HPLC-FD	[123]
Malawi	Maize based beer	$FB_1 + FB_2$	100%	1745µg/kg	LC/MS	[124]
		Aflatoxins	88.8%	90 µg/kg	LC/MS	[124]
DR.Congo	Maize	FBs	100%	2207.5ppb	TLC/HPLC	[36]
		Aflatoxins	95%	257.5ppb	TLC/HPLC	[36]
		OTA	45%	61.9ppb	TLC/HPLC	[36]
		ZEA	92.5%	247.8ppb	TLC/HPLC	[36]
	Beans	FBs	83.3%	243.5ppb	TLC/HPLC	[36]
		Aflatoxins	80%	215.5ppb	TLC/HPLC	[36]
	P.nut	AFB_1	70%	1.5-937 µg/kg	TLC	[125]
		Aflatoxins	100%	2.19-1258 µg/kg	HPLC	[80]
	Maize	AFB_1	50%	1.14-71.67ppb	ELISA	[91]
	Beans	OTA	82.4%	47.8ppb	TLC/HPLC	[36]
	Beans	ZEA	90%	185.2ppb	TLC/HPLC	[36]

Table 1. (Continued)

Country	Food/Commodity	Mycotoxin	Freq.	Conc.	Method	Ref.
	Peanuts	Aflatoxins	100%	368.7ppb	TLC/HPLC	[36]
		OTA	72.5%	11.6ppb	TLC/HPLC	[36]
Lesotho	Maize	DON	100%	1,469μg/kg	HPLC	[126]
		Moniliformin		5-320 μg/kg	HPLC	[126]
		Beauvericin		1.16 μg/kg	HPLC	[126]
		Fusaproliferin		128.5 μg/kg	HPLC	[126]
		AFB_1		0.43 μg/kg	HPLC	[126]
		FB_1		7-936 μg/kg	HPLC	[126]
		ZEA		10.2 μg/kg	HPLC	[127]
		Fusaric acid		56.8 μg/kg	HPLC	[126]
Uganda	Maize	AFB_1		1-1000ng/g	VICAM	[128]
	Maize	Aflatoxins		>20ppb	VICAM/TLC	[129]
	Stored maize	Fumonisins		5.7 mg/kg	VICAM	[130]
	Cassava flour	Aflatoxins	20%	16.0 ppb	VICAM	[131]
	Eshabwe soup	Aflatoxins	16%	18.0 ppb	VICAM	[131]
	G.nut	Aflatoxins	3%	11.5 ppb	VICAM	[131]
Mozambique	Peanut	AFB_1		1.2-5.5ng/g	HPLC	[132]
Benin	Cowpea seeds	Fumonisins		0.8-25.30 μg /g	HPLC	[72]
	Cowpea	AFB_1	0.03	0.03 μg /g	HPLC	[133]
		FB_1	0.06	3.58 μg /g	HPLC	[133]
	Dried yam chips	Aflatoxins	80%	Above 4ng/g	HPLC	[134]
	Cowpea seeds	AFB_1	6.52%	3.58 μg/kg	HPLC	[133]
		FB_1	3.26%	0.03 μg/kg	HPLC	[133]
	Cotton seed feed	AFB_1		26-164 μg/kg	VICAM	[78]
	Maize	Fumonisins		nd-12 mg/kg	VICAM	[135]
Angola	Maize	Fumonisins	100%	8-5929.0 μg/kg	HPLC	[136]
		Aflatoxins	81.7%	0.4-425.9 μg/kg	HPLC	[136]
		Ochratoxins	33.3%	0.5-40.3 μg/kg	HPLC	[136]
		ZEA	8.3%	0.7-143.0 μg/kg	HPLC	[136]
Rwanda	Peanut	Aflatoxins	75%	0.06-1755.9 μg/kg	HPLC	[137]
		Fumonisins	90%	0.2-5.0 μg/kg	HPLC	[137]
		OTA	95%	0.16-1155.8 μg/kg	HPLC	[137]
		DON	40%	2.5-254.2 μg/kg	HPLC	[137]
		ZEA	60%	0.2-5.2 μg/kg	HPLC	[137]

Country	Food/Commodity	Mycotoxin	Freq.	Conc.	Method	Ref.
	Maize	Aflatoxin	85%	0.2-3210 µg/kg	HPLC	[137]
		Fumonisins	85%	0.2-4591.0 µg/kg	HPLC	[137]
		OTA	5%	8.9-144.4 µg/kg	HPLC	[137]
		DON	85%	0.3-225.1 µg/kg	HPLC	[137]
		ZEA	35%	0.0-0.1 µg/kg	HPLC	[137]
	Cassava	Aflatoxins	65%	0.05-534.9 µg/kg	HPLC	[137]
		Fumonisins	30%	0.5-1.0 µg/kg	HPLC	[137]
		OTA	10%	33.8-44.2 µg/kg	HPLC	[137]
		DON	70%	0.3-6.1 µg/kg	HPLC	[137]
		ZEA	65%	0.03-3.0 µg/kg	HPLC	[137]
	Beans	Aflatoxins	70%	0.2-154.9 µg/kg	HPLC	[137]
		Fumonisins	100%	0.4-7.1 µg/kg	HPLC	[137]
		OTA	30%	8.9-144.4 µg/kg	HPLC	[137]
		DON	95%	0.2-35.4 µg/kg	HPLC	[137]
		ZEA	60%	0.02-2.6 µg/kg	HPLC	[137]
	Rice	Aflatoxin	80%	0.2-55.6 µg/kg	HPLC	[137]
		Fumonisins	70%	0.08-0.62 µg/kg	HPLC	[137]
		OTA	20%	7.1-36.9 µg/kg	HPLC	[137]
		DON	90%	0.11-419.7 µg/kg	HPLC	[137]
		ZEA	30%	0.02-0.07 µg/kg	HPLC	[137]
Togo	Okra	AFB_1		5.4 µg/kg	RF-HPLC	[138]
		AFB_2		0.6 µg/kg	RF-HPLC	[138]
	Hot chilli	AFB_1		3.2 µg/kg	RF-HPLC	[138]
Zambia	Maize	Fumonisins	96.4%	0.02-21.44 ppm	ELISA	[139]
		AFB_1	21.4%	0.7-108.39 ppb	ELISA	[139]
	Maize	Fumonisins	100%	3.7-192 ppm	ELISA	[140]
		Aflatoxins	100%	0.2-10 ppb	ELISA	[140]
Libya	Dairy milk	AFM_1	71.4%	0.03-3.113 ng/l	HPLC	[141]

Table 1. (Continued)

Country	Food/Commodity	Mycotoxin	Freq.	Conc.	Method	Ref.
	White soft cheese	AFM_1	75%	0.11-0.52ng/g	HPLC	[141]
	Breast milk	AFM_1	2.4%	0.015-0.343 pg/ml	HPLC	[142]
	Infant food	Aflatoxins	2.4%	44.5 μg/kg	HPLC	[127]
Tunisia	Barley	OTA	40%	11-940 μg/kg	HPLC	[143]
	Cereals	OTA	36%		HPLC	[143]
	Cereals	OTA	59.8%	0.8-36.4 ng/g	ELISA	[144]
		AFB_1	37%	0.7-50 ng/g	ELISA	[144]
	Wheat pasta	FB_1		184 μg/kg	LC-MS/MS	[109]
Mali	Okra	AFB_1		5.4 μg/kg	RF-HPLC	[138]
		AFB_2		0.6 μg/kg	RF-HPLC	[138]
	Hot chilli	AFB_1		3.2 μg/kg	RF-HPLC	[138]

3.1. Mycotoxins Produced by *Aspergillus* and *Penicillium* Species

Aspergillus and *penicillium* species are closely related and can produce the same mycotoxins. These fungi grow on a wide variety of food and feedstuffs worldwide especially cereals and oilseeds [145, 146]. Important *Aspergillus* and *penicillium* mycotoxins include Aflatoxins and Ochratoxins.

3.1.1. Aflatoxins

Aflatoxins (AF) are produced by fungi belonging to the family, *Aspergillus* on a variety of crops. The highest level of contamination has been recorded in peanuts, maize, Brazil nuts, pistachio nuts and cotton seeds. Lower levels are recorded in almonds, pecans, walnuts, raisins, spices and figures [145]. The four main naturally occurring aflatoxins consist of AFB_1, AFB_2, AFG_1 and AFG_2 with AFB_1 being the most important in terms of toxicity and occurrence [147]. Other members of the aflatoxin group include AFM_1 which is a hydroxylated metabolite found mostly in animal tissues and fluid (milk and urine) as a metabolic product of aflatoxin B_1 (AFB_1) [148].

AFB_1 is the most potent hepatocarcinogens in nature and is classified by IARC as Group 1 carcinogen [149]. The liver is the predominant target of aflatoxins. AFB_1 and its metabolites have been found in the heart, kidney, brain tissues and urine or feaces of humans. Aflatoxins have been implicated in human diseases including liver cancer, encephalopathy and fatty degeneration of the viscera in children, known as Reyes syndrome, Indian childhood cirrhosis, chronic gastritis, aflatoxic hepatitis in India and Kenya. Reports by different authors, have also shown aflatoxin to be present in blood, urine and livers of children with kwashiorkor [150, 151]. Positive correlation between aflatoxin content of food grains and occurrence of liver enlargement in children has been shown [19]. In Kenya, the average birth weight of the offspring of women exposed to aflatoxin prenatally was lower than those who had not been exposed [152]. There is also incidence of decreased vitamin A and assimilation of other nutrients. The chronic incidence of aflatoxin in diets is evident from the presence of Aflatoxin M_1 in human breast milk in Ghana, Nigeria, Sierra Leone and Sudan as well as in umbilical cord blood samples in Ghana, Kenya, Nigeria and Sierra Leone [153]. Epidemiological studies have shown a strong connection between exposure to aflatoxins and primary liver cancer [12].

Aflatoxin exposure in children is also associated with child stunting and child neurological impairment [154]. A cross- sectional study in Benin and Togo investigated aflatoxin exposure in children and it was revealed that exposure to this toxic contaminant increases markedly after weaning and associated with reduced growth [153, 155, 156]. Studies on Gambian children and human adults in Ghana have shown significant correlation between aflatoxin exposure and suppression of the immune system [157, 158]. Thus, this may likely increase the susceptibility of children to infections and cause failure of immunizations. Aflatoxins have been identified as a risk factor in neonatal jaundice (NNJ), although recent epidemiological studies show inconclusive evidence of the association between NNJ and exposure to aflatoxins [159]. Other concerns of dietary aflatoxin exposure include causing a variety of symptoms on various animal species. Aflatoxin B_1 (AFB_1) has been shown to cause acute toxicity at concentration (LD50 = 0.3-18) milligram per kilogram body weight to all studied animals, birds and fish species [160]. It was shown to have adverse effects on growth and antinutritional effects in animals, for example, AFB_1 was reported to reduce hepatic vitamin A significantly in a variety of animals, including chickens [161]. However, in all animals, aflatoxin damages the liver, decreases reproductive performance, reduces egg production, causes embryonic death, teratogenicity (birth defect), tumors and suppresses immune system functions [21] even when low levels are consumed. Exposure to Aflatoxins also causes general reduction in weight gain for a variety of production animals, including cattle, pigs and poultry. Also, in dairy animals milk production can drop with the consumption of aflatoxin contaminated foods [162]. Although many countries have regulatory limits for aflatoxins in foods/feeds, outbreaks of poisoning commonly occur. The 2004 outbreak of aflatoxin poisoning in eastern Kenya resulted in the death of 125 of the 317 cases of acute hepatic failure which was linked to consumption of contaminated maize [81, 84]. Furthermore according to [7] aflatoxin contaminated food was implicated in the death of over 100 people in Nigeria in 2005. The occurrence of Aflatoxins in foods and feeds in different countries have also been published; South African peanut butter [163], Ghanaian fermented maize dough [164], Peanut samples from Cameroon [165], Egyptian maize and maize products [166] Beans and wheat from Nigeria [167], Animal feed from South Africa [47, 168, 169], Sudan feed [47, 121].

3.1.2. Ochratoxins

Ochratoxin A (OTA) is a mycotoxin produced mainly by certain species of **Aspergillus and Penicillium** [170, 171]. Other members of the ochratoxin family include OTB and OTC, although OTA is the most commonly detected and most toxic among the three. Toxicological reports have shown that OTA is nephrotoxic, carcinogenic, teratogenic, hepatotoxic and immunotoxic and has been classified by the International Agency for Research on Cancer (IARC) as possibly carcinogenic to humans (Group 2B) [149]. Ochratoxin A has been found as a contaminant in a wide variety of commodities such as raisins, barley, soy products and coffee, but at relatively low levels [4]. Its occurrence in food and feedstuffs has been reported in many African countries, for example; Nigeria [167, 172, 173], Côte d'Ivoire [99, 100],

Kenya, Algeria, Sudan [47], Morocco [105], South Africa [174, 175], Tunisia [143], Cameroon [165].

Ochratoxin A has also shown to contaminate human milk and thus can cause kidney disorders in breast –fed infants. The kidney is the primary target organ for OTA toxicity but can also damage the liver at sufficiently high concentration [148]. There have been speculations that OTA is involved in the human disease, Balkan endemic nephropathy [2]. The condition is a progressive chronic nephritis that occurs among people living in the former Yugoslavia and it is also evident that it contributed to human renal and urinary tract tumors [148]. It is also suspected of playing a role in chronic interstitial nephropathy in North Africa. It has been detected in blood and serum and analysis of urine from children in Sierra Leone detected ochratoxin and aflatoxin [2]. The EU scientific committee recommended that OTA be reduced to TDI of below 5ηg/kg body weight. It has also been hypothesized that OTA might be a risk factor for testicular cancer [2]. Ochratoxin A has elicited great concern and has been previously and extensively reviewed by various authors [176].

Abortions have also been reported alongside other effects such as polyuria, depression, reduced weight gain, low specific gravity of urine as well as dehydration. Fortunately ruminants can breakdown and detoxify OTs to other less harmful metabolites and thus, mycotoxins are not usually a problem [26]. Exceptions may occur in high milk producing cows because they have increased rumen passage rate and thus faster processing of contaminated feed may overwhelm the rumen micro flora so that they may not be able to denature all the toxins [27, 28]. Also young stocks that have not developed functioning rumens may not be able to detoxify the toxins [21].

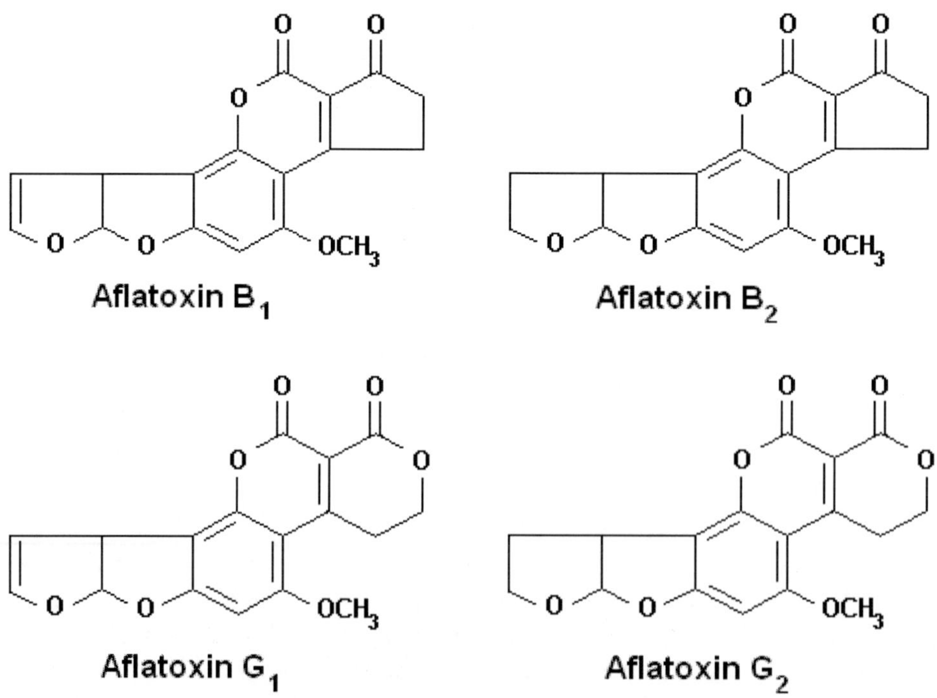

Figure 1. Chemical structure of aflatoxin B_1, B_2, G1 and G_2 [8].

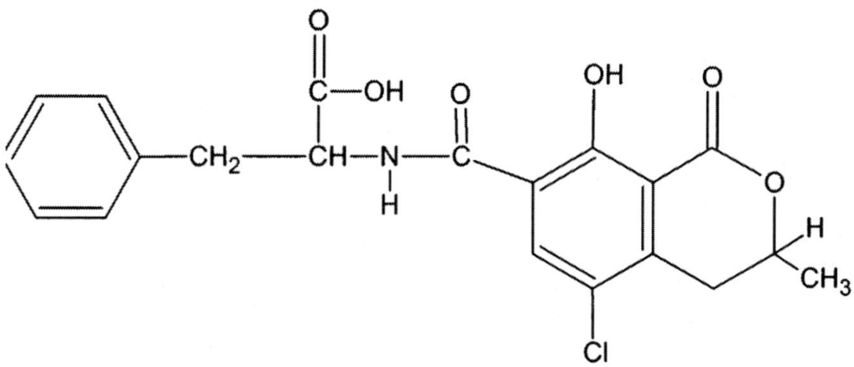

Figure 2. Chemical structure of Ochratoxin A [177].

3.2. Mycotoxins Produced by *Fusarium* Species

Fusarium species are pathogens on a wide variety of crop plants throughout the world. They cause mainly stem and root rots and frequently infect the grains of cereal crops. Several species of *Fusarium* produce mycotoxins on a range of cereal crops which include wheat, maize, rice barley and oats. Important *Fusarium* mycotoxins include fumonisins zearalenone, trichothescenes (t-2 toxin, deoxynivalenol (DON) and nivalenol) [148].

3.2.1. Fumonisins

Fumonisins were first discovered in 1988 and the most abundant member is the fumonisin B_1, [2]. They are known to be produced by *Fusarium verticillioides* (formerly known as *F. moniliforme*), a common fungal pathogen of maize. Also *Aspergillus niger* has now been found to form fumonisin B_2 (FB_2) [178]. FB_1 has been implicated in human esophageal cancer in the Eastern Cape, South Africa, linked to the consumption of contaminated maize, as well as in China and in North east Italy [2, 179]. According to Marasas et al., (2004), because FB_1 reduces uptake of folate in different cell lines, it is likely to be implicated in neural tube defects in human babies [180]. FB_1 has also been associated with high incidence of neural tube defects in babies of mothers consuming fumonisin-contaminated maize along the Texas-Mexico border [181]. The (IARC) designated FB_1 in Group 2B as ‘possibly carcinogenic to humans’ [182]. Its structural resemblance to the sphingoid bases explains the inhibition of the biosynthesis of sphingolipid complexes leading to cell destruction and subsequent cell death [47, 148]. Fumonisins are relatively heat stable, but in heat-processed foods FB_1 occurs in covalently bound forms [148]. Anyway, not like most well-known mycotoxins, which dissolve in organic solvents, fumonisins are water soluble which makes them challenging to be studied [2].

On the other hand, though fumonisin has a relatively simple chemical structure, its blockage of sphingolipid metabolism can have varied and complex effects on different animal systems [183]. Fumonisin impairs immune function, causes liver and kidney damage, decreased weight gains and increased mortality rates. Horses are the most sensitive animals to fumonisin toxicity and levels ranging from 0.2-126 ppm FB_1 were found in feed samples associated with outbreaks of leukoencephalomalacia, LEM (hole in the head syndrome), which is a necrosis of the brain, in North America, South America and South Africa [184,

185]. Porcine pulmonary edema in pigs is associated with the consumption of fumonisin-contaminated feed [186, 187]. Dairy cattle may be more susceptible to fumonisins than beef cattle due to greater production stress [26].

3.2.2. Zearalenone

Zearalenone, also known as F-2 is a metabolite produced by fungi of *Fusarium* genius. It is commonly found in maize but can also be found in other crops worldwide [4, 5, 148]. Several publications have reported its occurrence in foods as well as in livestock feeds in different African countries such as South Africa [168], Nigeria [50]. As an endocrine disruptor, they are widely believed to reduce male fertility in human and animal populations. It has been claimed that the high frequency of early menarche in Puerto Rico might be due to ZEA and related compounds in diet [2]. Due to its high estrogenic activity Zearalenone has been used to treat post-menopausal symptoms in women. Despite that, research has also shown the potential for ZEA to provoke the growth of human breast cancer cells containing estrogen response receptors [189]. However, the biological potency of this compound is high, but its actual toxicity is low [2] and was placed in Group 3, which means not classifiable as to their carcinogenicity to humans [149].

Since Zearalenone exhibits estrogenic activity and appears to bind to estrogen receptors [148], it has also been implicated in numerous mycotoxicoses in domestic animals. The ingestion of contaminated grains by animals can result in the development of feminine characteristics in males, premature sexual development of young females, and infertility in adults, abortion, stillbirths and birth of deformed offsprings, atrophy of the testes, and decreased libido in males. Cattle may have reddened, swollen vulvas and increased size of nipples while vaginal and rectal prolapse may follow. Also noticeable is a reduction in feed intake and perhaps feed refusal. The effects are clearly visible in pigs. Infact pigs appear to be the most susceptible spp. to ZEA [5, 148]. Recommended safe human intake of ZEA per day is in the estimate of 0.05 microgram per kilogram body weight [2].

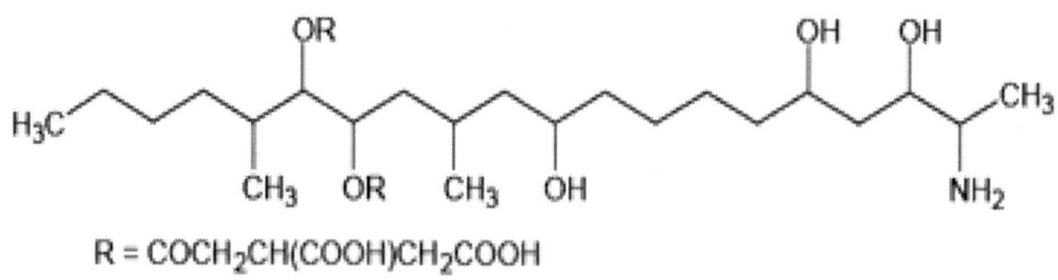

Figure 3. Chemical structure of fumonisin B_1 [188].

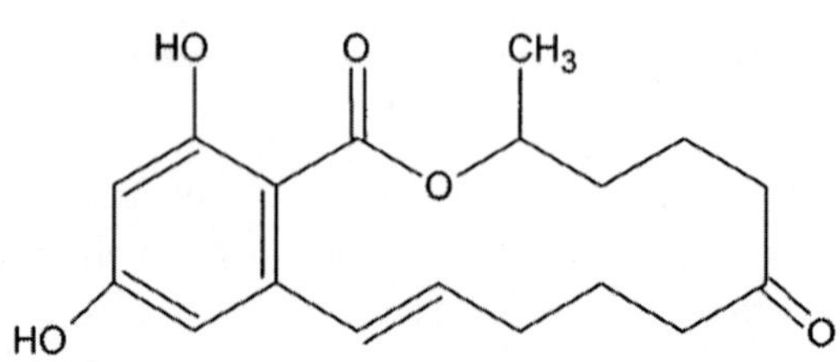

Figure 4. Chemical structure of ZEA [177].

3.2.3. Trichothescenes

Trichothecenes are the largest group of mycotoxins comprising of Type A e.g.: T-2 toxin and Type B e.g.: deoxynivalenol (DON) and nivalenol. They can contaminate commodities such as corn, wheat, barley and oats. The most abundant source of trichothecenes contamination in cereal grains today is as a result of *Fusarium* head blight, mostly caused by type-B trichothecene-producers [190]. Trichothescenes inhibit protein synthesis in the cell and induce apoptosis. A survey conducted by Rodrigues et al. (2011) on the occurrence of mycotoxins in agricultural products, feeds and feed materials obtained from the Middle East and Africa showed the following results for level of trichothescene contamination in different African countries: Sudan-33% (100μg/kg), Nigeria-58% (316μg/kg), Kenya-48% (422μg/kg), Ghana-50% (955μg/kg), Algeria-0% (0 μg/kg), Egypt-38% (643 μg/kg) and South Africa-87% (1469 μg/kg). In South Africa, B-trichothecenes were the main contaminant, with 87% of samples testing positive at a mean contamination level of 1469 μg/kg.

3.2.4. Deoxynivalenol (DON)

Deoxynivalenol (DON) also known as vomitoxin is produced principally by *Fusarium graminearum* and in some geographical areas by *F. culmorum* [148]. DON may co-exist with zearalenone, which is also produced by these organisms. Small grains such as oats, wheat and barley are the main crops affected but DON can be found in corn as well [4]. DON was placed in Group 3, which means not classifiable as to their carcinogenicity to humans [149]. DON has been associated with reduced milk production in dairy cattle, vomiting by swine consuming contaminated feed or their refusal to eat feed containing the toxin due to unpalatability of the feed. Also reduced feed intake results in severe weight loss and inhibits reproductive performance and immune function in several animal spp [191, 192]

Humans are thought to exhibit a similar vomiting syndrome when they consume DON-contaminated grains [148]. Also, increased incidence of upper respiratory tract infection was reported in children who consumed DON contaminated wheat bread for more than one week and the illnesses subsided when consumption of the bread was discontinued [193]. DON's toxicity action is understood to be by the modulation of innate immune system [194]. Chronic low dose exposure to DON has shown to cause anorexia, impaired weight gain, growth hormone deregulation and aberrant IgA production whereas acute high dose exposure evokes gastroenteritis, emesis and a shock-like syndrome [194]. DON does not appear to be significantly carried over into tissues or fluids of animals consuming toxic levels and also baking and malting using contaminated wheat and barley are adversely affected [148].

3.2.5. T-2 Toxins

T-2 toxin is produced by fungi of the genus *Fusarium*, on grains allowed to remain unharvested in the field. T-2 toxins are irritative, hemorrhagic and can cause necrosis throughout the digestive tract. They also slow down regeneration of cells in the bone marrow and spleen, impair the functions of the immune system and cause some changes in the reproductive organs. Affected animals show symptoms of weight loss, poor utilisation of feed, lack of appetite, vomiting, bloody diarrhea, abortion and (in severe cases) death [191]. T-2 toxin was responsible for Alimentary Toxic Aleukia (ATA) that caused the death of thousands of people in Russia. Alimentary Toxic Aleukia, (ATA) causes fever, bleeding from the skin, nose, throat and gums, necrosis and in addition to their cytotoxicity, they also have immunosuppressive effects [2].

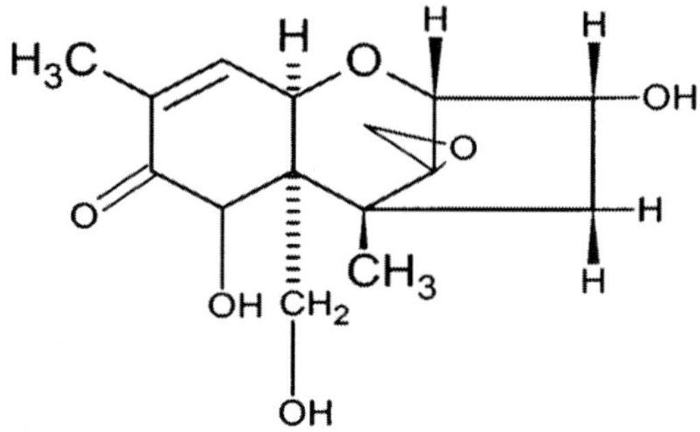

Figure 5. Chemical structure of DON [177].

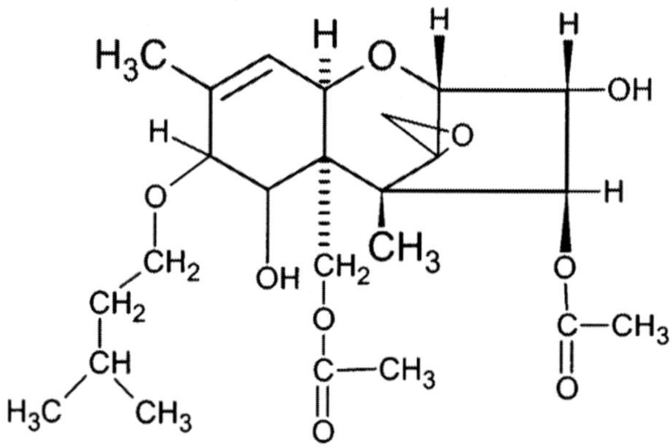

Figure 6. Chemical structure of T-2 toxin [177].

3.2.6. Nivalenol

Nivalenol with molecular formula $C_{15}H_{20}O_7$, is one of the trichothescenes toxins produced by *Fusarium* fungi. It is also a major problem for pig farmers. Nivalenol contamination of feed leads to feed refusal because of the unpalatability of the feed and may result in vomiting [191]. Maize containing more than 5% infected kernels should not be included in rations for pig. Cattle are more resistant to these effects and chickens do not seem to be affected. There are no reported effects on humans published so far. Nivalenol was also placed in Group 3, which means not classifiable as to their carcinogenicity to humans [149].

The co-occurrence of mycotoxins in agricultural commodities is a common phenomenon and indeed has been linked to chronic toxicity. However, their levels vary among commodities and also from region to region. Table 1 above summarizes mycotoxin contamination of different commodities in Africa as reported in the literature within the last decade.

4. Updates on Methods of Extraction, Detection and Quantitation of Different Mycotoxins

The detection of mycotoxins is crucial for food safety and food security in order to prevent outbreaks of mycotoxicosis which might be a public health concern. The presence of mycotoxins in a particular commodity is an indication that one or more fungi may have grown on the material at some stage [4]. The fact that a mycotoxin producing fungus is present does not mean that the associated toxin will also be present, as several factors are involved in its formation [195]. Similarly, the absence of any visible mold will not guarantee freedom from toxins as the mold may have died out while leaving the toxin behind [196].

Determination of mycotoxins in food and feeds as well as in biological samples is always challenging due to the great variety of sample matrices and also the different molecular species that need to be determined [197]. The fact that most mycotoxins are toxic in low concentrations requires sensitive and reliable methods for their detection [15, 25, 40]. Furthermore, given that these compounds have varied chemical structures [15, 24, 25, 198]. It is not possible to use one standard method to detect all mycotoxins thus; chemical analysis must be tailored to the individual mycotoxins, which invariably calls for different analytical methods [25, 37, 196].

What works well for some molecules may not be suitable for others of the same properties or for the same molecule in a different environment. Therefore, depending on the physical and chemical properties, of the mycotoxin of interest, procedures and analytical techniques should offer flexible and broad-based methods of detection [199]. A successful detection method should at least be sensitive and have a high degree of flexibility, over a wide range of compounds, but which can be very specific when required. All techniques should be reproducible to a high level and the results must be relevant and easy to analyse [198]. However, the methods of mycotoxin analysis may be broadly grouped into physicochemical analysis, biological assays and immunoassays [19]. Nevertheless, a number of recent reviews have been published on the methods of analysis of mycotoxins [11, 25, 196, 200, 201].

4.1. Physicochemical Analysis

Physicochemical/chemical assays are based on the differences in the physicochemical properties of different mycotoxins. Their solubility in certain organic solvents helps in their extraction from commodities. Their insolubility in some other solvents, for instance for aflatoxin in diethyl ether, petroleum ether and hexane provides a method for their separation from interfering substances and pigments. Their characteristic fluorescence and absorption under long wave UV light, aid their detection and estimation [19].

Generally, physicochemical assays involve sampling of the commodity (food and feed), sample extraction, sample clean-up and separation, detection and quantitation [2, 19, 24, 37].

Table 2. Physico-chemical properties of major Mycotoxins

Mycotoxin	Mol. Formula	Mol. Weight	Solubility	M.P	Storage	Physical state
Aflatoxin B_1	$C_{17}H_{12}O_6$	312.3	Soluble in DCM,DMSO or Methanol	268-269°C	4°C	Off-white powder, blue fluorescence
Fumonisin B_1	$C_{34}H_{59}NO_{15}$	721.8	Soluble in water, ACN. Insoluble in Chloroform, Petroleum ether, Hexane	147-149 °C	4°C	White to off-colour powder
DON	$C_{15}H_{20}O_6$	296.3	Soluble in ACN, Methanol, Ethyl acetate, water	151-153°C	4°C	White crystalline solid
T-2 toxin	$C_{24}H_{34}O_9$	466.6	Soluble in DCM, DMSO, Ethanol, Ethyl acetate, Petroleum ether, Water	151-152°C	-20°C	White crystalline solid
OTA	$C_{20}H_{18}ClNO_6$	403.8	Soluble in DMSO, Methanol, Ethanol	105-110°C	4°C	Yellow crystals, blue fluorescence
ZEA	$C_{18}H_{22}O_5$	318.4	Soluble in Benzene, ACN, DCM, Methanol, Acetone, Hexane, Water	159-163°C	-20°C	White powder
NIV	$C_{15}H_{20}O_7$	312.3	Soluble in DCM, DMSO, Methanol, Water	127-130°C	-20°C	White crystalline powder

[188, 202] (DMSO-Dimethylsulphoxide, DCM-Dichloromethane, ACN –Acetonitrile).

4.1.1. Sampling

The obvious reason for sampling a given lot of commodity is to obtain a portion for the estimation of the level of mycotoxin contamination of a particular lot [25, 203]. This may be done either for routine monitoring, surveillance or for quality control purposes [203] hence the method of sampling should be designed for the intended objective. Sampling is a very important stage in the analysis of mycotoxins since non-representative sample can invalidate the result of the analysis [196]. Nearly 90% of the errors associated with mycotoxin assay could be ascribed to how the original sample was collected. Mycotoxin contamination is usually heterogeneous within any given food commodity which constitutes a problem in obtaining representative sample for analysis [25, 196].

The problem is more severe with agricultural commodities such as grains and nuts. Within a given lot of such commodity, mycotoxin contamination can occur in pockets of high concentration and may not be randomly distributed [19]. A relatively small percentage (less than 0.5%) of the kernels can be contaminated and even within a single contaminated kernel, the distribution is not uniform, it may be concentrated in the heart and low in the outer portion while in others, it is concentrated in the outer portion and almost absent in the centre [19]. Individual peanut kernels have been reported to contain Aflatoxin B_1 at levels as high as 1.1mg/g. The presence of such kernels with extremely high concentrations would be sufficient to contaminate 10 kg sound kernels with high levels of Aflatoxin [160, 204-206].

Further processing of such a lot of peanut into peanut butter or of a lot of corn containing one highly contaminated kernel into meal would result in a finished product high in aflatoxin, yet the initial sample would have been negative for aflatoxin if it had not included the contaminated kernel. Hence, the importance of adequate sample selection for mycotoxin analysis cannot be overemphasised. Sampling plans for obtaining representative lot samples have been published by several authors as well as techniques for sample preparation [203, 205, 207-213].

It is usually felt that there is no error-free method of sampling, so it is suggested that whatever the quantity available, the sample should be ground and thoroughly mixed to obtain a homogeneous distribution of mycotoxins. The subsamples should be taken from as many points as possible [19]. It is immaterial whether a 5, 10 or 25g sample is analysed [214]. Analysed 5, 10 and 25g test portions of finely ground samples of contaminated corn and statistically compared them with 50g test portions for aflatoxin concentration. The results showed good agreement for all weights of the samples analysed.

In addition, after sampling, all possible precautions should be taken to avoid any alteration of samples by human error, spillage, contamination, loss of fine particles during handling or grinding, moisture uptake or loss, biodegradation and so on [203]. The selection of methods used for mycotoxin extraction will depend on the nature of the sample as well as on the nature of the toxin [2, 198].

4.1.2. Extraction

The purpose of extraction is to release as much of the mycotoxin from the food and feed commodity as possible into a solvent suitable for subsequent clean up and detection. The choice of extraction solvent used is determined by the structure of the toxin, the type of commodity, safety considerations and the analytical procedure. In this case, the polarity of the mycotoxins to be extracted is of major concern and thus, some solvents used for extraction are more suitable than others [37, 203]. Furthermore, mycotoxins with acidic functional groups such as fumonisins and ochratoxin A require an the addition of an acidic modifier inorder to enhance recoveries [25]. Ideally an extraction solvent is able to remove only the mycotoxin of interest from the sample matrix. Also, it should easily be recoverable, non-toxic and non-inflammable. Other aspects such as volatility, stability, transparency to UV light and environmental impacts of solvents are also important. Volatile organic compounds (VOC) may cause ozone depletion. Previously, chlorofluorocarbons (CFCs) were the great solvent choice due to their effectiveness, low cost and non- flammability. But today there are restrictions on CFCs and chlorinated solvent use due to environmental concerns [29, 197]. Extraction of mycotoxins from food and feed commodity is often performed with mixtures of water and relatively polar solvents such as acetone, acetonitrile, ethyl acetate, methanol and chloroform. Methanol and acetonitrile are by far the most common solvents used for extracting major mycotoxins, with the notable exception of patulin for which ethyl acetate is preferred [215, 216].

Furthermore, after evaluation of the extraction efficiency of some solvents, it has been reported that using either methanol-water or acetonitrile-water mixtures for the extraction of mycotoxin contaminated samples yielded different recovery rates. Possible reasons for the different recoveries are either interference with matrix compounds or possible absorption of water into the dry matter. However, in further evaluation using fumonisin (FUM) extraction solvents on corn-based samples, only methanol-water and acidified methanol water mixtures were used. A higher recovery for maize gluten using acidic extraction was shown [217]. For more polar metabolites than fumonisin such as deoxynivalenol (DON), extraction with only water is common [29] and other compounds need non-polar organic solvents to be extracted. Hydrophobic toxins such as aflatoxins rely on the use of organic solvents. Möller and Nyberg have evaluated the efficiency of four extraction solvent mixtures in the analysis of aflatoxins from a certified reference peanut meal. The extraction mixtures, chloroform-water (100 + 10), acetonitrile-water (60 + 40), acetone-water (85 + 15), and methanol-water (80 + 20) were tested. The recoveries for the reference material containing 206 μg/kg of aflatoxin were 211, 204, 160 and 123 μg/kg, respectively. Thus, solvents containing chloroform-water and acetonitrile-water mixtures showed good recoveries close the reference material [218]. For T-2 toxin, an acetonitrile-water extraction procedure was developed and tested to compare it with often suggested methanol-water mixtures. It showed that acetonitrile is more effective in the extraction of complex matrices. On the other hand, methanol is as suitable as acetonitrile in the extraction of simple matrices [219].

Sample to solvent ratio is also another aspect to consider in the extraction procedure. A rule of thumb is that 4-5ml of solvent/g of contaminated sample be used in the extraction of mycotoxins while sometimes higher sample to solvent ratios (e.g., 10ml/g) are proposed. According to the experiment on extraction efficiency of different extraction times, it was reported that it is easier to extract spiked samples than naturally contaminated ones [220]. This is because the added mycotoxins are not so strongly embedded in the matrix like the native ones, hence a longer extraction time is needed for naturally contaminated samples [25, 40]. In general, mycotoxins can be extracted from feed and food matrices either by shaking with mixtures of water or other polar solvents in a mechanical shaker or high-speed blending. Blending is more rapid than shaking when small numbers of samples are being analysed. However if large a number of samples are being analysed, the shaking may be preferable because modern shakers are capable of holding up to 20 samples concurrently. Then the liquid is separated from the solids either by filtration or centrifugation and will be ready for either clean-up or applied directly to the determinative step in the procedure.

In case of liquid samples such as fruit juices or milk, liquid-liquid extraction (LLE) can be applied, now being replaced by solid-phase extraction or immunoaffinity column clean-up. The LLE uses two immiscible liquid phases, which during the extraction, the analyte and any other compounds with similar properties will migrate into the extracting solvent until equilibrium is reached. This way, the target compounds can be condensed in a solvent and interfering substances removed [29]. The procedure is effective for several toxins and works well in small scale preparations. However it is time consuming, labour intensive requires large volumes of solvent and dependent on which matrix is being used, and which compounds are being determined. The disadvantage is the possible loss of sample by adsorption onto the glassware [29, 196]. Beside the commonly used extraction techniques, emerging extraction techniques such as pressurised liquid extraction (PLE), supercritical fluid extraction (SFE), microwave-assisted extraction (MAE) or ultrasonic extraction (USE), have been evaluated and established in the analysis of mycotoxin [25]. By comparing MAE, USE and PLE with conventional shaking and blending, it was shown that USE and MAE have higher extraction efficiency [221]. These extraction techniques usually are coupled to HPLC or LC-MS. Nevertheless, after extraction, additional clean-up steps are necessary in most cases to eliminate matrix interferences.

4.1.3. Clean-up

The goal of sample clean-up techniques is to remove co-extracted substances in the sample extract that may interfere with the detection of target analytes. In addition to removing interfering substances, the clean-up may further concentrate the analyte prior to the determinative step [29, 197, 215]. A variety of clean-up methods have been used, which include solid-phase extraction (SPE) columns, Ion-exchange columns and Immuno-affinity columns (IAC).

4.1.3.1. Solid-phase Extraction (SPE)

Formerly, solvent-consuming techniques such as LLE were used, which are now replaced by solvent-saving methods using SPE. SPE have gained enormous interest in the field of analytical chemistry including pharmaceutical, food and environmental analyses. This is because SPE is easy to perform, rapid and can simply be automated. There is no need to use chlorinated solvents and extraction time is reduced [196]. An SPE technique is based on partitioning analytes and interfering compounds between a mobile and stationary (immobile) phase. The stationary phase within the cartridge is composed of a solid adsorbent or an immobile liquid phase. C18, octadecy-silica is the most useful SPE column for mycotoxin detection [29, 222]. Solid-phase extraction (SPE) can either be used for sample clean- up, sample concentration and matrix elimination or applied as a reversed phase or a normal phase separation. A typical SPE sequence starts with a conditioning step, activating it with solvent. Then the aqueous sample extract is applied to the conditioned column and the analyte trapped alongside the matrix. After that, a rinsing procedure eliminates matrix compounds and finally the target analyte is eluted from the column with an organic solvent and a further preconcentration step is employed by evaporating excess solvent with nitrogen gas.

The disadvantages of SPE are the variations in reproducibility caused by sorbent properties and additionally often low capacities in retaining impurities, which interfere in subsequent determination [222]. Moreover, it is currently not likely to find a universal type of cartridge for extraction of all mycotoxins. Different types of cartridges function under certain conditions and performance can be affected by pH, solvent and ion concentration of the sample [196]. An important advantage of SPE is the ease of online coupling. Using online SPE techniques has made possible the development of faster methods by reducing sample preparation time and thus increasing sample throughput. The conditioning, rinsing and elution steps can be performed automatically [223].

4.1.3.2. Ion Exchange Columns

Ion exchange mechanisms are another type of clean-up employed if the analyte is present as an ion [29]. Anionic compounds isolate on strong anion exchange-bonded silica columns (SAX columns). The retention is based on the electrostatic interactions formed between the functional group of the analyte and charged groups on the silica surface of the column [29, 196]. The analyte is eluted by adding of a solution with high ionic strength because of its higher affinity to the sorbent. SAX (Strong anion exchange) columns can be used for the determination of mycotoxins such as ochratoxins and fumonisins [29]. Several researchers have applied this technology in analysis of mycotoxins, for instance in the analysis of Fumonisin B_1 from yam and cassava chips from Benin West Africa [224] and also Fumonisins from maize in South Africa [74].

4.1.3.3. Immunoaffinity Columns (IAC)

Immunoaffinity columns are easily used for the purification of samples contaminated with different mycotoxins. The mycotoxins are bound to the immobilised specific antibodies on the column [29, 37]. Since matrix components do not interact with the antibodies, a rinsing procedure eliminates almost all possible interferences and the toxin can be eluted by antibody denaturation by washing with organic solvents, such as methanol [29, 37].

Immunoaffinity columns feature high recovery rates and are commercially available columns for the clean-up of different mycotoxins. The advantages of IACS include clean extracts due to high specificity of the antibodies for one toxin or group of related toxins, rapidity of the purification step and reduction in the use of hazardous solvents [29] Nevertheless, the use of this technology, however, is constrained by cost due to lack of reusability [197, 225]. Application of Immunoaffinity columns has been reported in the clean-up of different mycotoxins such as OTA in cocoa products and chocolate [226, 227], OTA in cocoa powder [228], OTA and Aflatoxins in chocolate [229] and also in Aflatoxins in animal feeds [78]. A study conducted by Chen et al., (2005) [230] revealed that IAC clean-up was more sensitive than a Mycosep multifunctional column for milk samples.

However, the IAC are not absolutely selective for individual mycotoxins as mycotoxin analogues are usually bound to the material [25]. This could be due to the fact that antibodies often show cross reactivities with structurally similar compounds. However, when a separation technique is used for the subsequent determination, this does not present a problem. A detailed review on IAC clean-up techniques in food analysis has been done by [225].

4.1.4. Separation

In most mycotoxins assay methods, a further stage of separation is often needed to separate the toxins from other impurities still remaining in the purified extract. Moreover, most important mycotoxins such as aflatoxins and fumonisins occur as a number of different analogues thus cchromatographic separation is needed to quantify each analogue one by one. Several types of chromatographic techniques are used for mycotoxin separation in samples such as TLC, GC and HPLC. Although thin layer chromatography and gas chromatography are still in use, the separation technique that has come to predominate the analysis of mycotoxin is high-performance liquid chromatography (HPLC), coupled to a range of detectors.

Thin-layer chromatography (TLC) was the first chromatographic method to be applied to mycotoxin determination and is still in routine use in many laboratories especially in developing countries [231]. Its many advantages include being suitable for analyzing crude extracts, extensive choice of mobile and stationary phases, as well as assorted spraying agents used for the detection [25]. Even though used they are also for the more accurate determinations, TLC is now rarely used except for screening purposes [25].

After suitable extraction and clean up, aflatoxins are readily separated on silica gel coated TLC plates using various developing solvent systems. Chloroform: Acetone (9:1) is one of the most common developing solvent systems employed, although many alternatives are also available [19, 231]. After the chromatographic separation of aflatoxins on the silica gel plates, the fluorescence emitted when visualised under long wavelength UV light forms the basis for their estimation. The nomenclature of the toxins as ‘B’ and ‘G’ was derived from their blue and green fluorescence colours observed under such a light. Visual inspection of the plates and comparison with a similarly chromatographed set of authentic standards can yield semi-quantitative results [19, 25, 231]. This basic system of TLC has been improved in a number of ways, including the use of high-performance TLC plates, two-dimensional TLC and bi-directional TLC. Also, there has been improved quantification by the use of fluorodensitometric measurement which yields more precise, sensitive and accurate results [19, 25, 231].

4.1.5. Detection/ Quantitation

High-performance liquid chromatography (HPLC) is the most frequently and widely used method of mycotoxin detection in foods [232]. HPLC reference methods that are quite sensitive and have reasonably low levels of detection have been developed for most major mycotoxins; thus, these are good quantitative methods. The use of HPLC with fluorescence detector (HPLC-FD) has been used to detect and quantify mycotoxins such as ochratoxin, aflatoxin and citrinin among others [233]. However, some mycotoxins such as fumonisins, lack a suitable chromophore, for their determination requires chemical derivatisation using derivatising agents such as o-phthaldialdehyde [233]. It is sometimes helpful to use precolumn or postcolumn derivatization to assist in the sensitive detection the mycotoxin. For example, in the case of aflatoxins, on-line electrochemical bromination using a “Cobra” cell has become a well-established procedure in order to enhance fluorescence before passing through the detector [25, 215]. For many mycotoxins, the time for analysis following injection onto the column is less than 20 minutes, but the extract must be substantially cleaned up (e.g., through an immunoaffinity column) before injection.

HPLC has been a standard for mycotoxin detection in industries [196]. It has been employed in the analysis of mycotoxins in infant food and feed formulae retailed in North Africa using Kobra cell to derivatise the Aflatoxins [127]. Furthermore it was applied in the analysis of aflatoxin and fumonisin in yam and cassava chips from Benin, West Africa [224]. A more detailed and comprehensive review on HPLC has been conducted by [25].

4.2. Biological Assays (Bioassays)

Besides physicochemical tests, numerous biological systems have been proposed and used by several workers as bioassays for detection and quantification of mycotoxins [19]. Historically, the use of bioassay for detection of mycotoxins has proved pretty successful and in the discovery of Aflatoxin, which marked the birth of modern mycotoxicology, bioassay played a vital role. While chemical assays are generally preferred over biological assays for analysis of mycotoxins in food and feedstuffs, in that they are faster, more specific and more reproducible; they are limited to known mycotoxins with available standards and analytical methodology [234]. Obviously, a chemical assay cannot be used until a mycotoxin is chemically, characterised and the methodology developed. Therefore, chemical and bioassay methods complement each other. Hence, bioassay methods are generally for initial detection and isolation of unknown mycotoxin which then provides the basis for the development of subsequent chemical assays [234].

These factors should be considered before using bioassay methods for routine analysis of mycotoxins. The capability of the method to respond to diverse range of mycotoxins, sensitivity and repeatability must be acceptable, cost should be relatively low in terms of resources required and results should be obtained in a short time, less than 24hours. Finally, the assay should be unambiguous, which means that the occurrence of false positives must be minimal to none [234]. These assays employ various biological systems which include cell, tissue and organ cultures, intact animals and microorganisms. The assays are based on the toxic/carcinogenic effects on the test organism [19].

4.2.1. Cell, Tissue and Organ Culture

Various cells, tissues and organ cultures have been used such as, HeLa cell [235], human chang liver cell [236], human lymphocyte cell [237-239], rat kidney cells [240], rat hepatic tissue[241], pig jejunal epithelia cell [242-244] and kidney fibroblast cell [245].

Cytological assays involve exposure of various concentrations of mycotoxins or test materials to various cell cultures for various periods of time. Cytological, morphological and biochemical methods are used to evaluate general toxicity, morphological changes and effects on enzyme regulation. The advantages of these bioassays are that they are readily available, inexpensive and sensitive; however, they may suffer from lack of mycotoxin specificity [234].

4.3. Immunoassays

Immunoassays are highly specific and sensitive for the detection and quantification of mycotoxins and have been developed for various mycotoxins. Immunoassay methods are based on interactions between antigens and antibodies. In immunoassay methods for detecting mycotoxins, mycotoxins are the antigens. Antibodies are highly specific for structurally diverse compounds, but can exhibit significant cross-reactivity for structural analogs, since they can only recognise specific chemical groups- the epitope [25]. Mycotoxins are small, non-immunogenic molecules (haptens) and have to be bound to suitable carrier proteins to elicit adequate immune response and production of antibodies in an animal. Novel methods of recombinant antibody production can avoid the binding of hapten to protein and inherent risks that the free hapten would not be detected in samples [25].

Wide use has been made of immunoassays to detect mycotoxins in foods. The immunoassay methods used to analyse food commodities include enzyme-linked immunosorbent assay (ELISA), radioimmunoassay (RIA), fluorescence immunoassay (FIA), Surface-plasmon resonance-based immunoassays (SPR), chemical sensors and biosensors [25]. Markers, also known as tracers or labels are involved in the assay to aid detection. The marker could be radioactive in radioimmunoassay (RIA) – rarely used now, or a chromogenic reacting with enzyme in enzyme immunoassay (EIA, ELISA) or fluorogenic compound in fluorescence immunoassay (FIA) [25]. Marker- free immunoassays exploit the ability of some mycotoxins such as ochratoxins and aflatoxins to fluoresce naturally and also on conductometric or impedimetric measurements. In SPR, the change in the optical features observed when antibodies attach to surface- bound antigens is measured and the mycotoxin content is correlated to the quantity of free antibodies. Other immunoassay type measures fluorescence polarization induced by increased molecular mass of antibody-bound labeled mycotoxin [25].

The enzyme-linked immunosorbent assay (ELISA) has become very popular due to its relatively low cost and easy method of application [246]. ELISA is a heterogeneous procedure mostly performed in 96-well polystyrene plates. One of the immunoreagents is normally immobilised on the bottom and walls of the wells for separation of bound and free reagents. This heterogeneous nature of ELISA makes it easy to separate e the bound and free elements of an analytical system after the immunochemical reaction. Because mycotoxins are monovalent antigens, competitive ELISA techniques are usually used [247]. Direct competitive ELISA is the most commonly used approach where a known amount of enzyme labelled toxin competes with any possible toxin present in the sample to bind to specific amounts of antibodies on the wells. After incubation the compounds that are not bound are removed by washing and chromogenic substrate is added. The enzymatic activity in each well results in a coloured product that can be measured with colorimetric detection and is inversely proportional to the mycotoxin concentration [247]. Replacement of colorimetric detection by the chemiluminescent detection increases the sensitivity of mycotoxin detection, for example up to 10 times for FB_1 [248]. Since chemiluminescent signal_might be instantly read out after adding the substrate, the assay time is less than with colorimetric ELISA.

Factors such as structurally related mycotoxins as well as matrix constituents can interfere with conjugate attachments to the antibody. It may be by absorption of the conjugate or antibody, by denaturation of the antibody or by inhibition of the enzyme. This means that cross-reactivities and matrix dependence are the major disadvantages of ELISA methods and can lead to false positive results [215]. Other conditions influencing the results include temperature, sample viscosity, pH, and ionic strength [25]. If no sample clean-up or extraction is done before the test, matrix effects might well be expected [40] sometimes leading to significant overestimation of mycotoxin [215]. This has been applied in the analysis of aflatoxin in peanuts from Kenya [33].

5. Possible Intervention Strategies

The major challenge in controlling mycotoxin is its prevention. Since mycotoxins are formed both at pre and post-harvest stages of food and feed production, it is necessary that control measures be developed and adopted towards solving the problem at these stages[19]. The majority of mycotoxins are thermally stable [7] and tend to survive storage and processing, and even when cooked at pretty high temperatures such as those reached during bread making or breakfast cereal production [196]. Mycotoxins can equally be found in meat and dairy products and even temperature treatments (for example cooking and freezing) cannot destroy them. This shows that it is better to prevent their formation if possible, since it is likely not going to be easy to destroy them. Although it is said that "prevention is better than cure", no reliable preventive techniques have been developed so far, especially for preventing the formation of mycotoxin completely during pre-harvest stages of crop production [19].

5.1. Pre-Harvest Prevention

This involves the adoption of appropriate agronomic practices such as reduction of insect and fungal damage through the use of crop varieties that can resist insect and fungal infection. Crop rotation limits fungal inoculums in the field by breaking the chain production of these infectious agents. This therefore assists in the prevention of mycotoxin contamination. For example, it has been reported that a maize- soya bean rotation yielded a less extensive outbreak of *Fusarium* than did maize-maize planting operations [135]. Also proper and uniform irrigation reduces plant drought stress because maize grown under drought-stressed conditions had higher fumonisin concentration [18]. Destruction of weeds, grasses, debris, crop residues and alternate host plants, which serve as reservoirs to fungal inoculums also prevent mycotoxin contamination of crops in the field and consequently, a reduction in mechanical injury to crops to a minimum since incidence of *Aspergillus spp.* is generally higher in broken kernels. The use of appropriate and approved fungicides, pesticides and other chemicals, although could yield limited success in preventing aflatoxin contamination of grains, their judicious use should be considered_since they are toxic to mammalian cells [19].

5.2. Post-Harvest Prevention

This is primarily dependent on good management practices before and after harvest. Aflatoxins are potentially serious post-harvest problems unlike fumonisins, since *Fusarium spp.* does not grow in maize at less than 18-20% moisture [249]. Aflatoxin contamination is considered to be primarily a storage problem and this is very severe in many rural areas in Africa due to lack of infrastructure for drying and other appropriate storage conditions. Drying of food and feeds to safe moisture levels is perhaps the best method of choice. Keeping moisture of grains and ambient humidity, very low during storage will help prevent the formation of (aflatoxin). Any produce with a higher moisture level than that considered safe should not be stored. Cleaning and disinfection of storage containers and transportation equipment is recommended. In addition, during storage the initial grain condition is critical as good quality, clean, sound grain is easier to maintain in storage than physically damaged grains. Also Modified Atmosphere Packaging, (MAP) which is a useful technology involving O_2 absorbents, storage temperature and packaging film barrier can be used to prevent fungal growth and mycotoxin contamination on finished products [19].

However, various approaches such as physical and chemical decontamination of mycotoxins from commodities as well as biological methods that aid in destroying mycotoxins and their toxic effects have been developed. Several microorganisms, including bacteria and fungi, have the potential to degrade aflatoxin and or inhibit their formation in food and feed both *invivo* and *invitro* [250]. Also, other novel approaches such as addition of certain nutrients and adsorptive earth metals to the animal diet has helped in reducing the toxic effects or avoiding the bio-availability of mycotoxins respectively in animal systems [19].

Mycotoxin control strategies should also include approaches for the diversion of contaminated products to lower risk uses and laboratory safety measures to protect researchers from toxic effects of mycotoxins. There should be producer /consumer education and accurate and affordable detection methods of mycotoxin in developing African countries.

Discussion and Conclusion

In Africa, mycotoxin contamination of agricultural produce is a great concern with aflatoxins being the most prevalent in the last decade from the available data. The risk of contamination by mycotoxins is an important food safety concern for grains and other field crops. The amount of toxins produced and accumulated in food and feed depends on the influence of several intrinsic and extrinsic factors discussed earlier, such as rainfall, temperature and geographical location. Surveys on the occurrence of mycotoxins have been conducted in African countries, on various agricultural products such as cereals, oilseeds, vegetable oil, root and stem crops, spices, processed foods, herbal medicinal products, feeds and animal products. From the data, (Table 1), even though considerable research efforts have been made to survey the occurrence of mycotoxins in African food and feed commodities in the last decade, it is clear that surveillance reports from some of the countries are very few, scanty and outdated. There is still a need for continuous monitoring and surveillance of mycotoxins in agricultural crops, animal feedstuffs and human food. Dietary mycotoxins can cause adverse health effects on humans and animals linked with organ toxicity, carcinogenesis and immune suppression. Thus, even low levels of mycotoxins in staple foods should be viewed seriously, as their daily consumption is relatively high. A study conducted by [51] revealed a significant positive correlation between mothers' dietary exposure to AFB_1 and AFM_1 contamination of breast milk in Ogun State Nigeria. This indicates that even infants and young children are exposed to mycotoxins through breast milk. Moreover, during weaning other foods are often introduced which are additional sources of mycotoxins such as AFM_1 and OTA as in the cases of cereal- based foods, milk formula, meat, eggs and other edible products from animals that consume aflatoxin-contaminated feeds. The growing feotuses in the womb are not even safe; they are exposed to the toxic effects of these metabolites through the pregnant mother, as the mothers are generally exposed to mycotoxin-contaminated food especially in the African continent and in other developing countries. In order to minimise the risks of mycotoxins in the continent, the African population; farmers, traders and the consumers should be educated on mycotoxins, their health hazards and different ways to manage their occurrence. This can be done through seminars, workshops and media announcements. There is also a need to train personnel at all levels; scientists, technicians and extension workers, in sampling protocols and modern methods of analyzing mycotoxins. The adoption of GAP in the field and throughout the whole table chain, coupled with the best sampling practices and the use of validated and fit-for-purpose analytical methods will go a long way in solving the problem of mycotoxins. Besides, assistance by international organizations such as FAO may be necessary, particularly in developing countries such as those in Africa to motivate and put into operation the necessary HACCP approach of food control systems. Due to the various factors that trigger fungal and mycotoxin contamination of crops, threatening food security and food safety, constant monitoring and risk assessment of these mycotoxins is needed to prevent further outbreaks of mycotoxicosis.

REFERENCES

[1] Bräse, S. et al. (2009). Chemistry and biology of mycotoxins and related fungal metabolites. *Chem. Rev*, 903-3990.

[2] Bennett, J. W. & Klich, M. (2003). Mycotoxins. *Clinical Microbiology Reviews*, *16*, 497-516.

[3] Blount, W. P. (1960). Disease of turkey poults. *Veterinary Record*, *72*, 786.

[4] CAST. (2003). Potential economic costs of mycotoxins in the United States. In Mycotoxins: Risks in Plant, Animal and Human Systems. In *Council for Agricultural Science and Technology*, 136–142.

[5] EFSA. (2011). European food safety authority. Scientific opinion on the risks for public health related to the presence of zearalenone (ZEA) in food. *European Food Safety Authority.*

[6] Timbrell, J. A. (2002). *Introduction to Toxicology*. Taylor and Francis.

[7] Wagacha, J. M. & Muthomi, J. W. (2008). Mycotoxin problem in Africa: Current status, implications to food safety and health and possible management strategies. *International Journal of Food Microbiology*, *124*, 1-12.

[8] Hussain, S. H. & Brassel, M. J. (2001). Toxicity,metabolism and impact of mycotoxins in humans and animals. *Toxicology*, *67*, 101-134.

[9] Jacobsen, B. J. et al. (2007). Stored grain-Mycotoxins.

[10] Coppock, R. W. & Jacobsen, B. J. (2009). Mycotoxins in animal and human patients. *Toxicology and industrial health*, *25*, 637-655.

[11] Shephard, G. S. (2008). Determination of mycotoxins in food. *Chem. Soc. Rev*, *37*, 2468-2477..

[12] Bhat, R. V. & Vasanthi, S. (2003). Food safety in food security and food trade: Mycotoxin food safety risk in developing countries. In *International Food Policy Research Institute, (Brief 3).*

[13] WHO. (2006). Mycotoxins in African foods: implications to food safety and health. AFRO Food Safety Newsletter. WHO Food Safety.

[14] FAO. (2004). Food and Agricultural Organization. Worldwide Regulations for mycotoxins in food and feed in 2003. *FAO Food and Nutrition Paper 8.*

[15] Krska, R. et al. (2008). Mycotoxin analysis: An update. *Food Additives & Contaminants: Part A*, *25*, 152-163.

[16] Pascale, M. N. (2009). Detection methods for mycotoxins in cereal grains and cereal products. *Pro Nat. Sci.*, *117*, 15-25.

[17] Park, D. L. & Stoloff, L. (1989). Aflatoxin control - how a regulatory agency managed risk from an unavoidable natural toxicant in food and feed. *Regul. Toxicol pharmacol.*

[18] Miller, J. D. (2001). Factors that affect the occurrence of fumonisin. *Environmental Health Perspectives*, *109*, 321–324.

[19] Basappa, S. C. (2009). *Aflatoxins; formation, analysis and control.*, *66*, 103-104, Alpha Science International ltd. U.K..

[20] Dawlal, P. (2010). Resistance of maize cultivars against the infestation of mycotoxigenic fungi. University of Pretoria, Pretoria.

[21] Ncube, E. (2008). Mycotoxin levels in subsistence farming systems in South Africa Stellenbosch: University of Stellenbosch.

[22] Lee, H. B. & Magan, N. (2000). Impact of environmental and interspecific interactions between spoilage fungi and Aspergillus Ochraceus on growth and ochratoxin production in maize grain. *International Journal of Food Microbiology*, *61*, 11-16.

[23] Cardwell, K. F. et al. (2004). Mycotoxins: The cost of achievieing food security and food quality.

[24] Bennett, J. W. & Klich, M. (2009). *Mycotoxins*. Elsevier, 559-565.

[25] Cigić, I. K. & Prosen, H. (2009). An Overview of Conventional and Emerging Analytical Methods for the Determination of Mycotoxins. *Int. J. Mol. Sci.*, *10*, 62-115.

[26] Sultana, N. & Hanif, N. Q. (2009). Mycotoxin Contamination in Cattle Feed and Feed Ingredients. *Pakistan Vet. J.*, *29*, 211-213.

[27] Gremmels, J. F. (2008). The role of mycotoxins in the health and performance of dairy cows. *Vet. J.*, *176*, 84-92.

[28] Berge, A. C. (2011). Penicillium moulds in silage. How they affect rumen health In Dairy Solutions Newsletter.

[29] Rahmani, A. et al. (2009). Qualitative and quantitative analysis of mycotoxins. *Comprehensive reviews in food science and food safety.*

[30] Levetin, E. & McMahon, K. (2006). *Plant and Society*. McGraw Hill Publishers Inc.

[31] Egal, S. et al. (2005). Dietary exposure to aflatoxin from maize and groundnut in young children from Benin and Togo, West Africa. *Int. J. Food Microbiol.*, *104*, 215–224.
[32] Ngoko, Z., et al. (2008). Fungi and mycotoxins associated with food commodities in Cameroon. *Journal of Applied Biosciences*, *6*, 164-168.
[33] Mutegi, C. K. et al. (2009). Prevalence and factors associated with aflatoxin contamination of peanuts from Western Kenya. *International Journal of Food Microbiology*, *130*, 27-34.
[34] Ediage, E. N. et al. (2011). A validated multi-analyte LC-MS/MS method for the quantification of 25 mycotoxins in Rural Cooperation (CTA). Technical cassava, cassava floor, peanut cake and some maize samples. *Journal of Agriculture and Food Chemistry*, *59*, 5163-5180.
[35] Ezekiel, C. N. et al. (2012). Natural occurrence of mycotoxins in peanut cake from Nigeria. *Food Control*, *27*, 338-342.
[36] Mwanza, M. et al. (2013). Mycotoxins occurrence in selected staple food in main markets from Lubumbashi, DR Congo. *Journal of food, agriculture and environment*, 51-54.
[37] Jarvis, B. B. (2003). Analysis for mycotoxins: The chemist's perspective. *Achives of environmental health*, *58*, 8.
[38] Kumar, V. et al. (2008). Mycotoxin research and mycoflora in some commercially important agricultural commodities *Crop Protection*, *27*, 891–905.
[39] Muthomi, J. W. et al. (2009). The occurrence of Aflatoxins in maize and distribution of mycotoxin-producing fungi In Eastern Kenya. *Plant Pathology Journal*, *8*, 113-119.
[40] Krska, R. et al. (2007). Analysis of *Fusarium* toxins in feed. In *Fusarium and their toxins: Mycology, occurrence, toxicity, control and economic impact* (Morgavi, D.P. and Riley, R.T., eds), Anim. Feed Sci. Technol..
[41] Rai, M. K. et al. (2012). Mycotoxin: rapid detection, differentiation and safety. *J Pharm Educ Res.*, *3*.
[42] Mphande, F. A. et al.(2004). Fungi, Aflatoxins and Cyclopiazonic Acid Associated with Peanut Retailing in Botswana. *Journal of Food Protection*, *67*, 96-102.
[43] Nkwe, D. O. et al. (2005). Fungi, Aflatoxins, fumonisin B1 and zearalenone contaminating sorghum based traditional malt, wort and beer in Botswana. *Mycopathologia*, *160*, 177-186.
[44] Bankole, S. A. & Mabekoje, O. O. (2004). Occurrence of Aflatoxins and Fumonisins in Pre- harvest Maize from South-Western Nigeria. *Food Additives and Contaminants*, *21*, 251-255.
[45] Bankole, S. A. & Mabekoje, O. O. (2004). Mycoflora and Occurrence of Aflatoxin B1 in Dried Yam Chips from Markets in Ogun and Oyo States, Nigeria. *Mycopathologia.*, *157*, 111-115.
[46] Bankole, S. A. et al. (2004). Natural Occurrence of Moulds and Aflatoxin B1 in Melon Seeds from Markets in Nigeria. *Food Chem Toxicol*, *42*, 1309-1314.
[47] Rodrigues, I. et al. (2011). Mycotoxin occurrence in commodities, feeds and feed ingredients sourced in the Middle East and Africa. *Food additives and contaminants*, *3*, 168-179.
[48] Ogiehor, I. S. et al. (2007). The bioload and aflatoxin content of market garri from some selected states in southern Nigeria: public health significance. *African J. Health Sciences*, *7*, 223-227.

[49] Afolabi, C. G. et al. (2013). Maize contamination by zearalenone and T-2 toxin and human exposure in Nigeria. *Mycotoxins*, *63*, 143-149.
[50] Makun, H. A. et al. (2011). Natural multi-occurrence of mycotoxins in rice from Niger State, Nigeria. *Mycotox. Res.*, *27*, 97-104.
[51] Adejumo, O. et al. (2013). Correlation between aflatoxin M1 content of breast milk, dietary exposure to aflatoxin B1 and socioeconomic status of lactating mothers in Ogun State, Nigeria. *Food and Chemical Toxicology*, *56*, 171-177.
[52] Somorin, Y. M. et al. (2012). Aflatoxin and fumonisin contamination of yam flour from markets in Nigeria. *Food Control*, *25*, 53-58.
[53] Adebayo-Tayo, B. C. et al. (2006). Detection of fungi and aflatoxin in shelved bush mango seeds (Irvingia spp.) stored for sale in Uyo Eastern Nigeria. *J. Environ Agric Food Chem.*, *5*, 1569-1574.
[54] Atanda, O. et al. (2007). Aflatoxin M1 contamination of milk and ice cream in Abeokuta and Odeda local governments of Ogun State, Nigeria. *Chemosphere*, *68*, 1455-1458.
[55] Jonathan, G. et al. (2011). Nutritional compositions, fungi and aflatoxins detection in stored 'gbodo' (fermented Dioscorea rotundata) and 'elubo ogede' (fermented Musa parasidiaca) from South Western Nigeria. *Afr. J. Food. Sci.*, *5*, 105-110.
[56] Rubert, J. et al. (2013). A survey of mycotoxins in random street-vended snacks from Lagos, Nigeria, using QuEChERS-HPLC-MS/MS. *Food Control*, *32*, 673-677.
[57] Egbuta, M. A. et al. (2013). Co-contamination of Nigerian cocoa and cocoa-based powder beverages destined for human consumption by mycotoxins. *Ethno. Med.*, *7*, 187-194.
[58] Oluwafemi, F. & Ibe, N. I. (2011). Microbial contamination of seven major weaning foods in Nigeria. *J. of Health Popul. Nutr.*, *29*, 415-419.
[59] Okwu, G. I. et al. (2010). Quantification of aflatoxin B1 in ready to use food thickeners in South-east geo- political zone in Nigeria. *African Journal of Microbiology Research*, *4*, 1788-1793.
[60] Olufunmilayo, G. O. & Oyefolu, A. B. (2011). Natural occurrence of aflatoxin residues in fresh and sun-dried meat in Nigeria *Pan African Medical Journal*, *7*, 14.
[61] Adejumo, T. O. et al. (2007). Occurrence of *Fusarium* species and trichothecenes in Nigerian maize. *International Journal of Food Microbiology*, *116*, 350-357.
[62] Kayode, O. F. et al. (2013). Mycotoxins and fungal metabolites in groundnut- and maize-based snacks from Nigeria. *Food Additives & Contaminants*, *6*, 294-300.
[63] Jimoh, K. O. & Kolapo, A. L. (2008). Mycoflora and aflatoxin production in market samples of some selected Nigerian foodstuffs. *Research Journal of Microbiology*, *3*, 169-174.
[64] Baiyewu, R. A. et al. (2007). Survey of the post harvest diseases and aflatoxin contamination of marketed pawpaw fruit (Carica papaya L) in South Western Nigeria. *African Journal of Agricultural Research.*, *2*, 178-181.
[65] Odoemelam, S. A. & Osu, C. I. (2009). Aflatoxin B1 contamination of some edible grains marketed in Nigeria. *E-Journal of Chem.*, *6*, 308–314.
[66] Muhammad, S. et al. (2004). Survey of the market diseases and aflatoxin contamination of tomato (MILL) fruits in Sokoto, northwestern Nigeria. *Nutrition & Food Science*, *34*, 72 - 76.

[67] Diop, Y. et al. (2000). Contamination by Aflatoxins of local Peanut Oils Prepared in Senegal. *Annales Pharmacetiques Francaises*, *58*, 470-474.
[68] Odhav, B. & Naicker, V. (2002). Mycotoxins in South African traditionally brewed beers. *Food additives and contaminants*, *19*, 55-61.
[69] Sewram, V. et al. (2006). Mycotoxin contamination of dietary and medicinal wild plants in the Eastern Cape Province of South Africa. *Journal of agricultural and food chemistry*, *54*, 5688-5693.
[70] Dutton, M. F. et al. (2012). Mycotoxins in South African foods: a case study on aflatoxin M1 in milk. *Mycotoxin Res.*, *28*, 17–23.
[71] Burger, H. M. et al. (2010). Dietary fumonisin exposure in a rural population of South Africa. *Food Chem. Toxicol.*, *48*, 2103–2108.
[72] Kritzinger, Q. et al. (2003). Mycoflora and fumonisin mycotoxins associated with cowpea (Vigna unguiculata (L.) Walp seeds). *Journal of Agricultural and Food Chemistry*, *51*, 2188–2192.
[73] Shabangu, N. (2009). The occurrence of fungi and their mycotoxins in maize and bambara nuts and their effect on the health of the rural community in areas of Limpopo province. University of Johannesburg, South Africa.
[74] Chilaka, C. A. et al. (2012a). Fungal and mycotoxin contamination of South African commercial maize. *J Food Agric Environ*, *10*, 296–303.
[75] Chilaka, C. A. et al. (2012b). Comparative profiling of different analytical methods for fumonisin detection in maize. *Journal of Microbiology, Biotechnology and Food Science*, *2*, 833-849.
[76] Maenetje, P. W. & Dutton, M. F. (2007). The incidence of fungi and mycotoxins in South African barley and barley products. *Journal of Environmental Science and Health, Part B: Pesticides, Food Contaminants, and Agricultural Wastes*, *42*, 229-236.
[77] Mashinini, K. & Dutton, M. F. (2006). The incidence of fungi and mycotoxins in South Africa wheat and wheat-based products. *J. Environ. Sci. Health B*, *41*, 285-296.
[78] Dutton, M. F. et al. (2013). Analysis of a feed component imported into South Africa for Aflatoxin in relation to fungal and mycotoxin contamination. *Agricultural Science*, 101-14.
[79] Njobeh, P. B. et al. (2012). Estimation of multi-mycotoxin contamination in South African compound feeds. *Toxins*, *4*, 836–848.
[80] Kamika, I. et al. (2013). Mycological and aflatoxin contamination of peanuts sold at markets in Kinshasa, DRC and Pretoria, South Africa. *F Part B. Food additives and contaminants*, *7*, 120-126.
[81] Lewis, L. et al. (2005). Aflatoxin Contamination of Commercial Maize Products during an Outbreak of Acute Aflatoxicosis in Eastern and Central Kenya. *Environmental Health Perspectives*, *113*, 1763-1767.
[82] Mbugua, S. K. & Gathumbi, J. K. (2004). The contamination of Kenyan lager beers with Fusarium mycotoxins. *Journal of the Institute of Brewing*, *110*, 227-229.
[83] Okoth, S. A. & Kola, M. A. (2012). Market samples as a source of chronic aflatoxin exposure in Kenya. *Afr J Health Sci.*, *20*, 56-61.
[84] Azziz-Baumgartner, E. et al. (2005). Case control study of an acute Aflatoxicosis outbreak, Kenya, 2004. *Environ. Health Perspectives.*, *113*, 1779-1783.

[85] Muture, B. N. & Ogana, G. (2005). Aflatoxin levels in maize and maize products during the 2004 food poisoning outbreak in Eastern Province of Kenya. *East Afr. Med. J.*, *82*, 275-279.

[86] Daniel, J. H. et al. (2011). Comprehensive assessment of maize aflatoxin levels in EasternKenya, 2005-2007. *Environ. Health Perspect*, *119*, 1794-1799.

[87] Mwihia, J. et al. (2008). Aflatoxin levels in locally grown maize from Makueni District, Kenya. *East Afr Med J.*, *85*, 311-317..

[88] Muthomi, J. W. et al. (2008). The occurrence of Fusarium species and mycotoxins in Kenyan wheat. *Crop protection*, *27*, 1215-1219.

[89] Kang'ethe, E. K. & Lang'a, K. A. (2009). Aflatoxin B1 and M1contamination of animal feeds and milk from urban centers in Kenya. *Afr. Health Sci*, *9*.

[90] Ayalew, A. et al. (2006). Natural occurrence of mycotoxins in staple cereals from Ethiopia *Mycopathologia*, *162*, 57–63.

[91] Manjula, K. et al. (2009). Aflatoxin and fumonisin contamination of cassava products and maize grain from markets in Tanzania and republic of the Congo., *Toxin Rev.*, *28*, 63–69..

[92] Kimanya, M. E. et al. (2008). Co-occurrence of fumonisins with aflatoxins in homestored maize for human consumption in rural villages of Tanzania. *Food Addit. Contam. Part A Chem. Anal. Control Expo. Risk Assess*, *25*, 1353–1364.

[93] Njobeh, P. B. et al. (2010). Simultaneous occurrence of mycotoxins in human food commodities from Cameroon. *Mycotoxin Res.*, *26*, 47–57.

[94] Tchana, A. N. et al. (2010). Aflatoxin contamination in food and body fluids in relation to malnutrition and cancer status in Cameroon. *Int. J. Environ. Res. and Public Health*, *7*, 178-188.

[95] Mounjouenpou, P. et al. (2008). Filamentous fungi producing OTA during cocoa processing in Cameroun. *International Journal of Food Microbiology*, *121*, 234-241.

[96] Essono, G. et al. (2009). Aflatoxinproducing Aspergillus spp. and aflatoxin levels in stored cassava chips as affected by processing practices. *Food Control*, *20*, 648–654.

[97] Abia, W. A. et al. (2013). Determination of multi-mycotoxin occurrence in cereals, nuts and their products in Cameroon by liquid chromatography tandem mass spectrometry (LC-MS/MS). *Food Control*, *31*, 438-453.

[98] Kana, J. R. et al. (2013). Mycobiota and Toxigenecity Profile of *Aspergillus* flavus Recovered from Food and Poultry Feed Mixtures in Cameroon. *Journal of Animal and Poultry Sciences*, *2*, 98-107.

[99] Sangare-Tigori, B. et al. (2006b). Co-occurrence of aflatoxin B1,fumonisin B, ochratoxin A and zearalenone in cereals and peanuts from Coˆ te d'Ivoire. *Food Addit Contam.*, *23*, 1000–1007.

[100] Sangare-Tigori, B. et al. (2006a). Preliminary survey of ochratoxin A in millet, maize, rice and peanuts in Côte d'Ivoire from 1998 to 2002. *Hum. Exp. Toxicol.*, *25*, 211–216.

[101] Mupunga, I. (2013). A comparative study of natural contamination with aflatoxins and fumonisins in selected food commodities from Botswana and Zimbabwe. University Of South Africa.

[102] Weidenborner, M. (2011). *Mycotoxins and their metabolites in human and animals.* Springer Science, Business Media, LLC.

[103] Zinedine, A. et al. (2007b). Incidence of ochratoxin A in rice and dried fruits from Rabat and Salé area, Morocco *Food additives and contaminants*, *24*, 285–291.

[104] El-Mahgubi, A. et al. (2013). Distribution and toxigenicity of *Aspergillus* section flavi in spices marketed in Morocco. *Food Control, 32*, 143-148.

[105] Zinedine, A. et al. (2007c). Presence of aflatoxin M1 in pasteurized milk from Morocco. *International Journal of Food Microbiology, 114*, 25–29.

[106] Juan, C. et al. (2008). Ochratoxin A in rice on the Moroccan retail market. *International Journal of Food Microbiology, 126*, 83–85.

[107] Zinedine, A. et al. (2006). Natural occurrence of mycotoxins in cereals and spices commercialized in Morocco. *Food Control, 17*, 868-874.

[108] El Marnissi, B. et al. (2012). Occurrence of aflatoxin M1 in raw milk collected from traditional dairies in Morocco *Food and Chemical Toxicology, 50*, 2819–2821.

[109] Serrano, A. B. et al. (2012). Co occurrence and risk assessment of mycotoxins in food and diet from Mediterranean area. *Food Chem, 135*, 423–429.

[110] El-Tras, W. F. et al. (2011). Infants exposure to aflatoxin M1 as a novel foodborne zoonosis. *Food Chem. Toxicol., 49*, 2816–2819.

[111] Motawee, M. M. et al. (2009). Survey of Aflatoxin M1 in Cow, Goat, Buffalo and Camel Milks in Ismailia-Egypt. *Bull Environ Contam Toxicology, 83*, 766-769.

[112] Polychronaki, N. et al. (2006). Deterimation of Aflatoxin M1 in Breast milk in a selected group of Egyptian mothers. *Food additives and contaminants, 23*, 700-708.

[113] Tomarek, R. H. et al. (2011). Assessment of exposure of Egyptian infant to aflatoxin M1 through breast milk. *Journal of Egypt Public Health Association, 86*, 51-55.

[114] Hassan, A. M. et al. (2006). Study of ochratoxin A as an environmental risk that causes renal injury in breast-fed Egyptian infants. *Pediatr Nephrol., 21*, 102-105.

[115] Amer, A. A. & Ibrahim, M. A. E. (2010). Determination of aflatoxinM1 in raw milk and traditional cheeses retailed in Egyptian markets. *J. Toxicol. Environ. Health Sci., 2*, 50–53.

[116] Elzupir, A. O. & Elhussein, A. M. (2010). Determination of aflatoxin M1 in dairy cattle milk in Khartoum State, Sudan. *Food Control, 21*, 945-946.

[117] Elzupir, A. O. et al. (2012). Aflatoxin M1 in breast milk of nursing Sudanese mothers. *Mycotoxin Research, 28*, 131-134.

[118] Elshafie, S. Z. et al. (2011). Aflatoxin B1 contamination of traditionally processed peanuts butter for human consumption in Sudan *Mycopathologia, 171*, 435–439.

[119] Idris, Y. M. A. et al. (2010). Determination of aflatoxin levels in Sudanese edible oils. *Food and Chemical Toxicology, 48*, 2539-2541.

[120] Elzupir, A. O. et al. (2011). Aflatoxins in peanut butter in Khartoum State, Sudan *Mycotoxin research, 27*, 183-186.

[121] Elzupir, A. O. et al. (2009). Determination of Aflatoxins in animal feed in Khartoum State, Sudan. *Journal of Animal and Veterinary Advances., 5*, 1000-1003.

[122] Riba, A. et al. (2010). Aspergillus section flavi and aflatoxins in Algerian wheat and derived products. *Food and Chemical Toxicology*, 2772-2777.

[123] Nikiema, P. N. et al. (2004). Fumonisin contamination of maize in Burkina Faso. *Food additives and contaminants, 21*, 865-870.

[124] Matumba, L. et al. (2013). A limited survey of mycotoxins in traditional maize based opaque beers in Malawi. *Food control, 36*, 253-256.

[125] Kamika, I. & Takoy, L. L. (2011). Natural occurrence of Aflatoxin B1 in peanut collected from Kinshasa, Democratic Republic of Congo. *Food Control, 22*, 1760-1764.

[126] Mohale, S. et al. (2013). Mycotoxigenic fungi and mycotoxins associated with stored maize from different regions of Lesotho. *Mycotoxin Res.*, *29*, 209–219.

[127] Aidoo, K. E. et al. (2011). Occurrence of Fungi and Mycotoxins in Some Commercial Baby Foods in North Africa. *Food and Nutrition Sciences*, *2*, 751-758.

[128] Kaaya, N. A. Warren, H. L. (2005). A review of past and present research on aflatoxin in Uganda *African Journal of Food, Agriculture and Nutritional Development.*, *5*, 1-18.

[129] Kaaya, A. N. & Kyamuhangire, W. (2006). The effect of storage time and agroecological zone on mould incidence and aflatoxin contamination of maize from traders in Uganda. *International Journal of Food Microbiology*, *110*, 217-223.

[130] Atukwase, A. et al. (2012). Dynamics of Fusarium and fumonisins in maize during storage – A case of the traditional storage structures commonly used in Uganda. *Food Control*, *26*, 200-205.

[131] Kitya, D. et al. (2010). Aflatoxin levels in common foods of South Western Uganda: a risk factor to hepatocellular carcinoma. *Eur. J. Cancer Care Engl.*, *19*, 516–521.

[132] Conzane, R. S. et al. (2002). Detection and determination of aflatoxins B1, B2, G1 and G2 in peanuts from Mozambique using HPLC. *Deutsche Lebensmittel-Rundschhau*, *98*, 289-295.

[133] Houssou, P. A. et al. (2009). Natural infection of cowpea (Vigna unguiculata (L.) Walp.) by toxigenic fungi and mycotoxin contamination in Benin, West Africa. *Journal of Stored Products Research*, *45*, 40-44.

[134] Mestres, C. et al. (2004). Yam chip food sub-sector: hazardous practices and presence of aflatoxins in Benin. *Journal of Stored Products Research*, *40*, 575-585.

[135] Fandohan, P. et al. (2005). Natural occurrence of Fusarium and subsequent fumonisin contamination in preharvest and stored maize in Benin, West Africa. *International Journal of Food Microbiology*, *99*, 173-183.

[136] Panzo, J. D. (2013). The incidence of fungi and their mycotoxins in Angolan food and crops with particular reference to maize., University of Johannesburg.

[137] Nyinawabali. (2013). A survey of fungi and mycotoxins in selected food commodities from Rwanda. University of Johannesburg.

[138] Hell, K. et al. (2009). Mycoflora and occurrence of aflatoxin in dried vegetables in Benin, Mali and Togo, West Africa. *International Journal of Food Microbiology*, *135*, 99-104.

[139] Kankolongo, M. A. et al. (2009). Assessment of fungal, mycotoxin and insect spoilage in maize stored for human consumption in Zambia. *Journal of Science of food and Agriculture*, *89*, 1366-1375.

[140] Mukanga, M. et al. (2010). A survey of pre-harvest ear rot diseases of maize and associated mycotoxins in south and central Zambia. *International Journal of Food Microbiology*, *141*, 213-221.

[141] Elgerbi, A. M. et al. (2004). Occurrence of aflatoxin M1 in randomly selected North African milk and cheese samples. *Food Additives & Contaminants*, *21*, 592-597.

[142] Elgerbi, A. M. (2005). Studies on Occurrence of Aflatoxins M1in Milk Products and Effect of Lactobacilli and Related Genera on Toxin. Glasgow Caledonian University, Glasgow..

[143] Zaied, C. et al. (2009). Natural occurrence of ochratoxin A in Tunisian cereals. *Food control*, *20*, 218-222.

[144] Ghali, R. et al. (2008). Incidence of aflatoxins, ochratoxin A and zearalenone in tunisian foods. *Food Control*, *19*, 921–924.

[145] Jelinek, C. F. et al. (1989). Worldwide occurence of mycotoxins in foods and feeds-An update. *J. Assoc. Off. Anal. Chem.*, *72*, 223-230.

[146] Payne, G. A. (1998). Process of contamination by aflatoxin-producing fungi and their impact on crops. In *mycotoxins in agriculture and food safety* (Sinha, K.K. and Bhatnager, D., eds), 278-306, Dekker.

[147] Turcotte, A. M. et al. (2013). Analysis of cocoa products for ochratoxin A and aflatoxins *Mycotoxin Research*, *29*, 193–201.

[148] Richard, J. L. (2007). Some major mycotoxins and their mycotoxicoses—An overview. *International Journal of Food Microbiology*, *119*, 3-10.

[149] IARC. (1993). International Agency for Research on Cancer. Monographs on the evaluation of the carcinogenic risk of chemicals to human: Some naturally occurring substances. Food items and constituents, heterocyclic aromatic amines and mycotoxins. 397-444.

[150] Hatem, N. L. et al. (2005). Prevalence of aflatoxins in blood and urine of Egyptian infants with protein energy malnutrition. *Food Nutri. Bull.*, *26*, 49-56.

[151] Brabin, B. J. & Coulter, J. B. S. (2009). Nutrition-Associated Disease. In *Manson's Tropical Diseases* (22nd edn) (Cook, G.C. and Zumla, A.I., eds), 537-554, Elsevier Health Sciences.

[152] Cardwell, K. F. (1999). Mycotoxin in foods—Anti-nutritional factors. In *International institute of tropical Agricul.*, 11.

[153] Turner, P. C. (2013). The molecular epidemiology of chronic aflatoxin-driven impaired child growth. *Scientifica* DOI: doi:10.1155/ 2013/ 152879.

[154] Turner, P. C. et al. (2007). Aflatoxin exposure in utero causes growth faltering in Gambian infants. *International Journal of Epidemiology Advance*.

[155] Gong, Y. Y. et al. (2003). Determinants of aflatoxin exposure in young children from Benin and Togo, West Africa: the critical role of weaning and weaning foods. *International Journal of Epidemiology*, *32*, 556–562.

[156] Gong, Y. et al. (2004). Postweaning exposure to aflatoxin results in impaired child growth: a longitudinal study in Benin, West Africa. *Environmental Health Perspective*, *112*, 1334-1338.

[157] Turner, P. C. et al. (2003). Modification of immune function through exposure to dietary aflatoxin in Gambian children. *Environmental Health Perspectives*, *111*, 217–220.

[158] Jiang, Y. et al. (2005). Aflatoxin B1 albumin adduct levels and cellular immune status in Ghanaians. *Int Immunol*, *117*, 807-814.

[159] Galal, N. et al. (2006). Aflatoxin levels in Egyptian neonates with unexplainable neonatal jaundice: morbidity and correlation to maternal assays in breast milk. Egypt. *J. Neonatol.*, *7*, 161-169.

[160] Püssi, T. (2007). Analysis of natural toxins In *Methods of analysis of food components and Additives* (Otles, S., ed), 472-495, Tailor and Francis group.

[161] Williams, J. H. et al. (2004). Human aflatoxicosis in developing countries: a review of toxicology, exposure, potential health consequences, and interventions. *Am. J. Clin Nutr.*, *80*, 1106–1122.

[162] Phillips, T. D. et al. (2008). Reducing human exposure to aflatoxinthrough the use of clay: a review. *Food Additives & Contaminants: Part A*, *25*, 134-145.
[163] MRC. (2006). Medical Research Council. Aflatoxin in peanut butter: MRC policy brief.
[164] Kpodo, K. et al. (2000). Fusaria and fumonisins in maize from Ghana and their cooccurrence with aflatoxins. *Int. J. Food Microbiol.*, *61*, 147-157.
[165] Njobeh, P. B. et al. (2009). Contamination with storage fungi of human food from Cameroon. *Inter. J. Food Microbiol.*, *26*, 47.
[166] El-Sayed, A. et al. (2003). Occurrence of certain mycotoxins in corn and corn-based foods and thermostability of FB1 during processing. *Nahrung*, *47*, 224-225.
[167] Makun, H. A. et al. (2010). Fungal and aflatoxin contamination of some human food commodities in Nigeria. *Afr. J. Food Sci.*, *4*, 127-135.
[168] Mwanza, M. (2007). An investigation in South African domesticated animals, their products and related health issues with reference to mycotoxins and fungi. University of Johannesburg.
[169] Mnagadi, P. T. et al. (2008). Co-occurring mycotoxins in animal feeds. *Afr. J. Biotech.*, *7*, 2239-2243.
[170] Bayman, P. & Baker, J. L. (2006). Ochratoxins: a global perspective. *Mycopathologia*, *162*, 215–223.
[171] Clark, H. A. & Snedeker, S. M. (2006). Ochratoxin A: its cancer risk and potential for exposure. *J. Toxicol Environ Health B.*, *9*, 265–296.
[172] Makun, H. A. et al. (2007). Fungi and some mycotoxins contaminating rice (Oryza sativa) in Niger State, Nigeria. *Afr J Biotechnol.*, *16*, 99–108.
[173] Dongo, L. M. et al. (2008). Occurrence of ochratoxin A in Nigerian ready to sell cocoa beans. *Agric. J. 3*, 4-9.
[174] Shephard, G. S. et al. (2003). Quantitation of ochratoxin A in South African wines. *J. Agric Food Chem*, *51*, 1102-1106.
[175] Stoev, S. D. et al. (2010). Complex aetiology and pathology of mycotoxic nephropathy in South African pigs. *Ety. Mycotox. Res.*, 10, 31-46.
[176] Khoury, A. & Atoui, A. (2010). Ochratoxin A: General overview and actual molecular status. *Toxins*, *2*, 461-493.
[177] Zain, M. E. (2011). Impact of mycotoxins on humans and animals. *Journal of Saudi Chemical Society*, *15*, 129-144.
[178] Frisvad, J. C. et al. (2007). Fumonisin B2 production by Aspergillus niger. *J Agric Food Chem*, *55*, 9727–9732.
[179] Peraica, M. et al. (1999). Toxic effects of mycotoxin in humans. *Bull. WHO*, *77*, 754-766.
[180] Marasas, W. F. O. et al. (2004). Fumonisins disrupt sphingolipid metabolism, folate transport, and neural tube development in embryo culture and in vivo: a potential risk factor for human neural tube defects among populations consuming Fumonisin contaminated maize. *Journal of Nutrition*, *134*, 711–716.
[181] Missmer, S. A. et al. (2006). Exposure to fumonisins and the occurrence of neural tube defects along the Texas-Mexico border. *Environ Health Persp. 114*, 237–241.
[182] IARC. (2002). International Agency for Research on Cancer. Fumonisin B1. Monographs on the Evaluation of Carcinogenic Risks to Humans: Some Traditional Medicines, Some Mycotoxins, Naphthalene and Styrene. 301-366.

[183] Desjardins, A. E. (2006). Introduction to Fusarium mycotoxins. In *In Fusarium Mycotoxins Chemistry, Genetics, and Biology*, 1-9, The American Phytopathological Society Minnesota, USA.

[184] Wilson, T. M. et al. (1985). Linking leukoencephalomalacia to commercial horse rations *Veterinary medicine*, *80*, 63-69.

[185] Jackson, L. (2004). Fumonisins. In *Mycotoxins in food: detection and control* (Nagan, M. and Olsen, M., eds), 367–405, Woodhead.

[186] Harrison, L. R. et al. (1990). Pulmonary oedema and hydrothorax in swine produced by fumonisin B1, a toxic metabolite of fusarium moniliform *Journal of Veterinary Diagnostic Investigation*, *2*, 217-221.

[187] Marasas, W. F. O. (2001). Discovery and occurrence of the fumonisins: A historical perspective. *Environ. Health Perspect.*, 239-243.

[188] Biotechnology, F. (2013) Fermentek Biotechnology.

[189] Ahamed, S. et al. (2001). Signal transduction through the rass/Erk pathway is essential for the mycoestrogen zearalenone- induced cell-cycle progression in MCF-7 cells. *Mol. Carcinog.*, *30*, 88-98.

[190] Foroud, N. A. & Eudes, F. (2009). Trichothecenes in cereal grains. *International Journal of Molecular Sciences*, *10*, 147-173.

[191] Agag, B. I. (2005). Mycotoxins in foods and feeds.5-trichothecenes, A T-2 toxins. *Ass. Univ. Bull. Environ. Res.* Vol.*8* No2.

[192] Pestka, J. J. & Amuzie, C. J. (2008). Tissue distribution and proinflamatory cytokine gene expression following acute oral exposure to deoxynivalenol: comparison of weaning and adult mice. *Food Chem. Toxicol.*, *46*, 2826-2831.

[193] Hopton, R. P. et al. (2010). Urine metabolite analysis as a function of DON exposure: an NMR-based metabolomics investigation. *Food Addit. Conta. Part A Chem. Anal. Contr Expo. Risk Assessment*, *27*, 255-261.

[194] Pestka, J. J. (2010). Deoxynivalenol-induced pro-inflammatory gene expression: mechanisms and pathological sequelae. *Toxins*, *2*, 1300-1317.

[195] Atanda, O. et al. (2013). Fungal and Mycotoxin Contamination of Nigerian Foods and Feeds. In *Mycotoxin and Food Safety in Developing Countries*. Atanda, O (ed). 1-36, InTech Open.

[196] Turner, N. W. et al. (2009). Analytical methods for determination of mycotoxins: A review. *Analytica Chimica Acta*, *632*, 168-180.

[197] Razzazi-Fazeli, E. & Reiter, E. V. eds (2011). *Sample preparation and clean up in mycotoxin analysis: principles, applications and recent developments*. Woodhead Publishing.

[198] Rubert, J. et al. (2012). Analysis of mycotoxins in barley using ultra high liquid chromatography, high resolution mass spectrometry: Comparison of efficiency and efficacy of different extraction procedures. *Talanta*, *99*, 712-719.

[199] Turner, N. W. et al. (2009). Analytical methods for determination of mycotoxins: A review. *Analytica Chimica Acta*, *632*, 168-180.

[200] Maragos, C. M. & Busman, M. (2010). Rapid and advanced tools for mycotoxin analysis: a review. *Food Additives & Contaminants: Part A*, *27*, 688-700.

[201] Köppen, R. et al. (2010). Determination of mycotoxins in foods: current state of analytical methods and limitations. *Appl Microbiol Biotechnology*, *86*, 1595–1612.

[202] Cousin, M. A. et al., eds (2005). *Food borne mycotoxins: chemistry, biology, ecology and Toxicology. In Fratamico, P.M., Bhumia, A.K., Smith, J.L. (eds) Food borne Pathogens: Microbiology and Molecular biology*. Horizon Scientific Press.

[203] Maestroni, B. & Cannavan, A. (2011). Sampling strategies to control mycotoxins In *Determining mycotoxins and mycotoxigenic fungi in food and feed* (Saeger, S.D., ed), 3-28, Woodhead Publishing.

[204] Bullerman, L. B. (1987). *Methods of Detecting Mycotoxins in Food and Beverages. In Food and Beverage Mycology*.

[205] Whitaker, T. B. (2003). Standardisation of mycotoxin sampling procedures: an urgent necessity. *Food Control, 17*, 233–237.

[206] Blanc, M. (2006). Sampling: the weak link in the sanitary quality control system of agricultural products. *Mol Nutr Food Res, 50*, 473–479.

[207] Schatzki, T. F. & Toyofuku, N. (2004). Sampling and sample preparation of pistachios. In *Proceedings Meeting* (Barug, D., Van Egmond, H. P., Lopez-Garcia, R., Van Osenbruggen, W. A. and Visconti, A., ed), 221–235, Wageningen Academic Publishers.

[208] Whitaker, T. B. (2004). Sampling for mycotoxins. In *Mycotoxins in Food: Detection and Control* (Magan, N. and Olsen, M., eds),69–81, Woodhead Publishing.

[209] Whitaker, T. B. (2006). Sampling food for mycotoxins. *Food Additives Contaminants, 23*, 50-61.

[210] Whitaker, T. B. et al. (2006). Sampling almonds for aflatoxin, Part I: estimation of uncertainty associated with sampling, sample preparation, and analysis. *Journal of AOAC International, 89*, 1027–1034.

[211] Ozay, G. et al. (2007). Sampling hazelnuts for aflatoxin: effect of sample size and accept/reject limit on reducing the risk of misclassifying lots. *Journal of AOAC International, 90*, 1028–1035.

[212] Whitaker, T. B. et al. (2007a). Sampling almonds for aflatoxin, Part II: estimating risks associated with various sampling plan designs *Journal of AOAC International, 90*, 778–785.

[213] Whitaker, T. B. et al. (2007b). Evaluating the performance of sampling plans to detect Fumonisin B1 in maize lots marketed in Nigeria. *Journal of AOAC International, 90*, 1050–1059.

[214] Francis, O. J. J. et al. (1988). Use of ten-gram samples of corn for the determination of mycotoxins. *Journal-Association of Official Analytical Chemists, 71*, 41-43.

[215] Pittet, A. (2005). Modern methods and trends in mycotoxin analysis. *Mitt. Lebensm, Hyg., 96*, 424-444.

[216] Murillo-Arbizu, M. et al. (2009). Occurrence of patulin and its dietary intake through apple juice consumption by the Spanish population. *Food Chemistry, 113*, 420–423.

[217] Meister, U. (1999). Effect of extraction and extract purification on the measurable fumonisin content of maize and maize products. Tests on the efficiency of acid extraction and use of immunoaffinity columns. *Mycotoxin Research, 15*, 13–23.

[218] Möller, T. E. & Nyberg, M. (2004). Analysis of aflatoxin by an LC with activation using a photochemical reaction. *Food additive and contaminants, 21*, 781-785.

[219] Trebstein, A. et al. (2009). Acetonitrile: the better extractant for the determination of T-2 and HT-2 toxin in cereals using an immunoaffinity based cleanup? *European Food Research and Technology, 228*, 519–529.

[220] Scott, P. M. (1995). Mycotoxin methodology. *Food Additives and Contaminants:Part A: Chemistry, Analysis, Control, Exposure and Risk Assessment, 12*, 395–403.

[221] Pallaroni, L. & Von Holst, C. (2003). Comparison of alternative and conventional extraction techniques for the determination of zearalenone in corn. *Analytical and Bioanalytical Chemistry, 376*, 908–912.

[222] Poole, C. F. (2003). New trends in solid-phase extraction. *Tr AC Trends in Analytical Chemistry, 22*, 362–373.

[223] Rodriguez-Mozaz, S. et al. (2007). Advantages andlimitations of on-line solid phase extraction coupled to liquid chromatography-mass spectrometry technologies versus biosensors for monitoring of emerging contaminants in water. *Journal of Chromatography A – Advances in Sample Preparation – Part I, 1152*, 97–115.

[224] Gnonlonfin, G. J. B. et al. (2008). Mycoflora and natural occurrence of aflatoxins and fumonisin B1 in cassava and yam chips from Benin, West Africa. *International Journal of Food Microbiology, 122*, 140-147.

[225] Senyuva, H. Z. & Gilbert, J. (2010). Immunoaffinity column clean-up techniques in food analysis. *Journal of Chromatography B, 878*, 115–132.

[226] Bonvehi, J. S. (2004). Occurrence of ochratoxin A in cocoa products and chocolate. *Journal of Agriculture and Food Chemistry, 52*, 6347–6352.

[227] Brera, C. et al. (2011). Ochratoxin A in cocoa and chocolate products from the Italian market: occurrence and exposure assessment. *Food Control, 22*, 1663–1667.

[228] Brera, C. et al. (2005). Interlaboratory study for Ochratoxin A determination in cocoa powder samples. *J. of Liq.chrom. Rel. Technol., 28*,35-61.

[229] Copetti, M. V. et al. (2012). Co-occurrence of ochratoxin a and aflatoxins in chocolate marketed in Brazil. *Food control, 26*, 36-41.

[230] Chen, C. Y. et al. (2005). Determination of aflatoxinM1 in milk and milk powder using high-flow solid-phase extraction and liquid chromatography-tandem mass spectrometry. *J. Agric Food Chem 53*, 8474–8480.

[231] Shephard, G. S. (2011). Chromatographic separation techniques for determination of mycotoxins in food and feed In *Determining mycotoxins and mycotoxigenic fungi in food and feed* (Saeger, S.D., ed), 71-89, Woodhead Publishing.

[232] Xu, B.J. et al. (2006). Review on the qualitative and quantitative analysis of the mycotoxin citrinin., *271–285* 17, 271–285.

[233] Yang, J. et al. (2013). Natural Occurrence, Analysis, and Prevention of Mycotoxins in Fruits and their Processed Products. *Critical Reviews in Food Science and Nutrition., 54*, 64-83.

[234] Cole, R. J. et al. eds (2012). *Biological screening methods for mycotoxins and Toxigenic fungi. In Cole R.J (ed) Modern methods in the analysis and structural elucidation of mycotoxins* Elsevier publishers.

[235] Bouaziz, C. et al. (2009). Fusarial Toxin-induced toxicity in cultured cells and in isolated mitochondria involves PTPC-Dependent activation of the mitochondria pathway of apoptosis. *Toxicological Sciences, 110*, 363-375.

[236] Kang, C. et al. (2013). Evaluation of oxidative DNA damage using an Alkaline Single-Cell Gel Electrophoresis(SCGE), Comet Assay and the protective effects of N-Acetylcysteine Amide on Zearalenone-induced cytotoxicity in Chang Liver Cells. *Toxicol. Res., 29*, 43-52.

[237] Mwanza, M. et al. (2009). The cytotoxic effect of fumonisin B1 and ochratoxin a on human and pig lymphocytes using the Methyl Thiazol Tetrazolium (MTT) assay, *25*, 233–238.

[238] Maenetje, W. et al. (2008). The use of isolated lymphocytes in mycotoxin cytotoxicity testing. *Inter. J. Mol. Sci.*, *9*, 1515-1526.

[239] Al-Hammadi, S. et al. (2013). The cytotoxicity of Aflatoxin B1 in Human Lymphocytes. *Sultan Qaboos University Med. J*, *14*, 65-71.

[240] Dominjan, A. et al. (2006). Standard and Fpg-modified comet assay in kidney cells of 14 S.D ochratoxin A- and fumonisin B1 treated rats. *Toxicological Sciences*, *222*, 53-59.

[241] Vasatkova, A. et al. (2009). Effect of naturally moldy wheat or fungi administration on metallothioneins level in brain tissues of rats. *Neuroendocrinology letters*, *30*, 163-168.

[242] Goossens, J. et al. (2012). Porcine intestinal epithelial barrier disruption by the Fusarium mycotoxins deoxynivalenol and T-2 toxin promotes transepithelial passage of doxycycline and paromomycin. *BMC Veterinary Research*, *8*, 245.

[243] Wan, L. Y. et al. (2013). Individual and combined effects of fusarium toxins, deoxnivalenol, nivalenol, zearalenone and fumonisin on swine jejunal epithelial cells *Food chem. Toxicol.*, *57*, 276-283.

[244] Wan, L. Y. et al. (2013b). Individual and combined effects of fusarium toxins on the mRNA expression of pro-inflammatory cytokines in swine jejunal epithelial cells. *Toxicol. Letter*, *18*, 238-246.

[245] Stec, J. et al. (2007). Cytotoxicity of feed-borne mycotoxin to animal cell lines invitro using the MTT Assay. *Bull. Vet. Inst.Pulawy*, *51*, 679-684.

[246] Goryacheva, Y. et al. (2007). Immunochemical methods for rapid mycotoxin detection: evolution from single to multiple analyte screening: a review. *Food Additives and Contaminants A*, *24*, 1169-1183.

[247] Goryacheva, I. Y. & Saeger, S. D. eds (2011). *Immunochemical methods for rapid mycotoxin detection in food and feed. In Determining mycotoxins and mycotoxigenic fungi in food and feed.* Woodhead Publishing.

[248] Quan, Y. et al. (2006). A rapid and sensitive Chemiluminescence enzyme-linked immunosorbent assay for the determination of fumonisin B1 in food samples. *Anal Chim Acta*, *580*, 1-8.

[249] Munkvold, G. P. & Desjardins, A. E. (1997). Fumonisins in maize. Can we reduce their occurrence? *Plant Disease*, *81*, 556-565.

[250] Samarajeuwa, U. et al. (1990). Detoxification of Aflatoxin in food and feed. *J.Food Prot*, *53*, 489-501.

INDEX

#

A

B

C

D

E

F

S

T

Z